Perspectives on Asian Tourism

Series editors
Catheryn Khoo-Lattimore
Griffith University
Nathan, Queensland, Australia

Paolo Mura
Taylor's University
Subang Jaya, Selangor, Malaysia

While a conspicuous body of knowledge about tourism in Asia is emerging, Western academic ontologies and epistemologies still represent the dominant voice within tourism circles. This series provides a platform to support Asian scholarly production and reveals the different aspects of Asian tourism and its intricate economic and socio-cultural trends.

The books in this series are aimed to pave the way for a more integrated and multifaceted body of knowledge about Asian tourism. By doing so, they contribute to the idea that tourism, as both phenomenon and field of studies, should be more inclusive and disentangled from dominant (mainly Western) ways of knowing.

More specifically, the series will fill gaps in knowledge with regard to:

- the ontological, epistemological, and methodological assumptions behind Asian tourism research;
- specific segments of the Asian tourist population, such as Asian women, Asian backpackers, Asian young tourists, Asian gay tourists, etc;
- specific types of tourism in Asia, such as film-induced tourism, adventure tourism, beauty tourism, religious tourism, etc;
- Asian tourists' experiences, patterns of behaviour, and constraints to travel;
- Asian values that underpin operational, management, and marketing decisions in and/or on Asia (travel);
- external factors that add to the complexities of Asian tourism studies.

More information about this series at http://www.springer.com/series/15382

Sajad Rezaei
Editor

Quantitative Tourism Research in Asia

Current Status and Future Directions

Springer

Editor
Sajad Rezaei
University of Hamburg
Hamburg, Germany

ISSN 2509-4203 ISSN 2509-4211 (electronic)
Perspectives on Asian Tourism
ISBN 978-981-13-2462-8 ISBN 978-981-13-2463-5 (eBook)
https://doi.org/10.1007/978-981-13-2463-5

Library of Congress Control Number: 2018960917

This Springer imprint is published by the registered company Springer Nature Singapore Pte Ltd.
The registered company address is: 152 Beach Road, #21-01/04 Gateway East, Singapore 189721, Singapore

Reviewer Acknowledgement

We appreciate all reviewers for their constructive comments on each chapter and are indebted to the reviewers for providing insightful and thoughtful comments that tremendously improved the quality of chapters to be included in this book. We are grateful to all those who contributed with chapters or advise and following scholars for the time and effort spent to review the chapters.

Roya Rahimi, Department of Marketing, Innovation, Leisure and Enterprise (MILE), University of Wolverhampton Business School, UK
Fevzi Okumus, University of Central Florida, Orlando, USA
Osman Nuri Aras, Zaman University, Phnom Penh, Cambodia
Jung Wan Lee, Administrative Sciences Department, Metropolitan College, Boston University, USA
Sreenivasan Jayashree, Faculty of Management (FOM), Multimedia University, Malaysia
S.R. Nikhashemi, Department of Marketing, Sultan Qaboos University, Muscat, Oman
Saeed Pahlevan Sharif, Faculty of Business & Law, Taylor's University, Malaysia
Milad Kalantari Shahijan, PNU International Center of Qeshm, Qeshm, Iran
S. Mostafa Rasoolimanesh, School of Housing, Building, and Planning, Universiti Sains Malaysia, Penang, Malaysia
Mustafa Öztürk, Burch University, Sarajevo, Bosnia and Herzegovina
Ali Ihtiyar, Department of Marketing, State University of New York at Oswego
Kshitiz Sharma, ABBS School of Management, Acharya Bangalore Business School, Bengaluru, Karnataka, India
Naser Valaei, KEDGE Business School, Talence, France
Ree C. Ho, Taylor's University, Malaysia
Hassan Gholipour Fereidouni, Faculty of Business and Law, Swinburne University of Technology, Australia

Acknowledgment

I would like to express my sincere appreciation to all contributors including the accepted chapters' authors and many other participants who submitted their chapters that could not be included in the book due to space limits. I would like to express my sincere thanks to the authors of the chapters for reporting their thoughts and experiences related to their research and for patiently addressing reviewers' comments and diligently adhering to the hectic deadlines to have the book sent to the publisher in a timely manner. In addition, I also appreciate series editors of *Perspectives on Asian Tourism* (Catheryn Khoo-Lattimore and Paolo Mura) for their constructive comments on the book. Lastly, I am grateful to all those who believed in this project and contributed with chapters or advise.

Hamburg, Germany
July 2018

Sajad Rezaei

Contents

Contributors

Osman Nuri Aras Zaman University, Phnom Penh, Cambodia

Mihir Dash General Management, Alliance University, Bengaluru, Karnataka, India

Maryam Emami Organizational Change Management Consultant, Childress Business Communications, Houston, TX, USA

See-Kwong Goh Faculty of Business & Law, Taylor's University, Subang Jaya, Malaysia

Ree C. Ho Taylor's University, Subang Jaya, Selangor, Malaysia

Ali Ihtiyar Department of Marketing, State University of New York, Oswego, NY, USA

Mostafiz Imtiaz Faculty of Business & Law, Taylor's University, Subang Jaya, Malaysia

Mastura Jaafar School of Housing, Building, and Planning, Universiti Sains Malaysia, Penang, Malaysia

Michael Kwag School of Hospitality Administration, Boston University, Boston, MA, USA

Jung Wan Lee Administrative Sciences Department, Metropolitan College, Boston University, Boston, MA, USA

Madhumita Guha Majumder Business Analytic, Welingkar Institute of Management Development and Research, Bengaluru, Karnataka, India

Parahny Manorungrueangrat Administrative Sciences Department, Metropolitan College, Boston University, Boston, MA, USA

S. Mostafa Rasoolimanesh School of Hospitality, Tourism and Events, Taylor's University, Subang Jaya, Malaysia

Paolo Mura Faculty of Hospitality, Food & Leisure Management, Taylor's University, Subang Jaya, Malaysia

Shuhaida Md. Noor School of Communication, Universiti Sains Malaysia, Penang, Malaysia

Fevzi Okumus University of Central Florida, Orlando, FL, USA

Mustafa Öztürk Burch University, Sarajevo, Bosnia and Herzegovina

K. R. Pillai School of Management, Manipal Academy of Higher Education, Manipal, India

Sajad Rezaei University of Hamburg, Hamburg, Germany

Behrooz Sangani Department of Management, Payame Noor University, Tehran, Iran

Saeed Pahlevan Sharif Faculty of Business & Law, Taylor's University, Subang Jaya, Malaysia

Kshitiz Sharma Marketing, ABBS School of Management, Acharya Bangalore Business School, Bengaluru, Karnataka, India

Moutushi Ganguli Sharma Soft Skill, Hinduja Global Solutions, Bengaluru, Karnataka, India

Naser Valaei KEDGE Business School, Talence, France

Sarah N. R. Wijesinghe Faculty of Hospitality, Food & Leisure Management, Taylor's University, Subang Jaya, Malaysia

Islam Yeasmin Independent University, Dhaka, Bangladesh

About the Editor

Dr. Sajad Rezaei currently works at the School of Business, Economics and Social Sciences, University of Hamburg. He teaches Quantitative Research Methods and International Market Strategies postgraduate level. Previously, he worked as a Lecturer at Taylor's University (Malaysia) dual award programs with the Bristol Business School, University of the West of England (UWE), UK. His recent research has been published in refereed journals such as *Journal of Business Research*; *International Journal of Contemporary Hospitality Management*; *Journal of Travel & Tourism Marketing*; *Tourism and Hospitality Research*; *Journal of Hospitality and Tourism Technology*; *Current Issues in Tourism*; *Computers in Human Behavior*; *Journal of Retailing and Consumer Services*; *Marketing Intelligence & Planning*; *Journal of Internet Commerce*; *Technology Analysis & Strategic Management*; *The International Review of Retail, Distribution and Consumer Research*; and *International Journal of Retail & Distribution Management,* among other scholarly outlets. His research direction has extended from his background in BSc Business Administration, MBA dissertation, and PhD theses with two key research themes including e-commerce and consumer behavior. He serves as a reviewer for several refereed journals and a member of the Editorial Advisory Board for *Journal of Hospitality and Tourism Insights*, *Journal of Promotion Management*, and *International Journal of Electronic Marketing and Retailing* among others.

Chapter 1
Quantitative Methods, Applications, and Trends in Asian Tourism Research

Sajad Rezaei

Abstract This chapter states the purpose and goals of the entire book covering the status and future directions of quantitative tourism research in Asia. As an introductory part, this chapter describes the scope of the book and provides a brief explanation and summary of chapters. As such, this chapter highlights the research paradigm, philosophy and design, and other quantitative-specific dimensions before intruding on each chapter. The chapters of the book are divided into 3 main parts including understanding tourism industry in Asia (Part I), the current status of quantitative techniques (Part II), and future directions for Asian tourism researches (Part III). In fact, the introduction chapter implicitly discusses how tourism context might be different from the other settings and argues that the creation of knowledge even in quantitative data analysis to some extent is context dependent. Therefore, this chapter discusses an overview of data analysis strategies that is often overlooked by researchers.

Keywords Quantitative methods · Tourism · Asian perspective · Research

1.1 Introduction

Indeed, although Asian tourism researchers may use the same methods and techniques as the other researcher's practice in their empirical studies, there might be some differences in how and for what research questions they have used the methods. The purpose of this book is twofold. First, this book is an attempt to map the state of quantitative research in Asian tourism context and provide a detailed picture of design, implementation, application, and challenges of quantitative methods in tourism in Asia. Second, this book also contributes to the tourism literature by introducing past, current, and future quantitative data analysis methods to tourism researchers. In order to achieve these objectives, after the introduction chapter, there will be 14 chapters divided into 3 main parts including understanding tourism

S. Rezaei (✉)
University of Hamburg, Hamburg, Germany
e-mail: sajad.rezaei@uni-hamburg.de

S. Rezaei (ed.), *Quantitative Tourism Research in Asia*, Perspectives on Asian Tourism, https://doi.org/10.1007/978-981-13-2463-5_1

industry in Asia (Part I), the current status of quantitative data analysis (Part II), and future directions for Asian tourism researches (Part III).

The current book discusses how Asian tourism context is different from other countries' contexts, what the current state of quantitative research in Asian tourism is, and how and for which research questions the most common methods have been used in Asian tourism research. This book by critically reviewing published researches seeks to identify the importance and position of data analysis underlying assumptions for Asian tourism scholars (Part I). Indeed, this book raises the question of validity and generalizability of research findings in Asian tourism literature. Therefore, reviewing regression analysis as one of the most widely used first-generation data analysis methods in tourism literature offers new insights into regression model-related methodologies and discusses the effectiveness of regression models in conducting empirical tourism research in South Asia. Moreover, through an empirical assessment, this book enlightens the fact that theory-driven assumptions are essential to support or justify regression models and results. As such, this chapter highlights the research paradigm, philosophy and design, and another quantitative-specific realm before intruding on the chapters.

This book by acknowledging the limitations of the first-generation data analysis techniques goes beyond regression analysis and looks at the use of structural equation modeling (SEM) including both CB-SEM and VB-SEM in Asian tourism. SEM by giving the opportunity to tourism scholars to ask complex research questions (Weston and Gore 2006) have enormously contributed to the tourism literature during the past decades (Nunkoo and Ramkissoon 2012). Although there are several studies (e.g., Crowley and Fan 1997; Fornell and Larcker 1981; Golob 2003; Hair et al. 2012; Lowry and Gaskin 2014) and books (Byrne 2013; Hair et al. 2013; Kline 2015) on the SEM methods and how to conduct them in general, and specifically in tourism literature (do Valle and Assaker 2015; Nunkoo and Ramkissoon 2012), there are few studies that have critically been reviewed using and reporting SEM research papers in Asian tourism context and/or by Asian scholars. This book aims to fill this gap in the literature by critically reviewing the utilization of SEM and mapping its state in Asian tourism literature.

This book also discusses the dilemma of choosing between reflective and formative measurement models and how the current tourism research theories and models can be specified as formative indicators or reflective foundations. More specifically, several conceptual, empirical, statistical, and theoretical indication and measures are proposed to shed light on formative and reflective measures/constructs in tourism research. A lack of research on the model specification in tourism highlights the importance of such an investigation.

In Part II and Part III, after exploring the most widely used first- and second-generation data analysis methods (i.e., regression analysis, CB-SEM, and VB-SEM) from different aspects in Part I, this book by conducting an empirical analysis on tourists' behavioral intention in Asia compares these three methods by highlighting the advantages and shortcoming of each of them. The book also contributes to the tourism literature by dedicating two chapters (Part III) to two more new data analysis methods: data envelopment analysis (DEA) and fuzzy set/Qualitative

Comparative Analysis (fs/QCA). In contrast to regression models that produce an average line across all decision-making units, DEA as a mathematical programming technique gives the opportunity to the researchers to produce an efficient frontier which encompasses the best unit performers (Banker et al. 1984; Donthu et al. 2005). One of the main reasons for the paucity of the studies using DEA in tourism studies is the absence of formal methodologies and measurement tools in the field (Donthu, et al. 2005). Chapter 14 is the first attempt that intends to provide researchers with the necessary information and guidance to apply DEA in tourism research. Thus, this book provides a guideline to researchers who are interested in applying fs/QCA in a tourism context.

Apart from the data analysis methods, research ethics is an extremely important part of every study. However, ethical principles and materials have been embedded mostly in qualitative topics, and most of the ethical topics in quantitative data analysis on specific areas of concern to tourism researchers (e.g., measurement, assumptions, sampling, design, model specification, reliability and validity assessment, and model fitting evaluation) are vague and "are disseminated without an ethical imperative" (Panter and Sterba 2011). Thus, this book will provide a practical review of ethical issues in quantitative methods in tourism research which would benefit tourism researchers in Asia. Lastly, this chapter continues with on data examination as the first necessary step in any data analysis method which is often overlooked by researchers.

1.2 Research Paradigm, Philosophy, and Design

In general, a research paradigm is an approach which is holistic and is a basis or ground for the methodological approach in a research (Sekaran 2006). Research paradigm is defined as "the underlying assumptions and intellectual structure on which research and development in a field of inquiry is based" including two main aspects: qualitative paradigm (inductive process used to identify patterns, concepts, and relationships) and quantitative paradigm (deductive reasoning process, moving from the specific to deduce larger generalizations) (Raines 2013, p. 425). Whereas methodology and methods focus on the practicalities and process of achievement of knowledge, research paradigm refers to the knowledge philosophy (Sekaran 2006; Trochim 2006). The quantitative paradigm is also referred to as the traditional, positivist, or empirical tradition (Raines 2013). In addition, Saunders et al. (2009) categorized research philosophy into four distinct aspects which are positivism, interpretivism, pragmatism, and realism. Most of the tourism studies in Asia adopted the positivism philosophy which refers to verifying knowledge through the measurement of phenomena or direct observations (Krauss 2005). Positivist researches normally use quantitative methods for empirical investigations and testing of formulated hypothesis, which generally involve attaining data through surveys and questionnaires and analyzing the structured data using statistical methods (Buttery and Buttery 1991). Therefore, tourism studies in Asia are mostly considered deductive because a predetermined set of theories is derived from the literature, and those theories would be tested by the hypotheses and relative statistical analysis. Thus, these types of studies underway with data collection and analysis of data gathered from different sources, including empirical research and literature, then proceed to analyze the surveyed data using structural equation modeling, for example.

Moreover, after settling the research paradigm, studies should continue with developing the appropriate research design as a function of research objectives. The research design is classified into three categories, namely, exploratory, descriptive, and causal (Burns and Bush 2002; Churchill and Iacobucci 2004). Multiple research designs are commonly applied by most researchers in tourism studies in Asia. Burns and Bush (2002) defined research design as a set of decisions that leads to the specification of the procedures and methods for collecting information as well as for analyzing such information through a comprehensive plan. For any type of research, an appropriate research design is required to determine the data type, collection technique, sampling methodology, and schedule. Basing on the available literature, the following procedures are applied in tourism research, which include determining target population and its criteria, development of survey questionnaire, pretest and pilot test, revision of questionnaire based on pretest and pilot test, CMV/CMB, and, finally, analysis of the data collected using the questionnaire.

1.3 Overview of Chapters

The first part of this book focuses on understanding the tourism industry in Asia. The second chapter provides a systematic review in Asia and introduces the Preferred Reporting Items for Systematic Reviews and Meta-Analyses (PRISMA) protocol to tourism and hospitality scholars. More specifically, by considering the items of the Preferred Reporting Items for Systematic Reviews and Meta-Analyses, also known as PRISMA (Liberati et al. 2009), this chapter reviews whether and how they have been employed in the systematic reviews conducted in the field of tourism and hospitality. This chapter enhances researchers' understanding of the procedures followed by tourism scholars in conducting systematic reviews and paves the way for researchers to adopt PRISMA in their studies. Thus, the second chapter provides a systematic review of systematic reviews published in hospitality and tourism journals by scholars affiliated with Asian institutions.

The third chapter of the book is a panel data analysis for ASEAN member countries and empirically assesses the relationship between the tourism industry and economic growth. In this chapter, the relationship between the tourism industry and economic growth in ASEAN countries is studied. Panel data is the most available method to study on common characteristics of country groups; thus, logarithmic values of the variables are included in the analysis. Tourism receipts, export revenues, and foreign direct investment are the variables that were analyzed to determine how effective they are on the GDP of the ASEAN member countries using panel data analysis. In this chapter, stationarity of the variables was tested with Levin, Lin, and Chu; Im, Pesaran, and Shin W-stat; ADF-Fisher Chi-square; PP-Fisher Chi-square; and all variables are eliminated from the unit root, and then panel data analysis was performed via EViews 8 statistical program. The results of the analysis in this chapter show that there are statistically significant relations between GDP, export, and tourism; however, the effect of the foreign direct investment on the GDP is not significant as it was seen in many empirical studies before. Therefore, panel data analysis provided to overlook the common characteristics of ten member countries of ASEAN and provides tourism investors, tourism companies, and also governments to make plans regarding the characteristics of the whole region.

The fourth chapter of the book focuses on hospitality competitiveness index for Indian states and union territories using multi-criteria Technique for Order of Preference by Similarity to Ideal Solution (TOPSIS). Because the tourism sector has become an important contributor to the economic rise of India, multiple destinations in India offer them a bouquet of tourism and simultaneously confuse them in terms of suitable destination selection to visit. The chapter compares Indian states and union territories in terms of hospitality competitiveness index and considers Hemmington's framework on hospitality that includes five dimensions. TOPSIS is used for empirical analysis of host-guest relationship, generosity; performance, the small surprises, and the buying experience driven from data for the study pertained were collected from various available reliable sources. Therefore, the results facilitate easy selection of a safer, better, and hospitable destination for tourists and demonstrate an understanding toward tourism development in Asia.

The fifth chapter discusses the co-integration and causality analysis between the tourism sector and GDP in Cambodia. The objective of this chapter is to investigate and analyze the contribution of the tourism sector in economic development in Asia with the example of Cambodia. In this chapter, the relationship between tourism receipts and gross domestic product (GDP) growth rate is analyzed using Johansen co-integration test to identify whether there was a long-term relationship between the study variables. Moreover, vector error correction mechanism (VECM) was used to identify the existence of a short-run relationship between variables. Finally, residual tests were performed to check how the model is stable and desirable. According to the results of the study, Johansen co-integration test verifies that GDP, tourism receipts, and export revenues have a long-term association. The VEC Granger Causality/Block Exogeneity Wald Tests show that the causality from tourism receipts to GDP is statistically significant. Therefore, this chapter shows that the tourism policies and the tourism revenues are more effective on the GDP than the export policies and the revenues.

The second part of the book starts with an application of correlation and regression analysis in tourism research and discusses the concepts of product-moment correlation, partial and part correlation, regression, and regression with dummy variables. Chapter 6 examines the significance of the relationship between technological innovation and firm financial performance in the tourism industry in Thailand. The study employs multiple regression analysis with dummy variables, correlation analysis (i.e., the Pearson correlation), and Chi-square tests (i.e., the Pearson Chi-square test), and the associated statistics are explained. The study reveals that technological innovation has been played an important role in improving firm short-term profitability in the tourism industry and support incorporates technological innovation and develops technological innovation capabilities is likely well positioned for success in the market. Continuous technological innovation and greater technological innovation are crucial to superior firm performance and growth. As such, an application of regression analysis with dummy variables to estimate the economic effects of innovation on firm financial performance in the tourism industry of an economy in Asia is discussed.

In addition, Chap. 7 of the book provides a quantitative understanding of the dimension of tourism trajectory in the post-reform era using the longitudinal trends of various tourism-related indicators and corroborating policy levels. The chapter focuses on foreign tourists' arrival and its associated quantitative dimensions using data from Centre for Monitoring of Indian Economy's online repository. The database comprises foreign tourist arrival in gender and selected nationality dimensions, FOREX earnings from tourism, profitability, and industry return positions. The major analytical tools were CAGR, correlation, t-test, one-way ANOVA, and MANOVA. The results of the analysis confirm impressive strides in the variables taken for analysis. The chapter considers the tourism multiplier as an all-encompassing indicator to measure its economic essence owing to its impressive inter-sectoral linkages. Chapter 7 uses longitudinal examination that provides ample evidence to conclude that tourism has been an integral part of globalization initiatives and supports the robust CAGR in the arrival of foreign tourists and its consequent macroeconomic variables.

Furthermore, using questionnaires and targeting coffee stores in Phnom Penh, Cambodia, Chap. 8 provides an empirical understanding of the impact of intercultural communication and personality on customers satisfaction and, word of mouth, intention to revisit and pay more in coffee stores in Cambodia. The chapter reveals the unique context of intercultural communication to highlight several improvements and to encourage the advancement of intercultural communication in the hospitality and tourism literature. Based on the established theories, Chap. 8 assesses the role of customers' personality on intercultural communication competence and perceived cultural distance and its impact on inter-role congruence and interaction comfort empirically. The chapter uses structural equation modeling (PLS-SEM) and provides and reveals additional insights into some managerial and theoretical solutions for addressing the intercultural communication of service encounters in the Asian hospitality industry; thus, it proposed causal relationships and framework. In addition, Chap. 9 discusses an application of SEM and path analysis (PA) in tourism research. The chapter examines antecedents of corporate commitment to sustainable tourism and corporate environmental responsibility using 386 samples collected from tourism employees in South Korea. Furthermore, the chapter describes the related technique of path analysis and provides an example of SEM with a path model, of which path analysis assumes that all variables are measured without errors. The chapter discusses the basic concepts of SEM, followed by an explanation of the key statistics and terms associated with this procedure, and describes the procedure for conducting SEM, including second-order confirmatory factor analysis (CFA). Thus, the study employs factor analysis (i.e., exploratory factor analysis, confirmatory factor analysis, and internal consistency reliability tests) and structural equation modeling analysis and path analysis (i.e., the analysis of moment structures and regression analysis).

Chapter 10 furnishes the results of systematic review in tourism and hospitality researches with best-practice guidelines for conducting PLS-SEM. Because most of the tourism and hospitality researchers have used PLS-SEM, Chap. 10 addresses the question "to what extent Asian research in tourism and hospitality has followed the guidelines recommended by the most prominent literature on PLS-SEM." In doing so, 64 partial least squares structural equation modeling studies conducted by Asian researchers and/or on Asian contexts that were published in tourism and hospitality journals were systematically reviewed. The results identified some weaknesses in conducting the analysis which should be addressed in future empirical studies. Furthermore, Chap. 11 provides a robust application of SEM with latent variables in tourism research and discusses the basic concepts of SEM, an explanation of the key statistics and terms associated with procedure. Considering the critical factors influencing consumer behavior in medical tourism as an application, the chapter describes the procedure for conducting SEM, including second-order confirmatory factor analysis and related technique of SEM analysis. The chapter provides the explanation of latent variables that might be considered as hypothetical constructs that are invoked to explain observed covariation in behavior. The chapter identifies critical factors of consumer acceptance of medical tourism in Asia using the survey data collected from a sample of 486 international tourists who visited South Korea.

The third part of the book assesses the positive and negative perceptions of residents toward tourism development, vector autoregressive models with multivariate time series, fuzzy set/qualitative comparative analysis (fs/QCA), and data envelopment analysis (DEA) and finally proposes ethical considerations in quantitative research that provides a guide for tourism and hospitality researches. Chapter 12 investigates whether residents' positive and negative perceptions of tourism development, in the Asian context, are reflective or formative in nature and assumes positive perceptions (PP) and negative perceptions (NP) as both unidimensional and multidimensional constructs inclusive of economic, social, cultural, and environmental components. Using data from residents of the Lenggong World Heritage site in Malaysia, confirmatory tetrad analysis (CTA), and recently developed fit models, such as geodesic discrepancy, unweighted least squares discrepancy, and standardized root mean square residual, this chapter concludes that future quantitative researches in resident perception field should take into consideration the formative nature of NP and PP. The chapter confirms the formative nature of PP and NP and reveals that the PP and NP measurement model is best conceptualized as a reflective-formative second-order construct, as an application case. Chapter 13 reveals a presentation of vector autoregressive models with time series in tourism research and describes how to apply vector autoregression with time series in a multivariate setting to estimate the short-run and long-run effects of international tourist arrivals on a local economy and associated economic spillovers in the local Asian economy. Thus, Chap. 13 used quarterly time series data from the Bank of Korea to explain the nature and methods of vector autoregression and vector error correction models and describe the general model, estimation of parameters, standardized regression coefficient, significance testing, residual analysis, and the meaning of regression coefficients, including short-run and long-run dynamics and elasticities.

Furthermore, Chap. 14 discusses the methods of applying both the DEA and fuzzy set/Qualitative fs/QCA in tourism research. The authors argue that conventional quantitative methods in social sciences such as the system of regression and multivariate procedures are mostly based on frequency and consistency thresholds, while the basis of fuzzy set analysis is the fact that there is no "single correct answer." Moreover, this chapter discusses the DEA introduced as a nonparametric quantitative data analysis method and as a mathematical programming technique to develop and provide the best possible solutions. Thus, Chap. 14 shows that application of fs/QCA and DEA method in Asian tourism research would yield a fruitful contribution to the literature. Lastly, Chap. 15 highlights the research ethics as an important part of tourism researches. The chapter discusses a need in a structured quantitative context such as tourism and hospitality and explains that the choices related to research ethics are largely implicit and informal. This chapter focuses on several potential issues that might emerge in conducting research methods in tourism context such as data collection and questionnaire design as a practical example. Moreover, based on the principles of ethical issues in behavioral research, this chapter proposes and offers substantial guides in anticipating ethical in tourism and hospitality research. Thus, this chapter provides a practical review of ethical issues in quantitative methods in tourism and hospitality researches.

1.4 Advancing Our Understanding of Tourism Research

Considering the above, researchers should also explore the state of asking conditional research questions such as "how," "what," and "for whom" in quantitative tourism studies in Asia and provide a nontechnical guideline on mediation and moderation analysis to tourism researchers. While these techniques are among the most widely used statistical methods in the social sciences, there is a paucity of information regarding the application of these methods in tourism. More specifically, conditional researches contribute to the literature by answering the following questions: To what extent tourism studies in Asia have employed these methods? What are the most widely used mediation and moderation methods in tourism studies in Asia? How researchers have performed and reported these methods? And researchers in answering which type of research questions have used these methods.

Researchers also should consider partial least squares discriminant analysis (PLS-DA) as a new data analysis method to social sciences literature and of course to tourism literature. PLS-DA is applied for dimension reduction and sharpens the discrimination between different groups of samples simultaneously to recognize the maximum discrimination between them (Bassbasi et al. 2014; Dorigo et al. 2007; Peerbhay et al. 2013; Wolter et al. 2009). Although PLS-DA is a more powerful classification method than the conventional techniques such as PCA and discriminant analysis, there are few studies in social sciences that have used this technique.

Lastly, data examination also is an aspect of quantitative data analysis overlooked by tourism researchers. Indeed, statistical data analysis methods rely on a variety of assumptions about the nature of the underlying data, thus failure to meet the assumptions, questions, the validity, and the generalizability of the findings. Data examination in the symmetrical and asymmetrical distribution of data should be considered in order to achieve reliable research findings. Data examination including data cleaning and checking the assumptions that underlie a statistical test is a very first necessary step in any data analysis which is often overlooked by researchers. However, in many research papers, little information is reported on these assumptions; thus, further principles of data examination in conducting quantitative data analysis in tourism research should be considered.

References

Banker, R. D., Charnes, A., & Cooper, W. W. (1984). Some models for estimating technical and scale inefficiencies in data envelopment analysis. *Management Science, 30*(9), 1078–1092.

Bassbasi, M., De Luca, M., Loele, G., Oussama, A., & Ragno, G. (2014). Prediction of the geographical origin of butters by partial least square discriminant analysis (PLS-DA) applied to infrared spectroscopy (FTIR) data. *Journal of Food Composition and Analysis, 33*(2), 210–215.

Burns, A. C., & Bush, R. F. (2002). *Marketing research: Online research applications* (4th ed.). New Jersey: Prentice Hall.

Buttery, E. A., & Buttery, E. M. (1991). Design of a marketing information system: Useful paradigms. *European Journal of Marketing, 25*(1), 26–39.

Byrne, B. M. (2013). *Structural equation modeling with AMOS: Basic concepts, applications, and programming*. New York: Routledge.

Churchill, G. A., & Iacobucci, D. (2004). *Marketing research: Methodological foundations* (9th ed.). Ohio: Thomson South-Western.

Crowley, S. L., & Fan, X. (1997). Structural equation modeling: Basic concepts and applications in personality assessment research. *Journal of Personality Assessment, 68*(3), 508–531.

do Valle, P. O., & Assaker, G. (2015). Using partial least squares structural equation modeling in tourism research a review of past research and recommendations for future applications. *Journal of Travel Research, 55*, 695–708.

Donthu, N., Hershberger, E. K., & Osmonbekov, T. (2005). Benchmarking marketing productivity using data envelopment analysis. *Journal of Business Research, 58*(11), 1474–1482.

Dorigo, W. A., Milla, R. Z., De Wit, A. J. W., Brazile, J., Singh, R., & Schaepman, M. E. (2007). A review on reflective remote sensing and data assimilation techniques for enhanced agroecosystem modeling. *International Journal of Applied Earth Observation and Geoinformation, 9*(2), 165–193.

Fornell, C., & Larcker, D. F. (1981). Evaluating structural equation models with unobservable variables and measurement error. *Journal of Marketing Research, 18*(1), 39–50.

Golob, T. F. (2003). Structural equation modeling for travel behavior research. *Transportation Research Part B: Methodological, 37*(1), 1–25.

Hair, J. F., Sarstedt, M., Ringle, C. M., & Mena, J. A. (2012). An assessment of the use of partial least squares structural equation modeling in marketing research. *Journal of the Academy of Marketing Science, 40*(3), 414–433.

Hair, J. F., Hult, G. T. M., Ringle, C. M., & Sarstedt, M. (2013). *A primer on partial least squares structural equation modeling*. Thousand Oaks: Sage.

Kline, R. B. (2015). *Principles and practice of structural equation modeling*. New York: Guilford publications.

Krauss, S. E. (2005). Research paradigms and meaning making: A primer. *The Qualitative Report, 10*(4), 758–770.

Liberati, A., Altman, D., Tetzlaff, J., Mulrow, C., Gøtzsche, P., Ioannidis, J., Clarke, M., Devereaux, P. J., Kleijnen, J., & Moher, D. (2009). The PRISMA statement for reporting systematic reviews and meta-analyses of studies that evaluate healthcare interventions: Explanation and elaboration. *PLoS Medicine, 6*(7), e1000100.

Lowry, P. B., & Gaskin, J. (2014). Partial Least Squares (PLS) Structural Equation Modeling (SEM) for building and testing behavioral causal theory: When to choose it and how to use it. *IEEE Transactions on Professional Communication, 57*, 123–146.

Nunkoo, R., & Ramkissoon, H. (2012). Structural equation modelling and regression analysis in tourism research. *Current Issues in Tourism, 15*(8), 777–802.

Panter, A. T., & Sterba, S. K. (2011). *Handbook of ethics in quantitative methodology*. New York: Routledge.

Peerbhay, K. Y., Mutanga, O., & Ismail, R. (2013). Commercial tree species discrimination using airborne AISA Eagle hyperspectral imagery and partial least squares discriminant analysis (PLS-DA) in KwaZulu–Natal, South Africa. *ISPRS Journal of Photogrammetry and Remote Sensing, 79*, 19–28.

Raines, D. A. P. E. R. N. A. (2013). Research paradigms and methods. *Neonatal Network, 32*(6), 425–428 Retrieved from https://vpn.utm.my/docview/1463807316?accountid=41678.

Saunders, M., Lewis, P., & Thornhill, A. (2009). *Research methods for business students*. Harlow: Pearson Education.

Sekaran, U. (2006). *Research methods for business: A skill building approach*. New Delhi: Wiley.

Trochim, W. M. K. (2006). *The research methods knowledge base*. (2nd ed.). Retrieved Date Accessed, 2012 from http://www.socialresearchmethods.net/kb/.

Weston, R., & Gore, P. A. (2006). A brief guide to structural equation modeling. *The Counseling Psychologist, 34*(5), 719–751.

Wolter, P. T., Townsend, P. A., & Sturtevant, B. R. (2009). Estimation of forest structural parameters using 5 and 10 mSPOT-5 satellite data. *Remote Sensing of Environment, 113*(9), 1–17.

Part I
Understanding Tourism Industry in Asia

Chapter 2
Systematic Reviews in Asia: Introducing the "PRISMA" Protocol to Tourism and Hospitality Scholars

Saeed Pahlevan Sharif, Paolo Mura, and Sarah N. R. Wijesinghe

Abstract Reviews of the literature have been regarded as essential exercises to assess the nature of knowledge produced in a field of inquiry, its gaps, and possible future developments. Despite this, studies assessing the nature and quality of the systematic review papers published in the tourism literature are scarce. This chapter provides a systematic review of systematic reviews published in hospitality and tourism journals by scholars affiliated with Asian institutions. More specifically, by considering the items of the Preferred Reporting Items for Systematic Reviews and Meta-Analyses, also known as PRISMA (Liberati A et al, PLoS Med 6(7):e1000100, 2009), this chapter reviews whether and how they have been employed in the systematic reviews conducted in the field of tourism and hospitality. By doing so, this study enhances our understanding of the procedures followed by tourism scholars in conducting systematic reviews. It is expected that this study would provide researchers with best-practice guidelines for conducting a systematic review and pave the way for researchers to adopt PRISMA in their studies.

Keywords Systematic reviews · PRISMA · Tourism and hospitality knowledge · Asia

2.1 Introduction

Reviews of the literature have been regarded as essential exercises to assess the nature of knowledge produced in a field of inquiry, its gaps, and possible future developments (Grant and Booth 2009). As Tranfield et al. (2003, p. 208) have rightly pointed out, literature reviews play a pivotal role "to map and to assess the existing intellectual territory, and to specify a research question to develop the existing body of

S. Pahlevan Sharif (✉)
Faculty of Business & Law, Taylor's University, Subang Jaya, Malaysia
e-mail: saeed.sharif@taylors.edu.my

P. Mura · S. N. R. Wijesinghe
Faculty of Hospitality, Food & Leisure Management, Taylor's University, Subang Jaya, Malaysia

S. Rezaei (ed.), *Quantitative Tourism Research in Asia*, Perspectives on Asian Tourism, https://doi.org/10.1007/978-981-13-2463-5_2

knowledge further". More specifically, the importance of reviews lies on the recognition that "most research can only be understood in context – and a key part of that context consists of the results of other studies" (Petticrew and Roberts 2006, p. 3).

While the importance of literature reviews has been recognized by scholars operating within different disciplinary boundaries, such as management (Mays et al. 2005) and education (Pickering and Bryne 2014), it is in the medical field that they have received particular attention. The pronounced emphasis placed on reviews by scholars in the medical field has been mainly driven by the advent of evidence-based practices, which have reiterated the need to base clinical decisions on reliable and comprehensive assessments of the existing knowledge (Grant and Booth 2009). As such, there have been substantial efforts in the medical sciences to improve the quality of literature reviews by reducing biases and omissions. In this respect, the relatively subjective nature of reviews has not been neglected within academic circles (see Petticrew and Roberts 2006); yet, rigorous processes and procedures have been developed to minimize implicit assumptions and researchers' biases.

It is important to emphasize that different typologies of reviews exist. Grant and Booth (2009), for example, identify 14 types of reviews, based on the different methods employed for searching, appraising, synthesizing, and analysing the items constituting the body of knowledge. Among the various options available, systematic reviews have been regarded of particular value due to their propensity to reduce biases, increase reliability, and potentially improve the communication of the findings (Liberati et al. 2009). Petticrew and Roberts (2006, p. 9) define systematic reviews as "literature reviews that adhere closely to a set of scientific methods that explicitly aim to limit systematic error (bias), mainly by attempting to identify, appraise and synthesize all relevant studies (of whatever design) in order to answer a particular question (or set of questions)". Drawing from Klassen et al. (1998), Weed (2006) discusses the key features of systematic reviews, namely, comprehensiveness, transparency, and rigourosity. Furthermore, Mays et al. (2005) suggest four different approaches for synthetizing the findings of a systematic review: narrative, qualitative, quantitative, and Bayesian meta-analysis and decision analysis. As tourism knowledge has progressively expanded over the years, several systematic reviews have been conducted by tourism scholars to map the epistemological foundations of the field and its theoretical and methodological developments (see Airey 2015; Mura and Pahlevan Sharif 2015; Wijesinghe et al. 2017). Despite this, studies assessing the nature and quality of the systematic review papers published in the tourism literature are scarce. To our best knowledge, at the current time, the study conducted by Kim et al. (2018) is the only work that provides a systematic analysis of review papers in the hospitality and tourism literature. However, although their work represents a solid contribution to our understanding of the trends and impacts of the existing review studies, it only considers work published in *Web of Science* indexed journals (namely, 32 journals).

This chapter was conceived as an attempt to overcome this limitation by focusing on a larger number of tourism and hospitality journals. Moreover, an additional contribution of this work lies on the methodological approach selected, namely, the Preferred Reporting Items for Systematic Reviews and Meta-Analyses, also known as PRISMA (Liberati et al. 2009). This systematic approach has been widely recognized and valued in the health field; yet, it has not been used by tourism scholars. More specifically, by considering the items of the PRISMA checklist, this chapter reviews whether and how they have been employed in the systematic reviews conducted in the field of tourism and hospitality. By doing so, this study enhances our understanding of the procedures followed by tourism scholars in conducting systematic reviews. It is expected that this study would provide researchers with best-practice guidelines for conducting a systematic review and pave the way for researchers to adopt PRISMA in their studies. Importantly, this chapter provides a systematic review of systematic reviews published in hospitality and tourism journals by scholars affiliated with Asian institutions. The choice of focusing on scholars based in Asia was mainly driven by the idea that Asian tourism knowledge is relatively underrepresented within the global academic scenario due to crystallized postcolonial and neocolonial structures of power (Wijesinghe and Mura 2018). As such, this systematic review of review papers casts additional light on "other" forms of knowledge.

2.2 Methods

This systematic review was performed in accordance with the Preferred Reporting Items for Systematic Reviews and Meta-Analyses (PRISMA; Liberati et al. 2009). We systematically searched for systematic reviews conducted by scholars in Asian institutions and published in hospitality and tourism journals. This study used a dataset created for a broader project to assess the quality of systematic reviews in the field of tourism and hospitality (Pahlevan Sharif et al. Forthcoming). By following Liberati et al. (2009) and Shamseer et al. (2015)'s approaches, the current study provides "explanation and elaboration" of the reporting checklist items of the PRISMA that are followed to conduct systematic reviews in the social sciences.

To create the main dataset, a protocol was developed in advance to document the analysis method and inclusion criteria. Scopus, Google Scholar, Emerald, ProQuest, Tandfonline, and the journals' websites were utilized to search for papers published in 66 tourism and hospitality journals, included in the Australian Business Deans Council (ABDC) quality list, and containing "review" in their titles, abstracts, and/or keywords. The search was performed on 1 July 2017, and no date and language restrictions were imposed. This study did not use "systematic" as a search term to identify systematic reviews that did not mention "systematic".

The title, abstract, keywords, authors' names and affiliations, journal name, and year of publication of the identified records were exported to an MS Excel spreadsheet. The screening of the titles and abstracts was performed by two independent reviewers. Then, the full texts of the remaining papers were assessed based on the eligibility criteria. All reviews that identified and selected papers using an "explicit, reproducible, and without a priori assumption" method were included (Green et al. 2008; Petticrew and Roberts 2006; Pickering and Bryne 2014). More specifically, reviews that selected literature through searching preselected keywords in databases were included (Pickering and Bryne 2014). Disagreements between the reviewers were resolved by consensus. Reviews that only included papers published in one journal were excluded.

The bibliographic details of the included studies and essential items of the PRISMA checklist (with some extensions) were added to the data management excel file. Items related to risk of bias and combining results of meta-analysis studies (items 12–16 and 19–23) were omitted as they were related to meta-analyses. A pilot test on 40 randomly selected included papers was conducted to refine the extracted items and coding accordingly. Next, all included papers were carefully reviewed to extract and code the data. Finally, for the purpose of this study, the papers published by authors with Asian affiliations were included.

2.3 Results

Although the search of the electronic databases retrieved 2420 records, records were excluded in the screening of the abstracts as they were not systematic reviews. The remaining 572 records were assessed in more detail on the basis of their full texts. Of these, 378 records were discarded as they did not meet the eligibility criteria. Two additional studies were excluded as they claimed to be systematic reviews, yet they were traditional narrative reviews. Hundred ninety-two records remained. Then, non-Asian reviews were excluded on the basis of the authors' affiliations. In total, 48 reviews were included in this systematic review. The study selection process is shown in Fig. 2.1.

2.3.1 Title

Title

Identify the report as a systematic review, meta-analysis, or both.

Example "The genesis of a new body of sport tourism literature: a systematic review of surf tourism research (1997–2011)." (Martin and Assenov 2012, p. 257)

Explanation The paper should be identified as a systematic review. However, only three reviews used the term "systematic review" in their titles. Twenty-four reports mentioned "review", "critical review", or "literature review" in their titles, and the title of one of the reviews described the report as "meta-study". Eighteen papers did not mention any of these terms in their titles. Identifying a review that has been conducted systematically as a "systematic review", and choosing an informative title would provide more accurate information about the review to the readers and also would improve indexing and the process of identification by search engines.

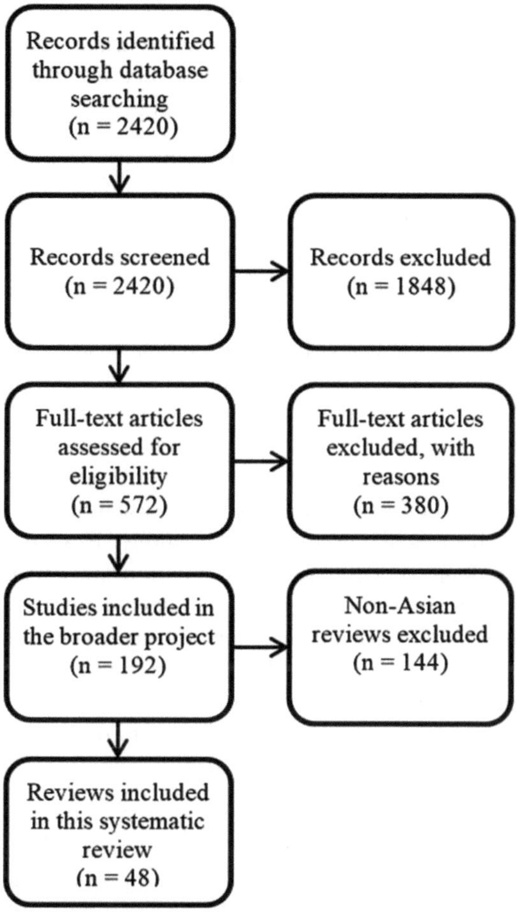

Fig. 2.1 Flow chart of study selection process

Interestingly, the word "review" was included in the keywords of only 25 papers, and only 2 of the papers used "systematic review" as 1 of their keywords. Although keywords are not part of the PRISMA checklist, researchers should not neglect the importance of choosing relevant keywords as they make papers more accessible through indexers and search engines.

2.3.2 Abstract

Structured Summary *Provide a structured summary including, as applicable, background, objectives, data sources, and study eligibility criteria, study appraisal and synthesis methods, results, limitations, conclusions and implications of key findings, funding for the systematic review, and systematic review registration number.*

Example "Purpose: This paper aims to examine the prevalence and trend of experimental research in hospitality and tourism. Hospitality and tourism researchers have long been encouraged to increase their use of experimental designs. However, a solid support for such advocacy is lacking, and the present paper fills in this research gap.

Design/methodology/approach: By using a systematic approach, this study reviews 161 tourism and hospitality articles and conducts content analysis based on certain criteria including journal outlets, Social Sciences Citation Index journals, years of publication, contexts, disciplinary foci, experimental designs, settings, number of independent variables, number of studies per article, manipulation methods, manipulation check, research subjects, sample size, subjects per experimental condition, statistical analyses and provision of effect size. The criteria between hospitality and tourism publications are also compared.

Findings: Findings show that the number of experimental publications has significantly increased over the past decade, especially in hospitality publications. Nonetheless, there is still room for improvement in applying the experimental design in hospitality and tourism research.

Research limitations/implications: Researchers in hospitality and tourism are recommended to report manipulation check results and the effect size of statistically significant results, as well as to devote more effort to knowledge accumulation and methodological advancement of experimental designs.

Originality/value: This study is the first to review experimental research in hospitality and tourism. The findings of this study provide significant implications and directions for hospitality and tourism researchers to conduct experimental research in the future." (Fong et al. 2016, p. 246)

Explanation Many electronic search engines and databases only provide access to the title, abstract, and keywords of a paper. A journal editor may desk reject a paper before sending it for review based on its abstract. A potential referee, invited by an editor, accepts or declines to review a paper only based on its abstract as it is the only part of a paper that he/she can access. In this regard, it has been claimed that for the vast majority of readers, the paper does not exist beyond its abstract. For the referees, and the few readers who wish to read beyond the abstract, the abstract sets the tone for the rest of the paper. It is therefore the duty of the author to ensure that the abstract is properly representative of the entire paper (Andrade 2011). An abstract that provides a concise summary of the major aspects of a paper, including its scope, objectives, methodology, findings, and conclusion, helps the audience decide whether to read the full paper.

PRISMA suggests a structured abstract consisting of background and objectives; methods in terms of data sources, eligibility criteria, and data extraction; results; limitations; as well as conclusions and implications of key findings. We acknowledge that different journals may require different formats of abstracts. However, an author could address these key information and at the same time tailor the abstract to the specific formatting requirements of the targeted journal. Among the reviews that we assessed, ten reviews published in the *International Journal of Contemporary*

Hospitality Management and one review published in the *Worldwide Hospitality and Tourism Themes* provided a structured abstract with clear information about the review methods. Also, there were several reviews with structured abstracts that did not mention anything about their review methods.

2.3.3 Introduction

Rationale

Describe the rationale for the review in the context of what is already known.

Example "the past three decades, a large number of studies were published that covered the issue of outbound tourism in China on different topics. For example, ...

Nonetheless, most previous studies investigated one destination in regards to China's outbound tourism. An overall picture of the changes of the destinations is still largely absent. Previous studies showed that different topics were investigated within China's outbound tourism context but the development and relationships among different topics was unclear. Furthermore, research problems and theories play vital roles in developing sound ideas. Additionally, findings of previous studies illustrated the strong scholarly interests in understanding this important market. Hence, the present study is intended to provide a systematic review of the 107 empirical research articles and 15 review/policy articles of China's outbound tourism regarding the change of researched destinations, research problems, topics, and research methods." (Law et al. 2016)

Explanation The authors need to explain the rationale for conducting a systematic review and how the study would contribute to the existing knowledge. To do so, the authors should cite and review previous reviews, discuss their strengths and limitations (see Law et al. 2012 and Goh and Law 2011), and explain why a new review should be conducted. If no review has been conducted, it should be highlighted as well. The results of our assessment show that 45 reviews provided the rationale for conducting their reviews. The PRISMA-Protocol Group suggests an introduction or background to (1) define the importance of the systematic review question from different perspectives (such as tourism policy, tourism and hospitality industry, tourists, etc.), (2) demonstrate awareness of the current state of knowledge in the subject and its limitations and connect the study with what has been published, and (3) state clearly the objectives of the systematic review and what it adds to the body of knowledge and perhaps to what extent the review addresses the mentioned limitations.

Objectives

Provide an explicit statement of questions being addressed.

Example "To fill the gaps and to provide an updated and comprehensive review of the latest development in website evaluation in hospitality and tourism driven by the rapid technological advancement in the new millennium, the present study reviews published articles on website evaluation in hospitality and tourism in the period 2000–2015. In particular, the primary objectives of this study are to categorize published website evaluation-related articles from consumers' perspective, suppliers' perspective and consumers and suppliers' perspectives [...] and to analyze and evaluate these studies from the above-mentioned perspectives using previous frameworks as basis to track the changes and indicate future research areas and directions". (Sun et al. 2017, p. 356)

Explanation The authors should state the objectives of the review clearly and precisely at the end of the introduction of the review. Research objectives may be stated broadly or narrowly, but a good research objective provides readers a clear picture of the scope of the study and what the study aims to achieve. The authors may also address some components of the methods and eligibility criteria in the objectives. The majority of the reviews (44 reports) considered in this study stated the objectives of their reports although many of them did not mention any information about the time frame, scope, or eligibility criteria of their reviews.

2.3.4 Methods

Protocol and Registration
Indicate if a review protocol exists and if and where it can be accessed (e.g. *web address*), *and, if available, provide registration information including registration number.*

Example "Based on the review aims, a review protocol was developed to guide the literature search; it contained information of the search terms, databases, and screening criteria". (Yang et al. 2017, p. 91)

Explanation Developing and registering a review protocol reduce the biases that may occur due to the post hoc modifications and changes in the review process by prespecifying the review objectives and review methods. Also, a protocol may improve the reliability of the reviews by enhancing consistency between the reviewers in a systematic review conducted by more than one reviewer. It is expected that authors document the objectives, eligibility criteria, information sources, study selection processes, and data collection processes of the review and how they will synthesize and summarize the results before conducting the review. Not all the modifications during the review process question the appropriateness of the review. For example, the authors may modify or extend the search terms, data sources, or time frame. The authors should justify the amendments in the protocol and explain the possible effects on the review results clearly. If there are several amendments, a table of amendments should be provided. None of the reviews conducted by Asian scholars mentioned details about the review protocol utilized, which calls for more transparency in conducting a systematic review.

Eligibility Criteria
*Specify study characteristics, and report characteristics (*e.g. *years considered, language, publication status) used as criteria for eligibility, giving rationale.*

Example "T[t]he article had to meet three pre-defined criteria. First, the article had to be published in a hospitality and/or tourism journal. [...] The second criterion for the selection of the papers was that the article had to be full-length and published in English. Thus, all other publications, such as research notes, editors' comments, readers' comment, and book reviews were excluded. [...] The final criterion was that the article had to be based on an economic theory and/or concept. In respect of this last criterion, it was not just enough for an article to mention an economic theory or concept in order to be selected, but, the theory and/or concept needed to be central to the research". (Mohammed et al. 2015, p. 102)

Explanation Eligibility criteria are predefined unambiguous guidelines for conducting a review systematically and consist of the characteristics of the studies that should be included or excluded from the review. Specifying and explaining inclusion and exclusion criteria in advance are fundamental prerequisites for performing an unbiased systematic review. These criteria vary from one study to another. However, authors are encouraged to justify the reason for imposing or lifting such restrictions. Eligibility criteria are necessary to appraise the validity, applicability, relevance, and comprehensiveness of a systematic review and eliminate the study selection bias. PRISMA suggests to report two types of eligibility criteria, including study characteristics and report characteristics if applicable. Inclusion or exclusion criteria related to the population, outcome, methodology, methods, design, and context are some examples of study characteristics. Report characteristics include the year of publication, publication status (such as original articles, research notes, editors' comments, book reviews, conference proceedings, and unpublished data), and language of publication.

Among the Asian reviews that were assessed in this study, 35 of them described inclusion and exclusion criteria for publication status. Also, 23 reviews mentioned the language of publications as one of the eligibility criteria.

Information Sources
*Describe all information sources (*e.g. *databases with dates of coverage, contact with study authors to identify additional studies) in the search and date last searched.*

Example "An extensive literature search was conducted using various databases such as the social science citation index (SSCI), Google Scholar, ScienceDirect, and EBSCOhost... Secondly, the citations from the identified articles were traced. Among the 120-plus publications identified, only those empirical studies related to the demand for air travel were selected for analysis. The literature collection was mostly conducted in December 2008 and an additional literature search was carried out in March 2009". (Wand and Song 2010, pp. 30–31)

Explanation PRISMA recommends a comprehensive review by including different information sources such as electronic databases, reference lists, contact with authors of included studies, and grey literature. For each database, authors should report the name of the database and the start and end dates of the search. Also, any supplement search should be reported clearly. For example, authors may hand search journals or use snowball methods, check the reference list of the included studies (e.g. Wang and Song 2010), download biographies, and contact authors.

Although a few of the Asian reviews did not report the name of all electronic databases which were used and provided a few examples (e.g. Loulanski and Loulanski 2011; Wang and Song 2010), the majority of them were transparent. In total, 45 reviews fully reported the information sources that they used to search. Among them, while some reviews limited their scope to only one journal (e.g. Law et al. 2012) or one electronic database (Leung and Law 2010; Farid et al. 2016), some assessed the eligibility of all papers found through searching several electronic databases (e.g. Law et al. 2016; Schuckert et al. 2015). Also, there are some reviews that searched only selected tourism and/or hospitality journals (e.g. Yang and Cheung 2010; do Valle and Assaker 2016) and some that searched all tourism and hospitality journals in a ranking system, such as ABDC quality journal list (e.g. Mura and Pahlevan Sharif 2015). Eight reviews referred to *rating tourism and hospitality journals* by McKercher et al. (2006) to select included journals. Reporting the exact date of the search was not common in the reviews. Only 3 reviews reported the date of their search, but 18 reviews reported the month and year the databases were searched.

Search

Present full electronic search strategy for at least one database, including any limits used, such that it could be repeated.

Example "the data were obtained from SCOPUS, on 8th January 2016. The search terms were 'climate change' and 'sustainable tourism' and these were combined with Boolean Operator 'AND' which explores 'article title, abstract, keyword' of every published document in this field". (Farid et al. 2016)

Explanation PRISMA requires a fully transparent search strategy so that readers could evaluate the comprehensiveness of the review and replicate it. Thus, authors should report all search terms that they used. If authors used different search strategies for different objectives of the review, they should report at least one example of each search strategy. Searching an electronic database is not limited to defining the search terms, and researchers also may impose other constraints (which should all be reported). For example, Scopus search engine allows researchers to restrict their search to title, abstract, and keywords of the papers, year of publication, publication status, source title, etc. Researchers may also use Boolean operators to combine or exclude keywords in their search. If authors are concerned about exceeding the word limit of the journal, they may provide the excluded information in a web-based appendix. Moreover, authors should report who performed the search. Among the papers that were reviewed in this study, 33 of them reported all search terms that they utilized against electronic databases, and 8 papers only provided some examples of the search terms.

Study Selection

*State the process for selecting studies (*i.e. *screening, eligibility, and included in systematic review).*

Example "All authors independently read through each article to select relevant papers to be included, and any disagreement on inclusion and classification was discussed until a consensus was reached". (Law et al. 2014, p. 145)

Explanation Authors should use the eligibility criteria defined in advance to decide about the exclusion of each identified study. The studies that do not meet the inclusion criteria should be excluded in this stage. Authors should report how they screened the identified studies. For example, they may first screen titles and abstracts of the papers and exclude those that despite mentioning the search terms do not meet the eligibility criteria and are not in the scope of the review. Then, authors may download the full text of the studies remained and carefully review them to identify those that should be excluded according to the eligibility criteria. PRISMA provides a flow diagram showing the different steps of the study selection (i.e. identification, screening, eligibility, and included) and number of studies identified, number of excluded studies in each step, and number of included studies (Fig. 2.1). Authors should use this flow diagram to summarize the study selection process. Also, authors should report whether the search was performed individually or by several reviewers (and in this case how they resolved disagreements). The study selection performed by more than two independent reviewers reduces the chance of excluding relevant studies.

In total, 17 tourism reviews by Asian scholars reported that they first screened titles and abstracts of the identified papers. Full texts of the papers were downloaded and assessed for eligibility in 24 reviews. One of the reviews reported that the study selection was performed by one author. Also, 11 reviews stated that 2 authors selected the studies, and among them 6 reviews explained how they resolved disagreements. None of the reviews used the PRISMA flow diagram. However, Mura and Pahlevan Sharif (2015) used a diagram to depict the different steps taken for conducting their review.

Data Collection Process

*Describe method of data extraction from reports (*e.g. *piloted forms, independently, in duplicate) and any processes for obtaining and confirming data from investigators.*

Example "The authors of this study then perused the articles and subsequently performed the content analysis. Any conflicts were resolved by discussions among the researchers until a consensus was achieved". (Fong et al. 2015, p. 163)

Explanation The authors should report all the steps taken to extract data from the included records. Providing detailed explanation is vital to establish reliability and objectivity of the content analysis process and reduce bias and mistakes during the data collection process. Authors may use an excel spreadsheet or other software packages for data management of the extracted data (e.g. "Every article was carefully read and the information on the research theme/focus, methods, and author/s was recorded on an Excel sheet". Musinguzi 2016). Also, the data collection process may be performed by one reviewer or by several reviewers independently. If more than one reviewer was involved in the data extraction process, authors should report how they resolved disagreements. If authors first pilot-test the data collection on a sample of included records to refine the process, it should be reported how and who conducted the pilot test. The pilot test may be performed by one or more than one reviewer. For example, Stepchenkova and Mills (2010) in their study, which was not included in our review, stated that "to refine the set of categories, 63 articles (~40% of the sample) were first speed read and tabulated. For representativeness of the issue scope, these 63 articles were taken from Annals of Tourism Research, Journal of Travel Research, and Tourism Management journals, as these tourism journals are, respectively, predominantly conceptually, methodologically, and practically oriented" (Stepchenkova and Mills 2010). Moreover, authors should describe any steps taken to identify duplicate and salami publications. Including duplicate publications may introduce bias to the findings of systematic reviews and meta-analysis studies.

Three tourism reviews conducted by Asian scholars reported the software package used for data management. None of the studies reported whether they pilot-tested the data extraction process. The data extraction process was conducted by more than 1 reviewer in 12 reviews, and 10 of them reported how they resolved the disagreements.

Data Items

List and define all variables for which data were sought and any assumptions and simplifications made.

Example "Every article was carefully read and the information on the research theme/focus, methods, and author/s was recorded on an Excel sheet". (Musinguzi 2016, p. 266)

Explanation All data items and definitions of the extracted variables should be listed and described. During the data collection process, authors may add new variables to the predefined items stated in the protocol. In this case, authors should justify adding the new extracted items. Among the included records in this review, 30 studies reported the extracted data items.

2.3.5 *Results*

Study Selection *Give numbers of studies screened, assessed for eligibility, and included in the review, with reasons for exclusions at each stage, ideally with a flow diagram.*

Example "Entering the keyword 'personality' as the search criterion resulted in 205 full-length articles published between 1967 and 2009 being retrieved. ... Among the 205 articles, 12 were 'editorial' prefaces or commentary articles, and 19 were not available as the authors' affiliated university does not subscribe these relevant journals. The remaining full-length articles were thoroughly read by the each author to confirm their appropriateness. A further 14 of the retrieved articles were excluded as they were unrelated to personality research. For instance, one article stated that 'Instead of trying to measure personality traits that might be important. .. [respondents were] simply asked what the term hotel loyalty meant to them' (Mason et al. 2006, p. 192). Another article analyzed wine personality (Johnston 1999). This left a total of 160 published articles pertaining to personality for further analysis". (Leung and Law 2010, p. 442)

Explanation Authors should report the total number of records identified through searching against electronic databases or other sources of information, number of records excluded in each step of the study selection process, and finally the total number of included records. In each step, the reason for excluding records should be explained, and some examples should be provided. For example, if applicable, the number of duplicates that were excluded from the records should be reported. Also, authors should report how many records were excluded during the screening process and how many excluded because of not meeting the prespecified eligibility criteria. It is recommended that authors use the PRISMA flow diagram (Fig. 2.1) to show the number of included/excluded studies in different steps of the study selection.

The results of our review show that 15 reviews reported the total number of identified records through searching information sources. While none of the reviews reported the number of excluded duplicates, only two reported the number of records that did not meet the inclusion criteria and were excluded. Among them, one review provided an explanation and example for the exclusion during the study selection process. In total, 46 reviews reported the number of included records. As it was mentioned, only Mura and Pahlevan Sharif (2015) used a diagram similar to the PRISMA flow diagram to show the different steps of the study selection process.

Study Characteristics
For each study, present characteristics for which data were extracted and provide the citations.

Example "Table 2.1 summarizes the main characteristics of the Delphi studies published in the top tourism and hospitality journals and other management and marketing journals. …The Delphi technique has been applied in a variety of locations, the most popular researched region being the USA (7), followed by the UK (6), Hong Kong (4), Australia (3), and Canada (3). In the 46 empirical studies published post-1970 reviewed in this study, the Delphi forecasting technique was mainly applied in three areas: (1) event forecasting, (2) forecasting tourism demand, and (3) forecasting future trends/market conditions (the most popular application)…". (Lin and Song 2015, p. 3)

Explanation Authors should report the citation of each included record and their main characteristics, such as the year of publication, population, outcome, methodology, methods, design, and context of the studies. This information allows readers to retrieve the records and also assess the validity of the systematic review's results. Authors may provide this information in a table and summarize them in the text. If the targeted journal does not allow the inclusion of a large table, authors may provide the table of characteristics in an appendix, as a supplementary document, or on a website. Readers of a systematic review can evaluate the relevance of the included studies through access to such characteristics. In the case of missing information, authors may contact the original authors of the document. If the required information was not obtained, it should be stated in the report.

Reviewing the reviews conducted by Asian scholars revealed that 15 reviews listed all included records and some of their characteristics. Among them, 11 reported the context, 6 reported the methods, and 1 reported the sample size of the included studies.

Syntheses of Results
Present the main results of the review.

Example "The findings […] show that the contextual aspects of people's accounts are relatively ignored. The majority of the articles analysed (25) provides information about the situational terrain, namely the specific physical and socio-cultural situation in which narratives are produced […]; yet, only a minority of the studies (2) employs analytical bracketing. In other words, most of the studies do not place emphasis on the importance of shifting intermittently between narrative work and narrative environments to understand how narrative reality is constructed. Likewise, the analysis of the interactional terrain is ignored by 41 of the 44 papers. Indeed, only in rare cases the processes that activate the narratives are reported. In most of the articles, for example, when the responses provided by the narrators/interviewees are presented there is no mention of the questions asked by the interviewer (which would probably explain why that specific narrative was activated). Moreover, the analysis of some important aspects of narrative work, such as silences, is not contemplated in any of the papers". (Mura and Pahlevan Sharif 2017, p. 9)

Table 2.1 The first two rows of summary of study characteristics: summary of Delphi forecasting applications in tourism and hospitality studies

ID	Study	Region	Panel components	Task(s)/purpose(s)	R	Panel size	Convergence	Feedback and analysis	Pretest
1	Dyck and Emery (1970)	Alberta, Canada	Six panels from different areas: social goals and values, the needs of the individual, political life, family life and child rearing, leisure and recreation, and intercultural relations	Predict future distribution of work and leisure time and the most likely uses of this leisure time in Alberta, Canada, over the years 1975–2005; forecast the probable dates and probabilities for the occurrence of events associated with leisure and recreation; and project trends using three 10-year time periods and their probabilities	3	305, 149, 126	Modal position (the average for the total panel)	Median dates, trend forecasts in graphical form, and reasons and arguments presented together in scenarios	
1	Shafer and Moeller (1974), Shafer et al. (1974), Moeller and Shafer (1983, 1994)	USA	Experts from public land management agencies, educational institutions, communications industries, public regulatory agencies, legislative bodies, quasi-public environmental organizations, and industry	Probe for social, managerial, and technological events that are likely to shape the future of park and recreation management to the year 2000	4	904 in r1, 405 in r4	Median, interquartiles	Individual estimates, median, interquartiles, graphic summary of the dates range, and reasons (of those with responses outside quartiles)	

Explanation Authors should describe the included studies' characteristics as explained above to provide general information about the population and their context. This information allows the readers to understand to what extent they can generalize the findings and assess the applicability of the results as well as potential biases. Also, a summary of the extracted variables and information of the included studies should be reported.

2.3.6 Discussion

Summary of Evidence
*Summarize the main findings including the strength of evidence for each main outcome; consider their relevance to key groups (*e.g. *healthcare providers, users, and policy makers).*

Example "One of the most evident findings of our analysis concerns the variety of different meanings attached by tourism scholars to narrative analysis. More specifically, while all the 44 papers refer to narrative analysis as their method of inquiry, multiple approaches are employed to operationalise the study of narratives. This is not surprising as the fragmented scenario existing in tourism studies reflects a wider lack of consensus among social scientists from other disciplines about the meanings and procedures to be followed in the analysis of people's accounts…". (Mura and Pahlevan Sharif 2017, p. 9)

Explanation A brief and a balanced summary of the findings of the reviewed studies should be provided by the authors. The findings should be synthesized narratively in a concise fashion. Depending on how broad the review question and inclusion criteria are, this part can be divided into different categories. The review should document the state of the art of a topic/field and identify and highlight areas of agreement and areas with discrepancies in the field.

Limitations
*Discuss limitations at study and outcome level (*e.g. *risk of bias) and at review level (*e.g. *incomplete retrieval of identified research, reporting bias).*

Example "this study also presents a number of limitations. One of them, for example, concerns the choice of focusing on journals to represent the tourism scholarly production. While journals are important platforms to disseminate research, it needs to be remembered that tourism knowledge can also be circulated in other channels, such as books, conference papers and book chapters. Also, it needs to be remembered that as tourism crosses disciplines and fields of inquiry, issues related to tourism are also published in non-tourism journals (e.g. sociology, anthropology, geography, marketing, etc.). Moreover, this study only focuses on three non-English-speaking systems (France, Iran, Italy), which obviously are not representative of other non-English-speaking countries". (Mura et al. 2017, pp. 8–9)

Explanation Authors should discuss the limitations of their systematic review as these may affect the validity of the results, representativeness of the included studies, as well as the comprehensiveness and applicability of the review. For example, the risk of bias due to the restrictions of the eligibility criteria (e.g. language, status, and year of publications), search process (e.g. search terms), variety of information sources, and missing extracting data should be addressed. In total, 32 reviews provided a discussion of the review process.

Conclusions

Provide a general interpretation of the results in the context of other evidence and implications for future research.

Example "[A summary of the results:] the current findings indicated that most of the previous studies investigated only one specific topic, such as travel motivation. Limited studies have provided a comprehensive overview of China's outbound tourism. Hence, this study contributes to hospitality and tourism literature by summarizing the applied theories of different topics and the detailed analytical methods. These aspects have not been discovered previously. The theories identified and summarized not only contribute to the theoretical contribution but also provide several enlightened ideas for academic researchers. Similarly, different types of data analysis methods can also diversify the investigation in future studies and enrich the results. Finally, a conceptual model is advocated for future studies.

[Implications for future research:] After analyzing 107 empirical and 15 review/ policy articles, this study found that several aspects could be improved in terms of research on China's outbound tourism. The study found that topics regarding China's outbound tourism were changing constantly. The combination of different topics may be good future research areas. In terms of topics and applied theories/ concepts, 26 articles lacked a theoretical foundation. Thus, future studies can consider adding a theoretical framework to enhance theoretical rigorous. Different theories have been applied to investigate the topics related to China's outbound tourism from the perspectives of both supply and demand. However, only one or two theories are generally applied to most of the topics. Hence, more theories can generally be added to investigate a certain topic. By contrast, the same theory can be applied to different topics. Therefore, future studies can also consider the enhancement of the theoretical part. In reference to research methods, innovative qualitative research methods are suggested for future studies. Considering the vital role of online information exchange platforms, such as social media, consumer's behavior can be tracked and retrieved easily (Bhatiasevi and Yoopetch 2015). The results also indicated that questionnaire survey and interview were the most common methods adopted by these studies. Nevertheless, the variety and innovativeness of the methodology are lacking. Thus, adopting different and innovative quantitative research methods can be considered. Moreover, the findings provided numerous data analysis methods that can be considered by academic researchers for adoption.

[Implications for practice:] Practically, considering the important elements identified for three travel stages, hospitality and tourism practitioners can develop better products to satisfy the needs of tourists and enhance revisit intention. For example, in the pre-trip stage, hospitality and tourism practitioners should solve the travel constraints of tourists. Moreover, they can develop travel products that are matched with tourist expectation. In the during trip stage, considering the information search behavior of tourists, hospitality and tourism practitioners can adopt user-friendly mobile-apps as distribution channels to deliver information on the six aspects that are commonly perceived by tourists (Wang et al. 2014). Additionally, mobile apps can also facilitate hotel reservations to enhance customer satisfaction and revisit intention in the post-trip stage. Enhancement of revisit intention is vital since the revenue of an organization is largely affected by the revisit of the consumers". (Law et al. 2016)

Explanation Authors should draw conclusions and make recommendations that are meaningful and related to the review objectives and based on the results of the review. If there are other similar reviews or if the review is an update of an existing review, authors should relate the findings to other evidence. Also, an important part of the conclusion of a review is the provision of implications for both future research and practice. The majority of the reviews that were reviewed systematically in this study provided a section for conclusion (39 reviews), and they discussed the implications of the review (46 reviews) although some ignored either the implications for future research or implications for practice.

2.3.7 Funding

Funding
Describe sources of funding for the systematic review and other support (e.g. *supply of data*); *role of funders for the systematic review.*

Example "I, Liang Wang, on behalf of Rob Law, Kam Hung, and Basak Denizci Guillet verify that we have no affiliations with or involvement in any organization or entity with any financial interest or non-financial interest in the subject matter or materials discussed in this manuscript". (Wang et al. 2014, p. 7)

Example "The authors acknowledge the financial support of Zhejiang Province Natural Science Foundation [grant number Q14G010016] and the Hong Kong Polytechnic University [grant number H-ZG1Z]". (Lin and Song 2015, p. 29)

Explanation PRISMA requires authors to report any source of funding used to conduct the review or they state that the author(s) received no specific funding for the review. As a systematic review may provide evidence for decision-making and policy development, as well as justification for research grants, disclosing funding and authors' financial relationships with individuals or organizations is vital as it may bias the results and conclusion of the review (Bekelman et al. 2003 and Peppercorn et al. 2007). Ten reviews assessed in this study declared whether the review was funded.

2.4 Conclusion

This chapter has assessed whether and how the items of the PRISMA procedure, developed by Liberati et al. (2009), have been employed in the systematic reviews conducted by scholars based in Asia in the field of tourism and hospitality. By doing so, this work was conceived as an attempt of suggesting best practices for conducting systematic reviews. Overall, our systematic review shows that not all the items contemplated by the PRISMA protocol have been considered by tourism and hospitality scholars. For example, the specific protocol employed in conducting the reviews, including a clear explanation of the inclusion and exclusion criteria to select/discard the papers assessed, was not provided in many of the reviews analysed. Also, other important items, such as sources of funding, were omitted in several reviews. As such, our analysis raises issues concerning the transparency of the processes underpinning the systematic reviews considered. This does not want to question the reliability of the existing reviews conducted by scholars based in Asian institutions. Indeed, our work certainly acknowledges the important contribution to knowledge represented by each review examined. Despite this, this chapter would like to serve as an opportunity to encourage scholars to place equal emphasis on all the items suggested by PRISMA. By doing so, future systematic reviews could be grounded on more transparent and reliable criteria, which are crucial to minimize implicit assumptions and researchers' biases.

References

Airey, D. (2015). 40 years of tourism studies – A remarkable story. *Tourism Recreation Research, 40*(1), 6–15.

Andrade, C. (2011). How to write a good abstract for a scientific paper or conference presentation. *Indian Journal of Psychiatry, 53*(2), 172.

Bekelman, J., Li, Y., & Gross, C. (2003). Scope and impact conflicts of interest in biomedical research: A systematic review. *JAMA, 289*(4), 454–465.

Bhatiasevi, V., & Yoopetch, C. (2015). The determinants of intention to use electronic booking among young users in Thailand. *Journal of Hospitality and Tourism Management, 23*, 1–11.

Dyck, H. J., & Emery, J. G. (1970). *Social futures alberta 1970–2005*. Edmonton: Human Resources Research Council of Alberta.

Do Valle, P., & Assaker, G. (2016). Using partial least squares structural equation modeling in tourism research: A review of past research and recommendations for future applications. *Journal of Travel Research, 55*(6), 695–708.

Farid, H., Hakimian, F., Nair, V., Nair, P., & Ismail, N. (2016). Trend of research on sustainable tourism and climate change in 21st century. *Worldwide Hospitality and Tourism Themes, 8*(5), 516–533.

Fong, L., Au, N., & Law, R. (2015). Recent research in Asia Pacific: The case of a professionally affiliated journal. *Journal of Travel & Tourism Marketing, 32*(3), 161–179.

Fong, L. H. N., Law, R., Tang, C. M. F., & Yap, M. H. T. (2016). Experimental research in hospitality and tourism: A critical review. *International Journal of Contemporary Hospitality Management, 28*(2), 246–266.

Goh, C., & Law, R. (2011). The methodological progress of tourism demand forecasting: A review of related literature. *Journal of Travel & Tourism Marketing, 28*(3), 296–317.

Grant, M., & Booth, A. (2009). A typology of reviews: An analysis of 14 review types and. *Health Information and Libraries Journal, 26*(2), 91–108.

Green, S., Higgins, J., Alderson, P., Clarke, M., Mulrow, C., & Oxman, A. (2008). What is a systematic review? In J. Higgins & S. Green (Eds.), *Cochrane handbook for systematic reviews of interventions: Cochrane book series* (pp. 1–9). London: The Cochrane Collaboration.

Johnston, P. A. (1999). Huc pater O Lenaee veni: The cultivation of wine in Vergil's georgics. *Journal of Wine Research, 10*(3), 207–221.

Kim, C., Bai, B., Kim, P., & Chon, K. (2018). Review of reviews: A systematic analysis of review papers in hospitality and tourism literature. *International Journal of Hospitality Management, 70*, 49–58.

Klassen, T., Jadad, A., & Moher, D. (1998). Guides for reading and interpreting systematic reviews: Getting started. *Archives of Pediatrics & Adolescent Medicine, 152*(7), 700–704.

Law, R., Leung, D., & Lee, H. (2012). An analysis of publications in the journal of China tourism research. *Journal of China Tourism Research, 8*(1), 37–60.

Law, R., Wu, J., & Liu, J. (2014). Progress in Chinese hotel research: A review of SSCI-listed journals. *International Journal of Hospitality Management, 42*, 144–154.

Law, R., Sun, S., Fong, D., Fong, L., & Fu, H. (2016). A systematic review of China's outbound tourism research. *International Journal of Contemporary Hospitality Management, 28*(12), 2654–2674.

Leung, R., & Law, R. (2010). A review of personality research in the tourism and hospitality context. *Journal of Travel & Tourism Marketing, 27*(5), 439–459.

Liberati, A., Altman, D., Tetzlaff, J., Mulrow, C., Gøtzsche, P., Ioannidis, J., et al. (2009). The PRISMA statement for reporting systematic reviews and meta-analyses of studies that evaluate healthcare interventions: Explanation and elaboration. *PLoS Medicine, 6*(7), e1000100.

Lin, V., & Song, H. (2015). A review of Delphi forecasting research in tourism. *Current Issues in Tourism, 18*(12), 1099–1131.

Loulanski, T., & Loulanski, V. (2011). The sustainable integration of cultural heritage and tourism: A meta-study. *Journal of Sustainable Tourism, 19*(7), 837–862.

Moeller, G. H., & Shafer, E. L. (1983). The use and misuse of Delphi forecasting. In S. Lieber & D. Fesenmaier (Eds.), *Recreation planning and management* (pp. 96–104). State College: Venture.

Moeller, G. H., & Shafer, E. L. (1994). The Delphi technique: A tool for long-range travel and tourism planning. In J. R. Ritchie & C. R. Goeldner (Eds.), *Travel, tourism, and hospitality research: A handbook for managers and researchers* (2nd ed., pp. 473–480). New York: Wiley.

Martin, S. A., & Assenov, I. (2012). The genesis of a new body of sport tourism literature: A systematic review of surf tourism research (1997–2011). *Journal of Sport & Tourism, 17*(4), 257–287.

Mason, D., Tideswell, C., & Roberts, E. (2006). Guest perceptions of hotel loyalty. *Journal of Hospitality & Tourism Research, 30*(2), 191–206.

Mays, N., Pope, C., & Popay, J. (2005). Systematically reviewing qualitative and quantitative evidence to inform management and policy-making in the health field. *Journal of Health Services Research & Policy, 10*(1), 6–20.

McKercher, B., Law, R., & Lam, T. (2006). Rating tourism and hospitality journals. *Tourism Management, 27*, 1235–1252.

Mohammed, I., Guillet, B. D., & Law, R. (2015). The contributions of economics to hospitality literature: A content analysis of hospitality and tourism journals. *International Journal of Hospitality Management, 44*, 99–110.

Mura, P., & Pahlevan Sharif, S. (2015). The crisis of the 'crisis of representation' – Mapping qualitative tourism research in Southeast Asia. *Current Issues in Tourism, 18*(9), 828–844.
Mura, P., & Pahlevan Sharif, S. (2017). Narrative analysis in tourism: A critical review. *Scandinavian Journal of Hospitality and Tourism, 17*(2), 194–207.
Mura, P., Mognard, E., & Sharif, S. (2017). Tourism research in non-English speaking academic systems. *Tourism Recreation Research, 42*(4), 436–445.
Musinguzi, D. (2016). Trends in tourism research on Qatar: A review of journal publications. *Tourism Management Perspectives, 20*, 265–268.
Peppercorn, J., Blood, E., Winer, E., & Patridge, A. (2007). Association between pharmaceutical involvement and outcomes in breast cancer clinical trials. *Cancer, 109*(7), 1239–1246.
Petticrew, M., & Roberts, H. (2006). *Systematic reviews in the social sciences: A practical guide/.* Oxford: Blackwell Publishing Ltd.
Pickering, C., & Bryne, J. (2014). The benefits of publishing systematic quantitative literature reviews for PhD candidates and other early career researchers. *Higher Education Research and Development, 33*, 534–548.
Shafer, E. L., & Moeller, G. H. (1974). Through the looking glass in environmental management. *Parks and Recreation, 9*(2), 20–23.
Shafer, E. L., Moeller, G. H., & Getty, R. (1974). Future leisure environments (USDA forest service research paper NE-301). Upper Dardy.
Schuckert, M., Liu, X., & Law, R. (2015). Hospitality and tourism online reviews: Recent trends and future directions. *Journal of Travel & Tourism Marketing, 32*(5), 608–621.
Shamseer, L., Moher, D., Clarke, M., Ghersi, D., Liberati, A., Petticrew, M., et al. (2015). Preferred reporting items for systematic review and meta-analysis protocols (PRISMA-P) 2015: Elaboration and explanation. *BMJ, g7647*, 349.
Stepchenkova, S., & Mills, J. (2010). Destination image: A meta-analysis of 2000–2007 research. *Journal of Hospitality Marketing & Management, 19*(6), 575–609.
Sun, S., Fong, D. K. C., Law, R., & He, S. (2017). An updated comprehensive review of website evaluation studies in hospitality and tourism. *International Journal of Contemporary Hospitality Management, 29*(1), 355–373.
Tranfield, D., Denyer, D., & Smart, P. (2003). Towards a methodology for developing evidence-informed management knowledge by means of systematic review. *British Journal of Management, 14*, 207–222.
Wang, M., & Song, H. (2010). Air travel demand studies: A review. *Journal of China Tourism Research, 6*(1), 29–49.
Wang, L., Law, R., Hung, K., & Guillet, B. (2014). Consumer trust in tourism and hospitality: A review of the literature. *Journal of Hospitality and Tourism Management, 21*, 1–9.
Weed, M. (2006). Sports tourism research 2000–2004: A systematic review of knowledge and a meta-evaluation of methods. *Journal of Sports & Tourism, 11*(1), 5–30.
Wijesinghe, S., & Mura, P. (2018). Situating Asian tourism ontologies, epistemologies and methodologies: From colonialism to neo-colonialism. In P. Mura & C. Khoo-Lattimore (Eds.), *Asian qualitative research in tourism: Ontologies, epistemologies, methodologies, and methods* (pp. 97–115). Singapore: Springer.
Wijesinghe, S., Mura, P., & Bouchon, F. (2017). Tourism knowledge and neocolonialism – A systematic critical review of the literature. *Current Issues in Tourism*, 1–17. https://doi.org/10.1080/13683500.2017.1402871
Yang, H., & Cheung, C. (2010). A review of English literature on ecotourism research in China. *Journal of China Tourism Research, 6*(4), 370–382.
Yang, E. C. L., Khoo-Lattimore, C., & Arcodia, C. (2017). A systematic literature review of risk and gender research in tourism. *Tourism Management, 58*, 89–100.

Chapter 3
The Relationship Between Tourism Industry and Economic Growth: A Panel Data Analysis for ASEAN Member Countries

Mustafa Öztürk, Ali Ihtiyar, and Osman Nuri Aras

Abstract The Association of Southeast Asian Nations (ASEAN) is one of the best performing regions in the world in terms of economic growth. Tourism industry is one of the sources for ASEAN region to realize higher economic growth. Tourism is a significant driver of economic growth and a potential strategic industry for economic growth in all of ASEAN member countries. In other words, tourism is a key economic sector that has implications for societies and cultures in this region.

In this chapter the relationship between tourism industry and economic growth in ASEAN countries is studied. Tourism receipts, export revenues, and foreign direct investment are the variables that were analyzed to determine how effective they are on the GDP of the ASEAN member countries. For that panel data analysis is used. Panel data is the most available method to study on common characteristics of country groups as it was in this case. Logarithmic values of the variables are included in to the analysis. First of all, stationarity of the variables was tested with Levin, Lin, and Chu; Im, Pesaran, and Shin W-stat; ADF-Fisher Chi-square, PP-Fisher Chi-square, and all variables are eliminated from unit root. After that panel data analysis was performed via EViews 8 statistical program. There are some studies used the method in tourism.

The results of the analysis show that there are statistically significant relations between GDP, export, and tourism. But the effect of the foreign direct investment on the GDP is not significant as it was seen in many empirical studies before. The coefficient of the export and the tourism are low maybe because of the economic diversity of the ASEAN countries. This means that the region is available for tourism investments and has potential to invite more tourists. The coefficient of the export and the tourism is low maybe because of the economic diversity of the ASEAN countries as well as the lack of enough investment and the lack effective tourism

M. Öztürk
Burch University, Sarajevo, Bosnia and Herzegovina

A. Ihtiyar · O. N. Aras (✉)
Zaman University, Phnom Penh, Cambodia
e-mail: aihtiyar@zamanu.edu.kh

S. Rezaei (ed.), *Quantitative Tourism Research in Asia*, Perspectives on Asian Tourism, https://doi.org/10.1007/978-981-13-2463-5_3

policies. For example, Singapore is export-oriented country, and the rate of the export on the GDP is so high, and Cambodia is tourism oriented, whereas Myanmar has no effective policies to invigorate the tourism and the export.

Highlights:

- First contribution analyzing ASEAN tourism receipts and its effect on the GDP.
- Panel data model which is the most available method to study on common characteristics of country groups.
- External demand to the economies of ASEAN countries as goods, services, and local assets.
- Tourism receipts and the export revenues seem effective on the GDP.
- Some of ASEAN countries are export oriented, while the others are tourism oriented.

Keywords Tourism industry · Tourism revenues · Economic growth · ASEAN countries · Panel data analysis

3.1 Introduction

Tourism sector, which has the largest and fastest-growing and place-dependent socioeconomic activities, provides significant opportunities to achieve comprehensive economic growth for almost each country all over the world.

The hypothesis of tourism-led growth has a spotlight from many economists because the tourism as an economic sector is capable to contribute to the economic growth with three different ways: direct, indirect, and induced ways. Directly, the tourism sector contributes to the foreign exchange revenue, with the number of foreign tourists arriving. The increase in foreign exchange revenues makes the economic growth process run more smoothly and steadily. Indirectly, the tourism sector decreases unemployment rate by creating many job opportunities in an economy. The induced way is called the impact of the tourism sector on the economy which measures the GDP and jobs supported by the spending of those who are directly or indirectly employed by the tourism sector.

Tourism sector has also an importance in the diffusion of technical knowledge, stimulation of research and development, and the accumulation of human capital. On the other hand, tourism has diverse and deep linkages with other economic sectors. For example, while investment in physical capital, such as transport infrastructure, is complementary to investment in tourism sector (Chou 2013), tourism activities also increase the economic efficiency of transport investments. In addition to transport, tourism activities build synergies for other sectors such as agriculture and natural resources, education, energy, transport, telecommunications, manufacture, water, construction, environmental services, and other urban infrastructure and services. Moreover, it has a wide range of forward linkages with sectors supplying services to tourists, including financial, retail, recreational, cultural, hospitality, and health services. These strong linkages catalyze a multiplier effect that creates broad-based economic benefits at the national level as well as employment opportunities and poverty alleviation at the local level. Without strong tourism linkages, such benefits do not materialize (UNCTAD 2010; Pleumarom 2012).

With strong linkages to other sectors and thematic issues, tourism (ADB 2017):

(a) Provides markets for raw and value-added agricultural products
(b) Creates millions of jobs – mostly women
(c) Catalyzes additional public and private investment for roads, ports, airports, and other infrastructures that are used by both tourists and the local population to access services and markets
(d) Generates taxes and fees to finance infrastructure maintenance and subsidize environmental services

As a rapidly growing sector, tourism is an important economic sector in most countries all over the world. Over the past six decades, the tourism sector has become an increasingly significant sector especially for developing countries as a source of revenue as well as a source of employment.

Comparing with other nonagricultural sectors, tourism (including travel) provides better labor-intensive and small-scale opportunities and employs a high proportion of female labor force. Tourism sector is a significant source of women's employment and economically empowers women (UNWTO 2002; ADB 2017).

Although the standard of living of an individual depends on many factors such as the ownership of assets, prices in various markets, taxes and subsidies, and distribution of income within the household, growth of tourism activities is an important influence on poverty alleviation.

Tourism, which has high interindustry and cross-country linkages, is one of the most powerful instruments in poverty alleviation in a country through employments, tax revenues, and other multiplier effects in other sectors, especially agriculture, handcraft and souvenir production, and construction.

Tourism sector creates noteworthy tax revenue for governments that can be used to support social development objectives and targets; alleviate poverty through education, health, and infrastructure development; advance environmental management; and protect cultural heritage.

It is not just the economic benefits that can be achieved, but a strong tourism industry can act as a mechanism for social change. It can be a reason for the incorporation of new and innovative ideas and values (Thirumaran and Arumynathan 2016).

The tourism sector generates many jobs that only need low skills, so that it provides a lot of jobs for the poor workers. Tourism industry has a significant role and importance on local economy development (Herman et al. 2017). The tourism brings advantages for the local economy, especially with the indirect multiplier effect. For example, according to a study (Mazumder et al. 2009), the tourism creates a number of multiplier effects, some of which are related to the increase in output, the number of jobs, incomes, additional value of various products, and imports.

The important and remarkable growth in tourism sector has been led to considerable interest in tourism sector as a tool for poverty reduction especially in developing countries, where widespread poverty exists. International development agencies have also turned to tourism as a way of alleviating poverty. A range of international development agencies put tourism directly on their agendas. Despite the lack of convincing empirical evidence to support the claim that tourism sector benefits the poor, tourism continues to be prioritized as a key development sector for struggling economies (Chok et al. 2007; Godfrey et al. 2001; Scheyvens 2015). Indeed, many international, government, nongovernmental, and private sector organizations are paying increasing attention to the argument that tourism sector can be used as a viable tool to alleviate poverty.

In fact, tourism's potential as a means of achieving poverty alleviation related to the fact that only some of the least developed countries in the world have significant levels of revenues (Blake et al. 2008).

The study shows that the tourism receipts and the export revenues of the region have positive effects on the GDP. But the effect of the FDI on the GDP is not statistically significant maybe because of weak foreign investment in the region. This means that the region has investment gap and available for investments especially on the tourism sector and has potential to invite more tourists. Also the governments in the region should focus on the tourism sector and direct their incentive policies on it. This is the easiest and the least costly way of economic growth. Otherwise it is required to invest on technology which is the longest and the costliest way of the development. Investing on the export sectors can be effective on the economic growth, but the labor incentive sectors and the outsource production in the region are competitive and give comparative advantage to countries of the region. Yet the profit rates of these sectors are lower than the tourism sector.

Panel data analysis is provided to overlook the common characteristics of ten member countries of ASEAN. This gives tourism investors, tourism companies, and also governments to make plans regarding the characteristics of the whole region. When advertising and making tourism projects, actors can move regarding the pull factors of whole region such as the exotic tastes, cultures, sea sides, and historical places, and also when making projects locally, actors can move regarding the push factors such as regional competitions and comparative advantages.

3.2 Economic Importance of Tourism Industry in the World

Tourism as an export sector generates a vital amount of foreign exchange revenues that contributes to the sustainable economic growth of developing countries. As such, tourism-generated proceeds have come to represent an important revenue source, increasing employment, household income, and government income in almost all countries (Oh 2005; Fayissa et al. 2008; Ekanayake and Long 2012).

The improvement and development of the tourism sector in the world can be traced after the end of the World War II (Lickorish and Jenkins 2007). At that time a number of European countries supported the development of the tourism sector and made it globally oriented. Before the era of the 1980s, the government of the rich countries of Europe began to release its intervention slowly and made the tourism really participate in the free market (Holik 2016).

Tourism industry has experienced continued growth and deepening diversification to become one of the largest and fastest-growing and most important economic sectors in the world. According to the World Tourism Organization (UNWTO 2017a), every day, more than 3 million tourists cross international borders. Every year, more than 1.2 billion people travel abroad. International tourist arrivals (overnight visitors) have increased from 25 million globally in 1950 to 278 million in 1980, 674 million in 2000, and 1235 million in 2016. In addition to over 1.2 billion international overnight visitors and many more domestic visitors, taking into account its wider economic impacts and implications, the total economic contribution of the sector was $7.6 trillion and 292 million jobs in 2016. With these data, the sector represents an estimated 10,2% of the world's GDP and 1 in 10 jobs globally. For the sixth successive year, growth in tourism outpaced that of the global economy (2.5%). Likewise, international tourism receipts earned by destinations worldwide have surged from $2 billion in 1950 to $104 billion in 1980, $495 billion in 2000, and $1220 billion in 2016. Considering these data, it can be said that money spent by foreign visitors to a country (or visitor exports) is a key component of the direct impact of tourism sector on the economy. International tourism receipts following the general tendency of international tourist arrivals include expenditures by international visitors on accommodation, food and drink, entertainment, shopping, and other goods and services in tourism destinations.

Moreover, tourism is a major category of international trade in services. International tourism also generates exports through international passenger transport services rendered to nonresidents. In addition to tourism receipts earned in destinations, international tourism also generated $216 billion in exports through international passenger transport services rendered to nonresidents in 2016, bringing the total value of tourism exports up to $1.4 trillion, or US$ 4 billion a day on average. International tourism (comprising both earnings in destinations and passenger transport) represents 30% of the world's service exports. It also accounts for about 7% of overall exports in goods and services. Considering the data, international tourism seems to grow faster than world merchandise trade. As a worldwide export category, tourism ranks third after chemicals and fuels and ahead of automotive products and food. In many developing countries, tourism is the top export category. For emerging economies as a whole, tourism represents 40% of services exports, well above the 30% world average. Tourism is increasingly an essential component of export diversification, both for emerging and advanced economies, and often shows a strong capacity to compensate for weaker export revenues in many commodity and oil-exporting countries (UNWTO 2017a, b; WTTC 2017a). Due to the growth in the tourism sector, the importance of the sector is increasing every year for the economies and communities. As a result, tourism has become one of the main pillars of economies, a passport to prosperity, and a transformative force for improving millions of lives.

Today, as a result of an ever-increasing number of destinations opening up and investing in tourism sector, the business volume of tourism equals or even surpasses

Table 3.1 Estimates and forecast about tourism and travel (world)

Worldwide	2016 USDbn(1)	2016 % of total	2017 Growth(2)	2027 USDbn	2027 % of total	2027 Growth(3)
Direct contribution to GDP	2306.0	3.1	3.8	3537.1	3.5	4.0
Total contribution to GDP	7613.3	10.2	3.6	11,512.9	11.4	3.9
Direct contribution to employment(4)	108,741	3.6	2.1	138,086	4.0	2.2
Total contribution to employment(4)	292,220	9.6	1.9	381,700	11.1	2.5
Visitor exports	1401.5	6.6	4.5	2.221.0	7.2	4.3
Domestic spending	3574.6	4.8	3.7	5414.1	3.9	3.9
Leisure spending	3822.5	2.3	3.9	5917.7	2.7	4.1
Business spending	1153.6	0.7	4.0	1719.9	0.8	3.7
Capital investment	806.5	4.4	4.1	1307.1	5.0	4.5

(**1**) 2016 constant prices and exchange rates; (**2**) 2017 real growth adjusted for inflation (%); (**3**) 2017–2027 annualized real growth adjusted for inflation (%); (**4**) 000 jobs *Source*: (WTTC 2017a)

that of oil exports, food products, or automobiles. Tourism has become one of the major players in international trade and represents at the same time one of the main income sources for many developing countries. The global and rapid development of tourism sector has produced diverse economic benefits (including employment) in many associated sectors such as construction, agriculture, transportation, and telecommunication.

On the other hand, the outlook for the tourism sector (including travel) remains robust and will continue to be at the forefront of wealth and employment creation in the global economy. The number of international tourist arrivals worldwide will reach to 1.4 billion by 2020 and to 1.8 billion by 2030, growing by about 3.3% each year, according to UNWTO's long-term forecast report Tourism Towards 2030. According to WTTC's forecast, this number (international tourist arrivals) will reach to 2.04 billion by 2027. According to WTTC's forecast, the total contribution of the sector to GDP (including wider effects from investment, the supply chain, and induced income impacts), which is 10.2% in 2016, will increase to 11.4% of GDP in 2027. Likewise, by 2027, tourism is forecast to support 381,700,000 jobs, an increase of 2.5% pa over the period. In other words, the contribution of tourism to total employment, which was 9.6% in 2016, will increase to 11.2% by 2027. Another important forecast is related to tourism's share of total national investment: According to WTTC's forecast, the sector's share of total national investment will rise from 4.5% in 2017 to 5.0% in 2027 (see Table 3.1).

The more important expectation is for the emerging economies: International tourist arrivals in emerging destinations (+4.4% a year) are expected to increase at twice the rate of those in advanced economies (+2.2% a year). The market shares of emerging economies increased from 30% in 1980 to 45% in 2016. The number of arrivals in emerging economies is expected to exceed those in advanced economies before 2020. By 2030, 57% of international arrivals will be in emerging economy destinations (versus 30% in 1980) and 43% in advanced economy destinations (versus 70% in 1980).

The most important expectation is for Asia and the Pacific: The strongest growth by region is expected to occur in Asia and the Pacific, where arrivals are forecast to increase by 331 million to reach 535 million in 2030 (+4.9% per year). Thanks to their faster growth, the global market shares of Asia and the Pacific will increase to 30% in 2030, up from 22% in 2010 (UNWTO 2017a, b; WTTC 2017a). In this way, over the longer term, growth of the sector will continue to be strong, and the role of tourism will become even more significant, as an important engine sector for economic growth and poverty reduction through job creation, increased income, and improving standard of living and as a vehicle for sharing cultures, creating peace, and building mutual understanding. Considering the UNWTO's long-term forecasts, countries, especially developing countries, should absolutely take into account the potential power of the tourism sector in achieving sustainable economic growth and reducing the poverty. In other words, countries should use the positive impact of well-managed tourism on inclusive and equitable economic growth, sustainable development, and lasting peace. Countries should take into consideration that tourism is indeed a big economic and social driver for growth, development, and peace. Policy makers should encourage all the public and private stakeholders and partners in the sector to work together to formulate more defined tourism policies, develop more relevant programs, implement institutional mechanisms, and optimize the economic, social, and environmental impacts and implications of tourism.

In fact, sustainable tourism development also requires that all relevant stakeholders share similar values and commitment to sustainable tourism (Athanasopoulou 2013).

Besides, countries' policy makers need to take account of the fact that the contribution of tourism to economic well-being depends on the quality and the revenues of the tourism offer.

According to UNWTO, the large majority of international travel takes place within travelers' own regions (intraregional tourism), with about four out of five arrivals worldwide originating in the same region. Therefore, policy makers need to focus more on regional countries in policies to develop the tourism industry.

3.3 The Importance of the Tourism Sector on Economic Growth in the ASEAN

The Association of Southeast Asian Nations (ASEAN) was established on 8 August 1967. The member countries of ASEAN are Brunei Darussalam, Cambodia, Indonesia, Lao PDR, Malaysia, Myanmar, the Philippines, Singapore, Thailand, and Vietnam.

ASEAN was established as a means of maintaining peace and stability in Southeast Asia by providing a forum for the discussion and resolution of regional issues which had the potential to destabilize the region (Wong et al. 2011).

ASEAN recognizes that regional economic integration is a dynamic, ongoing process as economies as well as domestic and external environments are constantly evolving (ASEAN 2015a). ASEAN has been using soft, informal, and voluntary mechanisms for bringing about regional cooperation and integration. This soft approach helps ASEAN to achieve increased integration among member countries (Thirumaran and Arumynathan 2016).

In the era of decentralization, the region is also required to improve the regional economic growth. The fiscal decentralization has a significant positive effect on the economic growth. The regional economic growth in ASEAN member countries can be realized through the utilization of its potential, and one of them is the tourism sector (Holik 2016). Tourism is one of the main priority industries for ASEAN economic integration. Tourism industry is a key industry and an important generator of income and employment for ASEAN countries.

The diversity of tropical Southeast Asia offers a rich palette of colors and tastes to tourists. From history, culture, and nature to shopping, spas, and luxury holidays, all of these and more are available in a region that is predominately peaceful and stable (Thirumaran and Arumynathan 2016).

One study (see Harrison 2015) lists the exogenous and endogenous factors that are thought to have affected the development of tourism in the ASEAN region. According to this study, there are two types of exogenous factors: global and regional. Those that are global have knock-on effects not only in the ASEAN region but also elsewhere and will include environmental, economic, and cultural features.

Current ASEAN tourism collaboration is undertaken based on the policy framework set out by the 2002 ASEAN Tourism Agreement. While the factors associated with the development of ASEAN tourism are much more complex, according to the agreement, the long-term objectives of ASEAN cooperation in the tourism industry are (ASEAN 2002; UNWTO 2010):

(a) To cooperate in facilitating travel into and within the region
(b) To enhance cooperation in the tourism industry among ASEAN countries in order to improve its efficiency and competitiveness
(c) To substantially reduce restrictions to trade in tourism and travel services among ASEAN countries
(d) To establish an integrated network of tourism and travel services in order to maximize the complementary nature of the region's tourist attractions
(e) To enhance the development and promotion of ASEAN as a single tourism destination with world-class standards, facilities, and attractions
(f) To enhance mutual assistance in human resource development and strengthen cooperation to develop, upgrade, and expand tourism and travel facilities and services in the region
(g) To create favorable conditions for the public and private sectors to engage more deeply in tourism development, intra-ASEAN travel, and investment in tourism services and facilities

The guiding principles that are taken into consideration in the development of the tourism industry and thus in achieving the tourism objectives in the region are as follows (Wong et al. 2011; Yousuf and Lal 2013): integrated and structured tourism development, sustainable and responsible development, wide-ranging stakeholder collaboration, quality tourism products, service excellence, and distinctive and interactive experiences.

On the other hand, in order to achieve these goals, the following activities are being carried out in ASEAN tourism industry (Yousuf and Lal 2013; Wong et al. 2011):

(a) Exchange of information and experiences; coordination and harmonization of tourism policies and programs; marketing, training, research, and information dissemination; facilitation of intra-ASEAN travel
(b) Promotion of tourism incentives to facilitate the development of tourism infrastructure and other related travel and tourism facilities
(c) Promotion of private sector participation and increasing collaboration between public and private sector
(d) Closer cooperation with ASEAN dialogue partners and other emerging markets to promote tourism to ASEAN
(e) Joint approaches in addressing international and regional tourism issues in areas of common interest
(f) Developing a critical pool of tourism human capital
(g) Marketing the ASEAN region as a single tourist destination with multifaceted attractions and world-class standards and facilities
(h) Encouraging tourism investments under a more competitive regime
(i) Promoting environmentally responsible and sustainable tourism development

ASEAN tourism efforts was limited to familiarization tours, official training exchange programs, and website promotion for industry partners without substantial regional integration of travel initiatives or programs until 2011. In 2011, the ASEAN Secretariat released a strategic plan for tourism development in the region. The ASEAN Tourism Strategic Plan (2011–2015) especially focused on changes in the global world order, future directions of ASEAN, and expected changes in the ASEAN tourism industry itself. This development plan has also targeted to involve residents through sustainability and responsible tourism. This strategic plan has intended to develop tourism standards with a certification process recognized by all member states, travel facilitation, and ASEAN connectivity. Of course, these approaches directly affect the tourism industry. The ASEAN has chosen tourism as an important industry to enhance the region's attractiveness and consolidate its dynamism to a developed level (UNWTO 2010; Thirumaran and Arumynathan 2016; Yonn 2017).

In 2016, the ASEAN Secretariat released a new strategic plan for tourism development in the region (see ASEAN 2015b). The new tourism strategic plan covers the period from 2016 to 2025. According to the ASEAN Tourism Strategic Plan 2016–2025, the vision for ASEAN tourism to 2025 is: "By 2025, ASEAN will be a quality tourism destination offering a unique, diverse ASEAN experience, and will be committed to responsible, sustainable, inclusive and balanced tourism development, so as to contribute significantly to the socioeconomic well-being of ASEAN people."

Toward achieving the ASEAN tourism vision, it will be necessary to complete existing and adopt new and innovative approaches to destination and product development and marketing and to work with other agencies controlling or influencing key elements of the development of tourism. In this context, the core strategic objectives of the Plan are (ASEAN 2015b):

1. To enhance the competitiveness of ASEAN as a single tourism destination
2. To ensure that ASEAN tourism is sustainable and inclusive

Within the framework of the Tourism Strategic Plan (2016–2025), ASEAN tourism aims to make a greater contribution toward the ASEAN integration goal during the post 2015 decade of realizing an economic growth that is more "inclusive," "green," and "knowledge-based."

If the strategic programs and projects for the tourism sector are fully implemented, the expectations for 2025 in the Tourism Strategic Plan (2016–2025) are as follows:

(a) The GDP contribution of ASEAN tourism could increase from 12% to 15%.
(b) Tourism's share of total resourced and employment could increase from 3.7% to 7%.
(c) Per capita spending by international tourists could increase from $ 877 to $ 1500.
(d) Increase the average length of stay of international tourist arrivals from 6.3 nights to 8 nights.
(e) The number of accommodation units could increase from 0.51 units per 100 head of population in ASEAN to 0.60 units per 100 head of population.
(f) The number of awardees for the ASEAN tourism standards could increase from 86 to 300.
(g) The number of community-based tourism value chain project interventions could increase from 43 to over 300.

3.3.1 The Effects of the AEC on Tourism in ASEAN Region

The ASEAN Economic Community (AEC) was established in 2015. The AEC is a major milestone in the regional economic integration among ASEAN countries. It is also expected to have an impact on labor and entrepreneurial mobility in the tourism industry. The overall effect of the AEC on tourism industry is expected to be positive.

One of the biggest challenges facing AEC is to present the community as a single destination. The nations have developed destination themes, in particular to encourage multi-country trips across the region. Different taglines are used for promotional activities, such as "The Tastes of Southeast Asia"; "ASEAN, a Tropical Paradise"; "World Class Cities"; "Experience Diverse Traditions"; "Sport and Relaxation"; and "Diverse Contemporary Creativity" (Thirumaran and Arumynathan 2016).

Another problem among ASEAN countries is related to visa. Aside from technical issues related to interoperability of a regional system, state security and immigration requirements remain inhibiting factors for a universal visa scheme for ASEAN (Thirumaran and Arumynathan 2016).

ASEAN has initiated different studies for the AEC Blueprint 2025. One of the envisions of the AEC 2025 is to create a deeply integrated and highly cohesive ASEAN economy that would support sustained high economic growth and resilience even in the face of global economic shocks and volatilities. On the other hand, the vision of AEC 2025 for tourism is for ASEAN to be a quality tourism destination offering a unique, diverse ASEAN experience and committed to responsible, sustainable, and inclusive tourism development, so as to contribute significantly to the socioeconomic welfare of ASEAN peoples. The proposed strategic directions and action programs address the core challenges facing the sustainable development of quality tourism and its integration within ASEAN countries: achieving a better balance in the distribution of benefits of tourism among ASEAN countries, decreasing concerns over safety and security, making cross-border formalities more appropriate and less costly, and declining transportation and destination infrastructure congestion (ASEAN 2015a).

In order to achieve the ASEAN tourism vision, AEC's activities are planned to focus on two main strategic measures (ASEAN 2015a):

1. Enhance competitiveness of ASEAN as a single tourism destination through the following key measures:
   (a) Intensify the promotion and marketing of ASEAN through its Southeast Asia campaign as a single destination
   (b) Diversify ASEAN tourism products
   (c) Attract tourism investments.
   (d) Raise capacity and capability of tourism human capital.
   (e) Implement and expand standards for facilities, services, and destinations.
   (f) Improve and expand connectivity and destination infrastructure.
   (g) Enhance travel facilitation.
2. Achieve a more sustainable and inclusive pattern of ASEAN tourism through the following key measures:
   (a) Mainstream local community and public-private sector participation in the tourism value chains at the destination level
   (b) Ensure safety and security, prioritizing protection and maintenance of natural and cultural heritage
   (c) Increase responsiveness to environmental protection and climate change

3.3.2 Tourism Performance in the ASEAN Region

ASEAN countries are heavily dependent on tourism as well as international trade, and future growth is expected to be fueled mainly by both these industries. The ASEAN region is a major world tourist attraction due to its rich endowment of natural beauty and cultures. The tourism resource base, coupled with government policies and measures to promote tourism, has resulted in phenomenal growth of tourist arrivals and revenues (Yousuf and Lal 2013).

Tourism industry contributes significant economic benefits to the ASEAN member countries.

The direct contribution of tourism (including travel) to GDP was $119.7 billion, 4.7% of total GDP in 2016. This contribution primarily reflects the economic activity generated by industries such as hotels, travel agents, airlines, and other passenger transportations (WTTC 2017b). According to the WTTC estimates, the direct contribution of tourism (including travel) to GDP will grow by 5.7% per annum to $222.8 billion, 5.3% of total GDP by 2027.

On the other hand, the total contribution of tourism and travel to GDP (including wider effects from investment, the supply chain, and induced income impacts) was $301.1 billon, 11.8% of total GDP in 2016. It is forecast to rise by 5.7% per annum to $563.0 billion, 13.5% of total GDP by 2027 (see Table 3.2).

Across the ASEAN region, there is wide variation in the total contribution of tourism and travel to GDP. Tourism and travel makes a total GDP contribution of below 10% in Myanmar, Brunei, and Indonesia but over 20% in Thailand and almost 30% in Cambodia. Even in higher-income Singapore with a sophisticated and diverse economy, the industry makes a substantively greater contribution to the economy than many other high-income economies in Europe or North America (WTTC 2016).

Table 3.2 Estimates and forecast about tourism and travel (ASEAN)

	2016		2017	2027		
ASEAN	USDbn(1)	% of total	Growth(2)	USDbn	% of total	Growth(3)
Direct contribution to GDP	119.7	4.7	7.3	222.8	5.3	5.7
Total contribution to GDP	301.1	11.8	6.9	563.0	13.5	5.7
Direct contribution to employment(4)	11,156	3.6	4.1	16,087	4.4	3.3
Total contribution to employment(4)	30,155	9.7	3.7	42,043	11.4	3.0
Visitor exports	123.7	8.6	8.5	245.5	10.8	6.2
Domestic spending	108.7	4.3	5.3	177.3	2.9	4.5
Leisure spending	167.3	3.4	6.1	314.7	3.9	5.9
Business spending	65.1	1.3	9.2	108.1	1.3	4.3
Capital investment	48.6	6.8	4.0	87.5	7.0	5.6

(**1**) 2016 constant prices and exchange rates; (**2**) 2017 real growth adjusted for inflation (%); (**3**) 2017–2027 annualized real growth adjusted for inflation (%); (**4**) 000 jobs *Source*: (WTTC 2017b)

Tourism generated 11,156,000 jobs directly, 3.6% of total employment in 2016. This number includes employment by hotels, travel agents, airlines, and other passenger transportation services (excluding commuter services). It also includes, for example, the activities of the restaurant and leisure industries directly supported by tourists (WTTC 2017b). According to the expectations, tourism and travel will account for 16,087,000 jobs directly, an increase of 3.3% per annum over the next decade, by 2027.

The total contribution of tourism and travel industry to employment (including wider effects from investment, the supply chain, and induced income impacts) was 30,155,000 jobs, 9.7% of total employment, in 2016. According to the WTTC forecasting, by 2027, tourism and travel industry will support 42,043,000 jobs, 11.4% of total employment, by increasing 3.0% per annum over the next decade.

The industry's growing GDP and job contribution depends on supportive infrastructure and investment, from air and ground transport and connectivity, health and hygiene improvements, safety and security developments, mobile phone penetration, reliable banking infrastructure, as well as tourist accommodation and other infrastructures necessary for the sector to thrive. There is a challenge for all the countries to ensure that capacity volume and quality expand in line with visitor expectations (WTTC 2016).

In 2016, as a key component of the direct contribution of tourism and travel industry, visitor exports realized $123.7 billion in the region. International tourist arrivals are forecast to total 215,117,000, generating expenditure of $245.5 billion, an increase of 6.2% per annum by 2027 (see Table 3.2). However, according to UNWTO expectations (UNWTO 2017a), in 2030, international tourist arrivals will increase by 180 million in the region.

3.4 The Relationship Between Tourism Revenues and GDP in the ASEAN Region

3.4.1 Review of the Literature

The academic studies on the relationship between tourism and economic growth can be divided into two main categories: single-country studies and country-group studies. On the other hand, the empirical results of these studies can be grouped three categories: First group studies support the tourism-led economic growth hypothesis. According to these studies, tourism sector leads to economic growth. Second group studies support the economic-driven tourism growth hypothesis or the growth-led tourism hypothesis. These studies maintain that the economic growth stimulates tourism sector. Additionally, there are also some studies as third group which put forward either a bidirectional relationship between tourism and economic growth or no relationship at all.

In this section, the results of some of country-group studies will be mentioned in more detail.

Othman and Salleh (2008) analyzed the pattern of relationship between the development of tourism industry and the economic growth in major ASEAN countries, namely, Malaysia, Thailand, Singapore, and Indonesia. In this study, the cointegration test and Granger causality analysis are employed. According to the results of the study, there is only one-way relationship between tourism development and economic growth in these countries. The economic growth is the leading factor to the tourism development in Thailand and Indonesia, while the tourism development is the leading factor to the economic growth in Malaysia and Singapore.

Tiwari (2011) examines the impact of tourism receipts, exports, and foreign direct investment on economic growth on four Asian countries, namely, India, China, Pakistan, and Russia. According to the results of this study, the revenues from the tourism create a positive impact on the economic growth in these four countries. These countries have an important potential for their tourism sector, and the tourism is proved to realize higher economic growth in these countries.

Another study by Aslan (2014) investigated the relationship between tourism development and economic growth in the Mediterranean countries using the newly developed panel Granger causality tests over the period 1995–2010. It is concluded that while there is bidirectional causal nexus between tourism development and economic growth for Portugal, unidirectional causal nexus from economic growth to tourism development is found for Spain, Italy, Tunisia, Cyprus, Croatia, Bulgaria, and Greece. Therefore, the growth-led tourism hypothesis is supported in case of these seven countries. On the other hand, there is no causal relation for Malta and Egypt. The study finds evidence to support the tourism-led growth hypothesis for a group of panel in Mediterranean countries.

Lean et al. (2014) analyzed the relationship between tourism industry and economic growth whether tourism contributes to the growth of economies (tourism-led economic growth) or it is impacted by growth of the economies (economic-driven tourism growth). This paper examines the impact of tourism on economic growth of Malaysia and Singapore. According to the results of the study, economic-driven tourism growth hypothesis is supported in Malaysia, while tourism-led economic growth hypothesis has been identified for Singapore.

Holik (2016) examines the impact of foreign tourists toward the economic growth in five ASEAN member countries (Indonesia, Malaysia, Thailand, the Philippines, and Singapore) from 1995 to 2012. As the quantitative method, the study uses one-way random effect of panel regression. In this study GDP is the dependent variable, meanwhile, the data of revenue from the foreign tourist visit, the number of foreign tourist arrival, and the exchange rate are the independent variables. According to the research results, there is evidence that international tourism can increase the economic growth in those countries. The three independent variables have a positive and significant impact to the dependent variables.

3.4.2 Methodology

In this study, panel data is the more appropriate model because it has the necessary mechanism to consider both inter-temporal dynamic movements and the individualistic behaviors of the actors. Baltagi (2008) listed the benefits of the panel data as follows:

- It provides controlling of individual heterogeneity.
- By using panel data, it is possible to provide more information about the variables, their degree of freedom and efficiency as well as less collinearity between the variables.
- This method is more useful in measuring and describing the effects than time series or horizontal cross-sectional data.
- Panel data models are more suitable for behavioral models than the others.

Panel data has individual and time dimensions which are symbolized as follows:

$$X_{it}, i = 1,2,3,\ldots\ldots\ldots N, \qquad t = 1,2,3,\ldots\ldots\ldots T$$

where i represents individual dimension and t represents time dimension.

Fixed effect and random effect estimators are generally used in estimation of static models. These two models are better estimators than the OLS model which ignores the panel structure. Pooled regression model doesn't consider the heterogeneity and individuality among the countries. However, each country has its individual characteristics and policies, an analysis neglecting the differences and the individuality of the countries may not be too healthy. So fixed effect model or random effect model may be more appropriate in the long run.

The formula of pooled regression model is as follows:

$$Y_{it} = \alpha + \beta X_{it} + u_{it}$$

Y_{it} is the dependent variable observed for individual i at time t.
α is the intercept or constant.
β represents coefficients of independent variables to be estimated.
u_{it} is the error term bases on the assumptions $E(u_{it}) = 0$ and $\mathrm{Var}(u_{it}) = \sigma_u^2$.

The pooled OLS estimator simply estimates α, β, and u_{it} without regarding the panel structure. But the error term is likely to have some specification problems.

When the unobserved cross-sectional error term u_i or the unobserved time-variant error term u_t are correlated with the observed variables, the observed variables may be biased to estimate coefficients. Regression equation of the OLS model has a composite error term ($u_i + u_t + u_{it}$) which is serially correlated because of the error terms that u_i is time-invariant error shared by all observations for the same entity and u_t is the time-variant error shared by all observation for the same year.

Serial correlation in the error term leads the standard errors not to be correct. Therefore, pooled OLS model is not the most efficient method.

Fixed effect or LSDV model allows for heterogeneity or individuality among the countries by allowing them to have their own intercept values which differs from one country to another but dont change over time. If there is a change in the dependent variable, not because of the unobserved variable that doesn't change over time, there must be some influences other than the fixed characteristics (Stock and Watson 2007).

The formula for fixed effect model is as follows:

$$Y_{it} = \beta X_{it} + \alpha_i + u_{it}$$

where α is the intercept or constant that represents unobserved time-invariant individual effect, X_{it} is a vector that represents the time-variant regressor matrix, β represents coefficients of independent variables to be estimated, and u_{it} is the error term which is correlated with the regressors in the fixed effect model.

Fixed effect is used when analyzing only the impact of variables that vary over time. It examines relationship between dependent and independent variables within an entity such as country, person, and company. Each entity has its own individual characteristics that have potential to influence the independent variables. Fixed effect removes the individualistic time-invariant effects to assess only the impact of the independent variables on the dependent variables.

Another assumption of the fixed effect model is that each entity is unique and has its own characteristics different from others; therefore error terms and the constants of the entities should not be correlated. Otherwise outcomes of the fixed effect model may not be correct, and random effect model may be more rationale to perform.

In random effect models, variations across the entities are assumed random and uncorrelated with the dependent or the independent variables. If there are some influences on the dependent variables stemming from the differences between the entities, random effect model should be used. Including the time-invariant variables into the model is an advantage of the random effect estimator.

The formula of the random effect model is as follows:

$$Y_{it} = \alpha_i + \beta X_{it} + u_{it} + \varepsilon_{its}$$

where u_{it} symbolizes individual time-invariant effects such as climate, geography, etc. which are fixed over time. However, ε_{it} is a symbol of time-varying random effects.

3.4.3 Data

This study uses annual time series of ten Asian countries as GDP, tourism receipts, export revenues, and foreign direct investment that were obtained from the World Bank Databank. The empirical period depends on the availability of data, where the time period used is 1995–2015. Logarithmic values of variables were used in the analysis. Table 3.3 presents the panel unit root tests at a 5% significance level, and the test statistics confirm that four series are nonstationary. Having first difference of each series, all variables were eliminated from unit root. Then a panel data model is performed in order to determine whether tourism receipts, export revenues, and the amount of foreign direct investment have an effect on GDP.

Table 3.3 presents the results of the panel unit root test, and all variables are stationary at level I (1) with an individual intercept, according to Levin, Lin, and Chu t; Im, Pesaran, and Shin W-Stat; ADF-Fisher χ^2; and PP-Fisher χ^2.

3.4.4 Estimation Outputs

There are three models, pooled regression, fixed effect, and the random effect models, studied based on the formula shown below:

$$\Delta\text{GPDLOG} = \alpha_1 + \alpha_2\ \Delta\text{TOURISMLOG} + \alpha_3\Delta\ \text{EXPLOG} + \alpha_4\ \Delta\text{FDILOG} + u_t$$

According to the outcomes of the three models shown in Table 3.4, tourism receipts and the export revenues have positive effects on GDP at 5% level of significance.

In the panel data outputs shown in Table 3.2, R^2 values for pooled regression and the random effect models are so low (0.201008 and 0.197691); however R^2 value for the fixed effect model is relatively better (0.494349) than the others.

Durbin-Watson statistic is used to detect the presence of autocorrelation in the residuals of the regression. If the value of it is 2, null hypothesis that the residuals of the regression are autocorrelated is rejected. This statistic for random effect and the fixed effect models are 1.68 and 1.83 which are close to 2, so it can be assumed that the residuals of the regressions are not correlated.

One of these three models is more appropriate to examine the relationship between variables, so in the next step, Hausman test will be performed to check whether fixed effect model or random effect model is more appropriate (see Table 3.5).

Hausman test (1978) checks the probability of null hypothesis based on the argument that the random effect model is appropriate. According to test results, the probability of null hypothesis is less than 5%, so the fixed effect model is more appropriate rather than the random effect model.

Table 3.3 Panel unit root tests

	GDP		Tourism		Export		FDI	
Method	Statistic	Prob.[a]	Statistic	Prob.[a]	Statistic	Prob.[a]	Statistic	Prob.[a]
Levin, Lin, and Chu t*	−5.7705	0.0000	−5.5311	0.0000	−4.2643	0.0000	−12.9543	0.0000
Im, Pesaran, and Shin W-stat	−5.7867	0.0000	−5.2530	0.0000	−4.8396	0.0000	−12.5117	0.0000
ADF-Fisher χ^2	72.9511	0.0000	65.9727	0.0000	59.3467	0.0000	153.561	0.0000
PP-Fisher χ^2	91.4897	0.0000	115.240	0.0000	107.809	0.0000	258.922	0.0000

[a]Probabilities for Fisher tests are computed using an asymptotic Chi-square distribution. All other tests assume asymptotic normality

Table 3.4 Estimation outputs of three models

Dependent variable: Δ GPDLOG				
Sample (adjusted): 1996–2015				
Variable		Pooled OLS	Fixed effects	Random effects
Δ TOURISMLOG	**Coefficient**	**0.034704**	**0.016877**	0.019622
	Std. Error	0.010108	0.008523	0.008479
	t-Statistic	3.433287	1.980330	2.314278
	Prob.	0.0007	0.0491	0.0217
Δ EXPOLOG	**Coefficient**	**0.083295**	**0.078810**	0.079526
	Std. Error	0.016647	0.013703	0.013681
	t-Statistic	5.003551	5.751503	5.812720
	Prob.	0.0000	0.0000	0.0000
Δ FDILOG	**Coefficient**	**0.001159**	**0.001262**	0.001246
	Std. Error	0.001388	0.001131	0.001131
	t-Statistic	0.834610	1.116126	1.101705
	Prob.	0.4050	0.2658	0.2719
Constant	**Coefficient**	**0.042551**	**0.044429**	0.044138
	Std. Error	0.002792	0.002291	0.005222
	t-Statistic	15.24218	19.39295	8.452640
	Prob.	0.0000	0.0000	0.0000
R-squared		**0.201008**	**0.494349**	**0.197691**
Adjusted R-squared		**0.188778**	**0.461901**	**0.185410**
Durbin-Watson stat		1.156303	**1.828263**	**1.675544**
Prob (F-statistic)		**0.000000**	**0.000000**	**0.000000**

Table 3.5 Correlated random effects – Hausman test

Test summary	Chi-Sq. Statistic	d.f	Prob.
Cross-section random	10.270629	3	0.0164

3.4.5 *Estimation of Fixed Effect Model Using Dummy Variables*

There are varieties of performing the fixed effect model. One of them is a regression model that the dummy variables are included into the model structure. For ten ASEAN member countries as the model entities, nine dummy variables (D) are included in the model. The formula of the regression model is as follows:

$$\begin{aligned}\Delta \text{GPDLOG} = {} & \alpha_1 + \alpha_2\, \Delta\text{TOURISMLOG} + \alpha_3\, \Delta\text{EXPLOG} + \alpha_4\, \Delta\text{FDILOG} + \alpha_5 D_2 \\ & + \alpha_6 D_3 + \alpha_7 D_4 + \alpha_8 D_5 + \alpha_9 D_6 + \alpha_{10} D_7 + \alpha_{11} D_8 + \alpha_{12} D_9 + \alpha_{13} D_{10}\end{aligned}$$

Table 3.6 Fixed effect model outcomes

Dependent variable: Δ GPDLOG				
	Coefficient	Std. Error	t-Statistic	Prob.
α_1	0.004665	0.006218	0.750244	0.4541
α_2	0.016877	0.008523	1.980330	0.0491
α_3	0.078810	0.013703	5.751503	0.0000
α_4	0.001262	0.001131	1.116126	0.2658
α_5	0.030403	0.008656	3.512241	0.0006
α_6	0.056444	0.008862	6.369311	0.0000
α_7	0.052665	0.008749	6.019657	0.0000
α_8	0.079767	0.008750	9.116014	0.0000
α_9	0.037155	0.008685	4.278316	0.0000
α_{10}	0.031755	0.008693	3.652846	0.0003
α_{11}	0.042320	0.008661	4.886089	0.0000
α_{12}	0.020338	0.008700	2.337671	0.0205
α_{13}	0.046795	0.008758	5.342907	0.0000

Coefficients of the tourism receipts (α_2 = 0.016877) and the export revenues (α_3 = 0.078810) are positive in appropriate with the theory and statistically significant at 0.05 level of significance. Also coefficient of foreign direct investment is positive but isn't statistically significant (see Table 3.6).

To check whether the pooled regression model or fixed effect model is more appropriate to the analysis, Wald test will be performed. If the coefficient of dummy variables is zero, pooled model is more appropriate; otherwise fixed effect model will be accepted.

Probability of the F-statistic which is less than 5% in the outcomes of the Wald Test verifies that the coefficients of the dummy variables are not zero and the fixed effect model is not rejected (see Table 3.7).

Conclusion and Policy Recommendations

Tourism industry, which exhibits powerful growth potential, contributes remarkable economic benefits to the ASEAN member countries. The interest of visitors remains large, and the ASEAN region remains attractive to be visited.

According to test results of this study, fixed effect estimator is the more appropriate model to determine how effective is the foreign demand on goods, services, and local assets of ASEAN economies. Tourism receipts and export revenues have positive effects on GDPs in the region. These results indicate that the region has potential to attract more tourists and to realize higher economic growth through opening the economies into the foreign demand. In many developing countries which are suffering from current account deficits, governments have been performing various policies to invigorate their tourism industry, and they have been getting positive results of these policies on their tourism yields. In the same way, ASEAN member countries may develop individualistic and common tourism policies to attract more tourists into the region.

Table 3.7 Wald Test

Null hypothesis: $\alpha_5 = \alpha_6 = \alpha_7 = \alpha_8 = \alpha_9 = \alpha_{10} = \alpha_{11} = \alpha_{12} = \alpha_{13}$			
Test statistic	Value	df	Probability
F-statistic	8.161526	8,187	0.0000
χ^2	65.29221	8	0.0000

Restrictions are linear in coefficients

Although relations between the dependent variable GDP and the independent variables tourism receipts and export revenues are statistically significant, the effect of the FDI on the GDP is not significant. This means that either there are no enough investments into the region or the effects of the foreign investments on ASEAN GDPs are weak. However, growing tourist numbers and the tourism receipts show that the region has tourism potential and available for tourism investments. Local and foreign investors individually or collectively could be benefitted from these opportunities.

On the other hand, regional competition between the ASEAN member countries is inevitable and has positive effects on the sector by triggering tourism investments and by providing better services.

The coefficient of the export and the tourism is low maybe because of the economic diversity of the ASEAN countries as well as the lack of enough investment and the lack of effective tourism policies. For example, Singapore is an export-oriented country, and the rate of the export on the GDP is so high, and Cambodia is tourism oriented, whereas Myanmar has no effective policies to invigorate the tourism and the export.

Increasing tourism investments invite foreign investors into the region and help the local urban development. And also as the number of tourists visiting the region increased, local people will become more open to the world and will have the connections from many different countries. These dynamics have potential to attract more investors into the region not just in tourism but in various sectors and have potential for better links in export sectors to enter into the new markets.

Awareness of the economic benefits of the tourism industry is quite high in the region. As a matter of fact, the possibilities of short-term earnings in terms of job creation and foreign exchange earnings are quite satisfactory. Therefore, in the policies for tourism industry, it is needed to focus on achieving long-term goals that can be defined as improving sociocultural understanding, ensuring environmental sustainability, and ensuring grassroots-oriented public participation.

The ecotourism areas of ASEAN, which have quite rich natural resources, are important attraction centers for visitors. However, unsustainable management of these areas beyond their carrying capacities can cause irreversible damages. Therefore, taking into account this issue, both the private and public sector should have a sustainable management approach and demonstrate such a management.

Though a number of initiatives were taken for the marketing of tourism in the region and the promotion of intraregional activities among ASEAN countries, there is more to be done to understand and deal with the obstacles in front of cooperation.

An integrated regional framework provides an avenue through the competiveness of the tourism industry in the region. However, at the same time, ASEAN member countries should spend more concerted efforts to cope with mutual interest and challenges.

ASEAN member countries should demonstrate a stronger political will to develop the tourism industry in the region. As a demonstration of strong political will, member countries need to allocate more resources on tourism projects in the region.

On the one hand, ASEAN member countries should consolidate previous gains; on the other hand, they should develop a more strategic approach to dealing with the single destination marketing, quality standards, human resource education and development, connectivity, investment, community participation, safety and security, and natural and cultural heritage conservation.

Additionally, developing a competitive tourism industry in the region requires a strong and entrepreneurial private sector. Governments of member countries should activate the industry by inviting private sector to join policy formulations, particularly in the areas of marketing, education, human resource development, and investment. In this way, in addition to public sector, the private sector will also have become more involved into the tourism industry.

Acknowledgment The research was supported by Zaman University in Cambodia, Research Office via Exploratory Research Grant Scheme of Zaman University, Grant No: ERGS/5/2017/ZURO/09/17

References

ADB (Asian Development Bank). (2017). *Tourism sector assessment, strategy, and road map for Cambodia, Lao People's Democratic Republic, Myanmar, and Viet Nam (2016–2018).* Philippines: Economics and Research Department, Asian Development Bank.

ASEAN. (2002). *ASEAN tourism agreement.* Jakarta: The Association of Southeast Asian Nations.

ASEAN. (2015a). *ASEAN economic community blueprint 2025.* Jakarta: The Association of Southeast Asian Nations.

ASEAN. (2015b). *ASEAN tourism strategic plan 2016–2025.* Jakarta: The Association of Southeast Asian Nations.

Aslan, A. (2014). Tourism development and economic growth in the Mediterranean countries: Evidence from panel granger causality tests. *Current Issues in Tourism, 17*(4), 363–372.

Athanasopoulou, A. (2013). *Tourism as a driver of economic growth and development in the EU-27 and ASEAN regions.* Singapore: EU Center.

Baltagi, B. (2008). *Econometric analysis of panel data.* Chichester: Wiley.

Blake, A., Arbache, J. S., Sinclair, M. T., & Teles, V. (2008). Tourism and poverty relief. *Annals of Tourism Research, 35*(1), 107–126.

Chok, S., Macbeth, J., & Warren, C. (2007). Tourism as a tool for poverty alleviation: A critical analysis of 'pro-poor tourism' and implications for sustainability. *Current Issues in Tourism, 10*(2–3), 144–165.

Chou, M. C. (2013). Does tourism development promote economic growth in transition countries? a panel data analysis. *Economic Modelling, 33*, 226–232.

Ekanayake, E. M., & Long, A. E. (2012). Tourism development and economic growth in developing countries. *The International Journal of Business and Finance Research., 6*(1), 51–63.

Fayissa, B., Nsiah, C., & Tadasse, B. (2008). Impact of tourism on economic growth and development in Africa. *Tourism Economics, 14*(4), 807–818.

Godfrey, M., Sovannarith, S., Saravy, T., Dorina, P., Katz, C., Acharya, S., Chanto, S. D., & Thoraxy, H. (2001, August). *A study of the Cambodian labour market: Reference to poverty reduction, growth and adjustment to crisis* (Working paper 18). Phnom Penh: Cambodia development resource institute.

Harrison, D. (2015). Development theory and tourism in developing countries: What has theory ever done for us? *International Journal of Asia-Pacific Studies:IJAPS, 11*(1), 53–82.

Hausman, J. A. (1978). Specification tests in econometrics. *Econometrica: Journal of the Econometric Society, 1*, 1251–1271.

Herman, G. V., Deac, A. L., Ciobotaru, A. M., Andronache, I. C., Loghin, V., & Ilie, A. M.. (2017). The role of tourism in local economy development. Bihor County Case Study. Urbanism. Arhitectura. Constructii. *8*(3):265.

Holik, A. (2016). Relationship of economic growth with tourism sector. *JEJAK: Jurnal Ekonomi dan Kebijakan, 9*(1), 16–33. https://doi.org/10.15294/jejak.v9i1.7184.

Lean, H. H., Chong, S. H., & Chee-Wooi Hooy, C.-W. (2014). Tourism and economic growth: Comparing Malaysia and Singapore. *International Journal of Economics and Management, 8*(1), 139–157.

Lickorish, L. J., & Jenkins, C. L. (2007). *Introduction to tourism*. London: Routledge.

Mazumder, M. N., Ahmed, E. M., & Al-Amin, A. Q. (2009). Does tourism contribute significantly to the Malaysian economy? Multiplier analysis using IO technique. *International Journal of Business and Management., 4*(7), 146–159.

Oh, C. O. (2005). The contribution of tourism development to economic growth in the Korean economy. *Tourism Management, 26*(1), 39–44.

Othman, R., & Salleh, N. H. M. (2008). Relationship between tourism industry development and economic growth in major ASEAN countries. *International Journal of Management Studies, 17*, 171–188.

Pleumarom, A. (2012). *The politics of tourism, poverty reduction and sustainable development*. Penang: Third World Network (TWN).

Scheyvens, R. (2015). Tourism and poverty reduction. In R. Sharpley & D. J. Telfer (Eds.), *Tourism and development: Concepts and issues* (pp. 118–139). Bristol/Toronto: Channel View Publications.

Stock, J. H., & Watson, M. W. (2007). *Introduction to econometrics*. Boston: Pearson Addison Wesley.

Thirumaran, K., & Arumynathan, P. (2016). Destination ASEAN, beyond 2015. In P. Mandal & J. Vong (Eds.), *Development of tourism and the hospitality industry in Southeast Asia* (pp. 1–12). Singapore: Springer. https://doi.org/10.1007/978-981-287-606-5.

Tiwari, A. (2011). Tourism, exports and FDI as a means of growth: Evidence from four Asian countries. *The Romanian Economic Journal., 14*(40), 131–151.

UNCTAD (United Nations Conference on Trade and Development). (2010). *Contribution of tourism to trade and development*. Geneva: United Nations http://www.unctad.org/en/docs/cid8_en.pdf. Accessed 22 Sept 2017.

UNWTO (World Tourism Organization). (2002). *Tourism and poverty alleviation*. Madrid: World Tourism Organization.

UNWTO (World Tourism Organization). (2010). *ASEAN integration and its impact on tourism*. Madrid: World Tourism Organization.

UNWTO (World Tourism Organization). (2017a). *UNWTO tourism highlights*. Madrid: World Tourism Organization.

UNWTO (World Tourism Organization). (2017b, June 21–23). *Measuring sustainable tourism: A call for action report of the 6th international conference on tourism statistics*. Manila, Philippines.

Wong, E. P., Mistilis, N., & Dwyer, L. (2011 Apr 30). A framework for analyzing intergovernmental collaboration–the case of ASEAN tourism. *Tourism Management, 32*(2), 367–376.

WTTC (World Travel and Tourism Council). (2016 October). *Travel and tourism investment in ASEAN*. London: WTTC.

WTTC (World Travel and Tourism Council). (2017a). *The economic impact of travel and tourism – world*. London: WTTC.

WTTC (World Travel and Tourism Council). (2017b). *Travel and tourism economic impact 2017 – South East Asia*. London: WTTC.

Yonn, R. (2017). The effect of Cambodia economy on ASEAN economic moving forward. *Journal of Management, Economics, and Industrial Organization, 1*(2), 1–16.

Yousuf, A., & Lal, R. (2013). The impact of ASEAN on tourism. *International Journal of Networking and Parallel Computing, 2*(3), 38–44.

Chapter 4
Hospitality Competitiveness Index for Indian States and Union Territories Using Multi-Criteria TOPSIS Model

Kshitiz Sharma, Mihir Dash, Madhumita Guha Majumder, and Moutushi Ganguli Sharma

Abstract The tourism sector has become an important contributor for the economic rise of India in the global scenario, with both domestic and foreign tourists wanting to engage in tourism activity more than ever. Multiple destinations in India offer them bouquet of tourism and simultaneously confuse them too in terms of suitable destination selection to visit. The present study compares Indian states and union territories in terms of hospitality competitiveness index. The study considers Hemmington's framework on hospitality that includes five dimensions: host-guest relationship, generosity, performance, the small surprises, and the buying experience. The multi-criteria decision method Technique for Order of Preference by Similarity to Ideal Solution (TOPSIS) is used for the empirical analysis. Data for the study pertained to the year 2014 and was collected from various available reliable sources.

The proposed methodology will help the Ministry of Tourism, India, to give a ranking measurement of Indian states and union territories on hospitality competitiveness every year. The results can facilitate easy selection of a safer, better, and hospitable destination for tourists.

Keywords Hospitality · Indian tourism · Competitiveness index · TOPSIS

K. Sharma (✉)
Marketing, ABBS School of Management, Acharya Bangalore Business School, Bengaluru, Karnataka, India

M. Dash
General Management, Alliance University, Bengaluru, Karnataka, India
e-mail: mihir@alliance.edu.in

M. G. Majumder
Business Analytic, Welingkar Institute of Management Development and Research, Bengaluru, Karnataka, India

M. G. Sharma
Soft Skill, Hinduja Global Solutions, Bengaluru, Karnataka, India

S. Rezaei (ed.), *Quantitative Tourism Research in Asia*, Perspectives on Asian Tourism, https://doi.org/10.1007/978-981-13-2463-5_4

4.1 Introduction

Hospitality has emerged as a strong economic force for development. Asia represents a bouquet of culture, traditions, language, and practices across the region. This is magnanimous part of its diverse hospitality. India has a unique position in Asia. Every Indian state and union territory offers its own color for tourists. In spite of its rich attractions, every state and union territory has different bearings on customers' mind. Some states and union territories are better placed than others in terms of attractions, safety, value for money, ease of travelling, and so on. Through hospitality competitiveness, the objective here is to apply an analytical approach to compare all Indian states and union territories on hospitality grounds and then to identify gaps in hospitality industry and distances among these states and union territories on various attributions that lead to hospitality competitiveness. This effort certainly will help governments and leadership to take right steps toward the industry development. The rest of the chapter is divided into five sections: Sect. 4.2 provides an overview of hospitality industry; Sect. 4.3 explains the fundamental building blocks of performance index; Sect. 4.4 describes the methodology of analysis; Sect. 4.5 brings out the results and discussion; and finally Sect. 4.6 draws the conclusion of the study.

4.2 An Overview of Hospitality Industry

The hospitality industry is one of the most important, robust industries of the world. According to Oxford English Dictionary, hospitality is defined as the "friendly and generous reception of guests or strangers." It is difficult to make a clear distinction between the fields of hospitality, tourism, and leisure. Activities in these areas are merged with each other so they overlap. Tourism, which is a significant component of the hospitality industry, has become the largest industry in the world; and projections of long-term growth and development worldwide for tourism are always higher than those of other industries (Slattery 2002).

According to 2015–2016 report of the Ministry of Tourism, Govt. of India, "Tourism is accepted as the potent engine for inclusive social economic progress at universal level through its forward and backward linkages and ability to create employment in the economy." Tourism has direct role in employment generation and poverty eradication sustainable for all segments of society. The Government of India seems oriented in this direction. The gradual increases in foreign tourist arrivals (FTAs) and tourism revenue are indicators of Tourism India being in forefront. In 2015, FTAs increased 4.5% in India, and the corresponding foreign exchange earnings (FEEs) from tourism in rupees witnessed a growth of 9.6%. Government initiative like Tourist Visa on Arrival with Electronic Travel Authorization (e-Tourist Visa), 24x7 multilingual tourist help, is now witnessing the result; in 2015–2016, 445,300 tourists used e-Tourist Visa facility (Annual Report 2015–17), Ministry of

Tourism. The facility is available at 16 airports for 150 nationalities. Two tourism schemes are also launched, Swadesh Darshan and Pilgrimage Rejuvenation and Spiritual Augmentation Drive (PRASAD). Tourism India is a 365-day tourist destination.

Tourism in India is reconstructing itself with the help of digital technology. The Ministry of Tourism has set up a Web-based public delivery system with the objective of receiving application and giving approval online. More so, grant approvals for hotel projects, classification/reclassification, etc. have also gone online with integrated payment gateway. A mobile application called Swachh Paryatan ("Clean Tourism") is also launched to assist people in reporting unattended garbage piles at various tourist destinations across the country.

The World Economic Forum has considered the following dimensions while creating the Travel and Tourism Competitiveness Index (The Travel and Tourism Competitiveness Report 2015).

Enabling Environment, which includes the business environment, safety and security, health and hygiene, human resources and labor market, and information and communications technology readiness

Travel & Tourism Policy and Enabling Conditions, which includes aspects such as prioritization of travel and tourism, international openness, price competitiveness, and environmental sustainability

Infrastructure, which includes air transport infrastructure, ground and port infrastructure, and tourist service infrastructure

Natural and Cultural Resources that includes World Heritage natural sites, known species, protected areas, natural tourism demand, and quality of natural environment termed as natural resources, and World Heritage cultural sites, oral and intangible cultural expression, sports stadium, international association meetings, and cultural and entertainment tourism digital demand termed as cultural resources

In recent times, some of the Asian nations including Japan, Korea, India, Vietnam, and Bhutan have gained considerable attentions in the global tourism scenario. According to the Travel and Tourism Competitiveness Index 2017, Japan ranked 4th, Korea 19th, India 40th, and Vietnam 67th, while Bhutan ranked 78th. It is worth noting that India improved its position in the index by 12 ranks since 2015 (The Travel & Tourism Competitiveness Report 2017). Further, it reveals that Southeast Asian and South Asian nations are more price-competitive destinations.

The Travel and Tourism Competitiveness Index is a robust index and covers the travel and tourism industry very well; however, hospitality industry requires a fresh look for a new index, as it functions in a completely different way. The hospitality industry operates and works 365 days a year and 24 h a day. Employees work in different shifts, but generally are working more than in some other industries. Weekends, holidays, and festivals are the periods when workers in this sector are very busy (Zagade et al. 2011), while it is a lean working period for other industries.

The hospitality industry can be divided into two categories: *entertainment* and *accommodation*. Accommodation is in the form of tourist resorts, hotels, guest-houses, inns, motels, and campgrounds. Entertainment category includes clubs, fast food, and other restaurants. The industry includes commercial support activities such as travel agencies and the airline staff. Workforce in hospitality and tourism, along with technology, are other key components of the industry.

The Higher Education Funding Council for England's Hospitality Review Panel described hospitality as having a core that covers food, beverages, and/or accommodation in a service context (In Search of Hospitality 2001). The hospitality industry is dedicated to catering and accommodation, from hotels and restaurants to leisure centers; tourism is oriented to hotels, restaurants, travel agencies, and tour operators. In short, tourism means moving people from one place to another; hospitality means providing temporary accommodation, food, and entertainment for tourists (Rusu et al. 2014).

4.3 Fundamentals of Hospitality Index

The present study considers Hemmington's (2007) framework of hospitality. It is based on five building blocks of hospitality industry: *host-guest relationship*, *generosity*, *performance*, *the small surprises, and buying experience*. The essence of the thought is that the customers do not buy delivery of services, they actually acquire *experiences*; they do not buy the quality of services, they actually retain the *memories*; and customers in this industry do not procure food and drink, they actually look for *flavor*.

1. The *host-guest relationship* is not similar to that of managers and customers. Host-guest relationship is more of a cultural and social phenomenon. Lashley and Morrison (2000) state that "hospitality is essentially a relationship based on hosts and guests." The paper use the following metrics to measure the subjective relationship between host and guest in India based upon the criteria identified in State Ranking Survey 2013:

    (i) *Literacy Rate*: It has indirect bearing on the hospitable position of any state or union territory. It is assumed that educated societies are more open for people from different cultures and regions. Literacy rate has been assigned least weightage in all the parameters.

    (ii) *Tourist Security*: Security is a paramount interest for any visitor for the selection of the place to visit. Incidence of crimes against foreigners and total cognizable crimes reflect the safety position of the state or union territory. As tourist security is a vital issue, we have assigned weight as 20.

    (iii) *Political Stability*: India has a unique combination of governance. At the center, there is one party ruling, while states are ruled by many other parties. This kind of political situations in states and at the center is sometimes unstable, as every political party tries to pull down others, and when there is a shift in ruling party, it results in reversal or cancellation of decisions taken by the previous party. The prospective investors are highly skeptical about the unstable political scenario in India. Therefore, ease of doing business is an important index to understand the political stability of a specific market place. This definitely affects the movement of people.

(iv) *Human Resource Index*: Availability of qualified human resource improves the standards of service. A qualified human resource can better deliver services to the customers. The human resource index improves host-guest relationship.

Generosity and economic relationship in commercial hospitality are two opposites that create tension among hospitality suppliers and buyers. States make economic progress by charging taxes on hospitality engagement, whereas the states' expenditure on tourism activities (annual tourism budget) improves to tourism infrastructure, enhancing guests' experiences from the hospitality industry. The variables considered for this dimension include the tourism budget and the luxury tax on hotels.

Performance is the third key indicator, which includes foreign tourist arrivals (FTAs), domestic tourist arrivals (DTAs), tourism revenue, average length of stay, and visiting months.

The small surprises is another important aspect of hospitality. Hospitality is an experience that the guests undergo, irrespective of the duration of services; it is the quality of attraction and quality of service that really surprise the guests, and for these surprises, they travel far and wide. The parameter considered for the small surprises is the effectiveness of marketing campaign.

The buying experience in hospitality is derived from physical products (i.e., artifacts), service environment (urbanization, mobile and Internet connectivity), service as a product (bars and restaurants, hotels, and tourist attractions), and service delivery infrastructure (roads, railways, ports, and airports).

Various reports, both from government and nongovernmental sources, have identified the best and worst tourist state/union territory as per these criteria. The present study has collated all the reports, and the summary report is presented below in Table 4.1.

Further information is available in Appendix 1.

4.4 Methodology

The study is a descriptive study of the hospitality industry, using 19 criteria listed under Hemmington's 5 dimensions of hospitality. Table 4.2 describes the dimensions and the criteria along with their respective weights assigned under present study.

Table 4.1 Best (B) and worst (W) states/union territories on single-criterion basis[a]

States	1	2	3	4	5	6	7	8	9	10	11	12	13	14	15	16	17	18	19
Andaman and Nicobar Islands						**B**				**B**		**B**			**W**			**B**	
Andhra Pradesh				**B**															
Arunachal Pradesh			**W**			**B**													
Assam																			
Bihar	**W**			**W**					**W**		**W**								
Chandigarh											**W**		**B**				**B**		
Chhattisgarh													**W**			**W**		**W**	**B**
Delhi		**W**				**W**								**B**	**B**	**B**			
Goa																			
Gujarat			**B**																
Haryana																			
Himachal Pradesh											**B**			**W**					
Jammu and Kashmir						**B**													
Jharkhand																			
Karnataka																			
Kerala	**B**																		
Madhya Pradesh										**W**									
Maharashtra																			
Manipur																	**W**		
Meghalaya						**W**													
Mizoram						**B**	**W**												
Nagaland		**B**				**B**		**W**											
Odisha						**B**													
Puducherry						**B**													
Punjab					**W**				**B**										
Rajasthan																			
Sikkim					**B**	**B**													**W**
Tamil Nadu							**B**	**B**											
Tripura																			
Uttar Pradesh																			
Uttarakhand											**B**								
West Bengal												**W**							

[a]Criteria

No	Criteria	No	Criteria	No	Criteria
1	Literacy rate (Literacy Rate in India 2016)	7	Foreign tourist arrivals (Foreign Tourist Arrival in India [Internet] 2017)	13	Listed bars and restaurants (Read Reviews 2015)

(continued)

Table 4.1 (continued)

No	Criteria	No	Criteria	No	Criteria
2	Tourist security (Crime in India 2015; Incidence of Crime/ crime committed by foreigners 2016)	8	Domestic tourist arrivals (State-wise comparison of characteristics of domestic trips in India 2009–2010[Internet] 2016)	14	Urbanization (CENSUS 2011)
3	Political stability (Assessment of State 2016)	9	Tourism revenue (GSDP 2013)	15	Road and railway infrastructure (Annual Report 2014–15 [Internet] 2016)
4	Human resource index (2011 Census Data [Internet] 2016)	10	Average length of stay [compiled]	16	Mobile and Internet connectivity (State wise Mobile Phone users in India 2012)
5	State tourism budget (State Wise Budget [Internet] 2016)	11	Visiting months [compiled]	17	Branded hotels (Read Reviews 2015)
6	Luxury tax on hotels (Thandani and Debjani 2015)	12	Marketing campaign (Keyword Research n.d.)	18	Tourism attractions in state (Statistics [Internet] 2016)
				19	Airport presence (Traffic News-Annexure IIIC [Internet] 2013)

Table 4.3 provides the list of Indian states and union territories, which have been included in the present study. However, as data for Telangana and Andhra Pradesh is not available separately, we have considered them as a single state.

The study has adopted the census method, as all the Indian states and union territories are included. The data for the study pertained to the year 2014 and was collected from various available reliable sources such as government publications and websites.

4.4.1 TOPSIS Modeling

The study uses the multi-criteria decision-making model, TOPSIS (Technique for Order Preference by Similarity to Ideal Solution), to combine the performance measures into a single composite score. It is based on relative distance to the ideal solution, the point with the best values for all criteria, and the anti-ideal solution, the point with the worst values for all criteria (Yoon and Hwang 1981; Hwang et al. 1993).

Table 4.2 Dimensions and criteria used for the study

Dimensions	Criteria no.	Criteria	Impact	Assigned weight for criteria	Dimension weight
Host-guest relationship	1	Literacy rate	Low	5	45
	2	Tourist security	High	20	
	3	Political stability/ease of doing business	Low	10	
	4	Human resource index	Low	10	
Generosity	1	Tourism budget	High	25	40
	2	Luxury tax on hotels	High	15	
Performance	1	Foreign tourist arrivals	High	20	95
	2	Domestic tourist arrivals	High	20	
	3	Tourism revenue	High	20	
	4	Average length of stay	High	20	
	5	Visiting months	Medium	15	
Small surprises	1	Effectiveness of marketing campaign	Low	10	10
Buying experience	1	Listed bars and restaurants	Low	15	90
	2	Urbanization	Low	10	
	3	Road and railway infrastructure	Low	10	
	4	Mobile and Internet connectivity	Low	10	
	5	Branded hotels	Low	15	
	6	Tourism attractions in state	High	20	
	7	Airport presence	Low	10	
Total	19			280	280

Table 4.3 States and union territories: India

1	Andhra Pradesh	12	Karnataka	23	Sikkim
2	Arunachal Pradesh	13	Kerala	24	Tamil Nadu
3	Assam	14	Madhya Pradesh	25	Tripura
4	Bihar	15	Maharashtra	26	Uttarakhand
5	Chhattisgarh	16	Manipur	27	Uttar Pradesh
6	Goa	17	Meghalaya	28	West Bengal
7	Gujarat	18	Mizoram	29	*Andaman and Nicobar Islands*
8	Haryana	19	Nagaland	30	*Chandigarh*
9	Himachal Pradesh	20	Odisha	31	*Delhi*
10	Jammu and Kashmir	21	Punjab	32	*Puducherry*
11	Jharkhand	22	Rajasthan		

The TOPSIS scores are calculated as follows. Let x_{ij} denote the value of the jth parameter for the ith bank and w_j denote the weight for the jth parameter. These values are first normalized so that their sum of squares is united by taking

$$y_{ij} = \frac{x_{ij}}{\sqrt{\sum_{i=1}^{m} x_{ij}}}$$

The ideal point $y^*$is defined as the point with the best values for each criterion, and the anti-ideal point y_* is defined as the point with the worst values for each of the criteria. The distance from each point to the ideal point and anti-ideal point is calculated, respectively, as follows:

$$d_i^* = \sqrt{\sum_{n}^{j=1} w_j \left(y_{ij} - y_j^*\right)^2} \text{ and } d_{i*} = \sqrt{\sum_{n}^{j=1} w_j \left(y_{ij} - y_{j*}\right)^2}$$

The TOPSIS score is then calculated as

$$S_i = \frac{d_{i*}}{d_{i*} + d_i^*}$$

The TOPSIS score S_i indicates how close each state/union territory is to the ideal and anti-ideal points. The closer the TOPSIS score is to zero, the closer the state/union territory is to the anti-ideal point, while the closer the TOPSIS score is to one, the closer the state/union territory is to the ideal point.

4.5 Results and Discussion

The present study has adopted the TOPSIS method in order to identify the ranks of individual states considering all the 19 criteria stated in Table 4.2. The TOPSIS scores and ranking index for individual states and union territories are presented in Table 4.4. The results reflect that Chandigarh is the only ideal state and the rest of the states/union territories are close to anti-ideal situation while considering all the 19 criteria to evaluate the performance score. Further, we observe that even though the TOPSIS scores of Puducherry (0.0834), Delhi (0.0319), West Bengal (0.0026), and Bihar (0.0021) are very close to anti-ideal scenario, relatively these states along with Chandigarh are performing far better than the rest of the states. The TOPSIS scores indicate the hospitality index for various states and union territories. Ranking of these units, based on the hospitality index, clearly identifies Chandigarh, Puducherry, Delhi, West Bengal, and Bihar as the top five performers against the bottom five states including Sikkim, Himachal Pradesh, Chhattisgarh, Andhra Pradesh, and Jammu and Kashmir.

Table 4.4 TOPSIS scores for the states and union territories

State/union territory	TOPSIS score	Rank
Chandigarh	1.00000	1
Puducherry	0.08337	2
Delhi	0.03190	3
West Bengal	0.00263	4
Bihar	0.00207	5
Tripura	0.00206	6
Punjab	0.00180	7
Assam	0.00167	8
Goa	0.00101	9
Andaman and Nicobar Islands	0.00084	10
Odisha	0.00047	11
Haryana	0.00044	12
Uttarakhand	0.00030	13
Madhya Pradesh	0.00027	14
Uttar Pradesh	0.00027	15
Tamil Nadu	0.00026	16
Gujarat	0.00025	17
Manipur	0.00024	18
Meghalaya	0.00024	19
Kerala	0.00022	20
Nagaland	0.00019	21
Jharkhand	0.00017	22
Rajasthan	0.00016	23
Arunachal Pradesh	0.00015	24
Maharashtra	0.00015	25
Karnataka	0.00014	26
Mizoram	0.00012	27
Jammu and Kashmir	0.00010	28
Andhra Pradesh	0.00008	29
Chhattisgarh	0.00005	30
Himachal Pradesh	0.00004	31
Sikkim	0.00000	32

4.6 Conclusion

Hospitality index is a tool to demonstrate the strength and/or weakness of the tourist destination from hospitality perspectives. As per the assigned weights of all the criteria, the importance of various dimensions may be prioritized as follows. Performance of the destination is the most important dimension (95 points), followed by buying experience (90 points). The third most important dimension is host-guest relationship (45 points) followed by generosity (40 points); and the last is small surprises (10 points). Each dimension has some criteria best identified with. Hospitality index provides overall picture of the destination.

The best state/union territory on hospitality index is Chandigarh. Chandigarh is the first planned city of India and maintains union territory status. Some strengths of Chandigarh in this index are literacy rate, ease of doing business, workforce participation, luxury tax, length of stay of tourists, restaurants and bars, urbanization, railway network, Internet connection, branded hotels, and airport connectivity. The least competitive state on this scale is Sikkim. Sikkim is located in Eastern Himalayas and one of the most beautiful states of the Indian Union. Some of the major weaknesses of Sikkim are low domestic tourist arrivals, ineffective marketing campaign, low presence of restaurants and bars, weak railway network, and absence of airport.

The proposed methodology can be helpful for the Ministry of Tourism for hospitality development in India. The scale can indicate the hospitality development for each state and union territory and the competitive position in the hospitality industry over a period of time. The results can facilitate easy selection of a safer, better, and hospitable destination for tourists. The results can also help states and union territories to develop and upgrade their facilities for effortless, unhindered tourism, so that they will be in a better position to compete with other states and union territories.

Appendix 1 Position of States/UT on Individual Criteria

- Literacy Rate: The best state is Kerala with 93.90% literacy rate. The worst performer is Bihar with 63.80% literacy rate (Literacy Rate in India 2016).
- Tourist Security: Nagaland is found the safest destination among all with the lowest crime incidence per capita. Delhi is again proven as crime capital with top spot in crime incidence per capita (Crime in India 2015; Incidence of Crime/crime committed by foreigners 2016).
- Political Stability/Ease of Doing Business: The top spot here is attained by Gujarat with a score of 71.06 as per 2015 KPMG report. The bottom score is 1.23 achieved by Arunachal Pradesh (Assessment of State 2016).
- Human Resource Index: Workforce participation per capita is best in Andhra Pradesh with a score of 0.39, whereas the score is poorest in Bihar with score of 0.2058. Clearly, more people are participating in the production in Andhra Pradesh than any other state per capita (2011 Census Data [Internet] 2016).
- State Tourism Budget: In terms of budget fraction, expenditures on tourism are best in Sikkim with 2.84% of the total budget. The least fraction is spent by Punjab which is 0.004% only (State Wise Budget [Internet] 2016).
- Luxury Tax on Hotels: In this parameter many states are occupying the same place, Andaman and Nicobar, Arunachal Pradesh, Jammu and Kashmir, Manipur, Mizoram, Nagaland, Odisha, Puducherry, and Sikkim with 0% luxury tax. The highest tax is in Delhi and Meghalaya with 20% rate (Thandani and Debjani 2015).

- Foreign Tourist Arrivals: Foreign tourist arrival is a key indicator of hospitality health in the state if international customer is concerned. FTA was reported maximum in Tamil Nadu and minimum in Mizoram in 2014 (Foreign Tourist Arrival in India [Internet] 2017).
- Domestic Tourist Arrivals: The scenario for domestic tourist arrival is also similar where Tamil Nadu tops the chart again. The minimum domestic tourist movement was reported in Nagaland in 2014 (State-wise comparison of characteristics of domestic trips in India 2009-2010[Internet] 2016).
- Tourism Revenue: Minimum earning from tourism was reported in Bihar which is reported only INR 100,000 for 2014. Maximum revenue was earned by Punjab which is 1.1bn rupees (GSDP 2013).
- Average Length of Stay: Andaman and Nicobar Island is a picturesque place, unadulterated environment which attracts nature lovers and tourists. A tinge of adventure makes it more interesting. Andaman and Nicobar Island tops the chart with 22 days average length of stay. The lowest number is 4.58 days in Madhya Pradesh [compiled by researcher].
- Visiting Months: Himachal Pradesh and Uttarakhand are open for 12 months, which makes them available to the tourists more than any other spot in India. Bihar and Chandigarh are suitable for visits for one-third of the year and hence occupy the bottom place [compiled by researcher].
- Effectiveness of Marketing Campaign: Andaman and Nicobar Tourism campaigns were ranked as the best marketing campaign, while West Bengal Tourism campaign was ranked as the worst (Keyword Research n.d.).
- Listed Bars and Restaurants: Chandigarh has the highest density of restaurant and bars that help improve ranking of Chandigarh on hospitality index. It has 7912 restaurants and bars per 1000 sq. km of Chandigarh. The lowest density is in Chhattisgarh (Read Reviews 2015).
- Urbanization: It is best in Delhi. With 97.30 score, Delhi is the most urbanized union territory. Himachal Pradesh is the least urbanized state with a score of 10. Delhi being India's capital receives major urban facilities and upliftment (CENSUS 2011).
- Road and Railway Infrastructure: Road and railway infrastructure is best in Delhi and poorest in Andaman and Nicobar. Andaman and Nicobar is an island and is connected through water route only (Annual Report 2014–15 [Internet] 2016).
- Mobile and Internet Connectivity: This connectivity also is best in Delhi and least in Chhattisgarh state. Delhi score is 85.80 and Chhattisgarh score is 28.40 (State wise Mobile Phone users in India 2012).
- Branded Hotels: Branded hotel concentration as per TripAdvisor India is maximum in Chandigarh, viz., 1667 hotels per 1000 sq. km. The least concentration is in Manipur with a score of 0.36 hotels sq. km (Read Reviews 2015).
- Tourism Attractions in State: Tourism attraction concentration is maximum in Andaman and Nicobar and minimum in Chandigarh. Score of Andaman and Nicobar is 134 and Chandigarh score is 12 only (Statistics [Internet] 2016).
- Airport Presence: Concentration of airport per 1000 sq. km is maximum in Chandigarh (8.77) and minimum in Sikkim (0.00) (Traffic News- Annexure IIIC [Internet] 2013).

References

Al Mamun, A., & Mitra, S. (2012, September). A methodology for assessing tourism potential: Case study Murshidabad District, West Bengal, India. *International Journal of Scientific and Research Publications., 2*(9), 1–8.

Annual Report 2015–16. (n.d.). Retrieved September 11, 2017, from http://tourism.gov.in/annual-report-2015-16

Annual Report 2015–17. Retrieved September 20, 2017, from Annual Report 2015–16, Ministry of Tourism.

Bhattacharya, M., Smyth, R., & Vicziany, M. (2004). *South Asia in the era of globalization: Trade, industrialization and welfare*. New York: Nova Publishers.

Brotherton, B., & Wood, R. C. (2001). *Hospitality and hospitality management. In search of hospitality* (pp. 134–156). Oxford: Butterworth-Heinemann.

Grant, D. A. (2016). *Managers', employees', and students' perceptions of ethnic minority career advancement in the hospitality industry*. Arizona: Northcentral University.

Hemmington, N. (2007, September 1). From service to experience: Understanding and defining the hospitality business. *The Service Industries Journal., 27*(6), 747–755.

Huang, J. J. (2011). *Multiple attribute decision making: Methods and applications*. Boca Raton: Chapman and Hall/CRC.

Hwang, C. L., Lai, Y., & Liu, T. Y. (1993). A new approach for multiple objective decision making. *Computers and Operational Research., 20*, 889–899.

Mihaela, S. C. (2014 Jan 1). Perception of managers on performance-empirical study conducted in the hospitality industry entities in Romania. *Annals of Faculty of Economics., 1*(2), 165–174.

Rusu, S., Isac, F., & Cureteanu, R. (2014, May 31). The dimensions of hospitality industry. *Lucrări Ştiinţifice Management Agricol, 16*(4), 28.

Slattery, P. (2002 Jan). Finding the hospitality industry. *Journal of Hospitality, Leisure, Sport and Tourism Education., 1*(1), 19–28.

Thandani, M., & Debjani, D. (2013). State ranking survey [Internet]. HVS | Home. [cited 2015 Dec 10]. Retrieved from http://www.hvs.com/Accessed 10 December 2015.

World Economic Forum. (n.d.). *The travel and tourism competitiveness report 2015*. Retrieved June 11, 2017, from The Travel and Tourism Competitiveness Report 2015.

Yoon, K., & Hwang, C. L. (1981). *Multiple attribute decision making: Methods and applications*. Berlin: Springer.

Zagade, S., Jarad, R., Jadhawar, J., Gawade, S., & Sinha, D. *Challenges for the indian hospitality industry*. InEDULEARN11 Proceedings 2011 (pp. 6524–6528). IATED.

Data Sources

Census Data [Internet]. Census of India Website: Office of the Registrar General & Census Commissioner, India. [cited 2016 Apr 11]. Available from: http://www.censusindia.gov.in/2011-Common/CensusData2011.html

2011 Census Data [Internet]. Census of India Website: Office of the Registrar General & Census Commissioner, India. [cited 2016 Apr 11]. Available from: http://www.censusindia.gov.in/201census/population_enumeration.htmll

Annual Report 2014–15 [Internet]. Ministry of Road Transport and Highways. [cited 2016 Nov 11]. Available from: http://www.bing.com/cr?IG=8354C009BDFA4BED9F-6D7B0718D28444&CID=283FEFE5D00E615F04AAE4D4D1086081&rd=1&h=-dAcN_3dYbDgMqCMc0a-NW_2poHfnxwA-thi0ljmO14&v=1&r=http%3a%2f%2fmorth.nic.in%2f&p=DevEx,5063.1

Assessment of State Implementation of Business Reforms [Internet]. Press Information Bureau. 2015 [cited 2016 Aug 2]. Available from: http://www.pib.nic.in/newsite/mainpage.aspx

Author. RBI – State Finances: A Study of Budgets 2013–14 [Internet]. Reserve Bank of India. [cited 2016 Oct 12]. Available from: https://www.rbi.org.in/

CENSUS 2011 (PCA – Final Data). (2014, December 22). Retrieved from http://planningcommission.nic.in/data/datatable/data_2312/DatabookDec2014 307.pdf

Competitiveness of Tourism Sector in India with Selected Countries of World [Internet]. Home | Ministry of Tourism. [cited 2017 Nov 5]. Available from: http://tourism.gov.in/

Crime in India 2015 [Internet]. National Crime Records Bureau. [cited 2016 Sept 20]. Available from: http://ncrb.nic.in/

Dadar and Nagar Haveli Census of India Website: Office of the Registrar General & Census Commissioner, India. [cited 2016 Apr 11]. Available from: http://www.census2011.co.in

Foreign Tourist Arrival in India [Internet]. Bureau Of Immigration. [cited 2017Nov5]. Available from: http://www.boi.gov.in/

GSDP of 2011–12 at Current Prices [Internet]. Directorate of Economics And Statistics, Ministry Of Agriculture, Government Of India. 2013 [cited 2017 Nov 5]. Available from: http://eands.dacnet.nic.in/

Incidence of Crime/crime committed by foreigners [Internet]. India Crimestat: Socio-Economic Statistical Information about Crime in India. [cited 2016 Sept 11]. Available from: http://www.indiacrimestat.com/

India Brochure [Internet]. Incredible India. [cited 2016 Aug 3]. Available from: http://incredibleindia.org/

India Tourism Statistics 2014 at a Glance [Internet]. Home | Ministry of Tourism. [cited 2017 Nov 5]. Available from: http://tourism.gov.in/

ixigo – Flight Booking, Train Reservation, Hotels, Cheap Air Tickets [Internet]. ixigo – Flight Booking, Train Reservation, Hotels, Cheap Air Tickets. [cited 2017 Apr 4]. Available from: https://www.ixigo.com/

Kerala tourism clocks record 24k cr revenue [Internet]. The Hindu Business Line. [cited 2016 Dec 14]. Available from: http://www.thehindubusinessline.com/news/

Keyword Research, Competitor Analysis, & Website Ranking | Alexa. (n.d.). Retrieved from http://Alexa.com/

Literacy Rate in India [Internet]. Indiaonlinepages.com. [cited 2016 Aug 6]. Available from: http://www.indiaonlinepages.com/population

Punjab emerged top tourism destination across country [Internet]. CITY AIR NEWS. 2014 [cited 2015 Sept 6]. Available from: https://cityairnews.wordpress.com/

Read Reviews, Compare Prices & Book [Internet]. TripAdvisor. [cited 2015 July 9]. Available from: https://www.tripadvisor.in/

State Wise Budget [Internet]. Empowered Committee of State Finance Ministers. [cited 2016 Sept 12]. Available from: http://empcom.gov.in/

State wise Mobile Phone users in India Census 2011. (2012, July 21). Retrieved from http://updateox.com/india/state-wise-mobile-phone-users-in-india-census-2011/

State Wise Report 2015 [Internet]. Home | Ministry of Tourism. [cited 2016, July 7]. Available from: http://tourism.gov.in/

State-wise comparison of characteristics of domestic trips in India 2009–2010 [Internet]. Home | Ministry of Tourism. [cited 2016 July 10]. Available from: http://tourism.gov.in/

Statistics [Internet]. Department of Tourism, Govt of Goa. [cited 2016, June 10]. Available from: http://goatourism.gov.in/

Thandani M, Debjani D. 2013 India, State Ranking Survey [Internet]. HVS | Home. [cited 2015 Dec 10]. Available from: http://www.hvs.com/

Traffic News- Annexure IIIC [Internet]. Airports Authority of India. 2013 [cited 2017 Nov 5]. Available from: http://www.airportsindia.org.in/

Dr. Kshitiz Sharma is a marketing professor at Acharya Bangalore Business School. His past association has been with Alliance School of Business (Alliance University), UEI Global (Berggruen Education Pvt. Ltd.), IIAS School of Hotel Management and Tourism, Radisson (Carlson Hospitality Worldwide), etc. Dr. Sharma is an author of *Introduction to Tourism Management* published by McGraw-Hill Education, ISBN-13: 978-1-25-902680-5, and many book chapters. He has published papers including research papers, articles, and case studies in national and international conferences, journals, and magazines. His research areas of interests are marketing management, consumer behavior, services marketing, marketing research, marketing strategy, and hospitality marketing.

Prof. Mihir Dash is a specialist in mathematical and statistical modeling. Prof. Dash is actively involved in teaching a variety of courses, including Business Mathematics and Business Statistics, Operations Research, Econometrics and Mathematical Economics, and Operations Management, and in research activities and projects. Prof. Dash's research interests are diverse, touching several areas of management and involving applications of quantitative methods in business decision problems.

Dr. Madhumita Guha Majumder an accomplished professional with national and international experience, is currently working as a professor at Welingkar Institute of Management Development and Research, Bangalore. She is a postgraduate in Economics and received UGC Research Fellowship to pursue her Ph.D. in the area of Development Economics. Dr. Madhumita has passion to develop econometrics and statistical models in social science research. Her research is based on diversified areas of management including knowledge management, financial management, and consumer analytics. Her current research focuses on analyzing unstructured information generated on social media and its impact on policy-making using text mining techniques.

Moutushi Ganguli Sharma has a combination of academic and corporate experience. She has been in academics for 7 years and then joined the corporate world as corporate trainer-soft skill with Hinduja Global Solutions. Simultaneously, Moutushi is enrolled as research scholar with Jain University and pursuing her Ph.D. in Management.

Chapter 5
Co-integration and Causality Analysis Between Tourism Sector and GDP in Cambodia

Osman Nuri Aras, Mustafa Öztürk, and Ali Ihtiyar

Abstract Tourism sector has emerged as one of the most important service sectors in the global economy including Cambodia especially in recent 25 years, benefiting destinations and communities worldwide.

The objective of this chapter is to investigate and analyze the contribution of tourism sector in the economic development in Cambodia.

After the cessation of conflict and the start of establishing political stability in 1993, the tourism sector has started to be a significant and very dynamic engine of economic growth and generator of foreign exchange revenues in Cambodia. For this reason, this study covers the period after the early 1990s when Cambodia opened its door to the world.

In this study, the relationship between Cambodian tourism receipts and gross domestic product (GDP) growth rate is analyzed. Johansen co-integration test was performed to identify whether there was a long-term relationship between the variables. Then vector error correction mechanism (VECM) was used to identify the existence of short-run relationship between variables. Finally, residual tests were performed to check if the model is stable and desirable. E-views 8 statistical program is used for data analysis.

According to the results of the study, Johansen co-integration test verifies that GDP, tourism receipts, and export revenues have long-term association. The VEC Granger Causality/Block Exogeneity Wald Tests show that the causality from tourism receipts to GDP is statistically significant. And also this study shows that the tourism policies and the tourism revenues are more effective on the GDP than the export policies and the revenues.

Keywords Tourism sector · Tourism revenues · Tourism policies · Sustainable tourism development · Economic development · Gross domestic product · Johansen co-integration test · Cambodia

O. N. Aras (✉) · A. Ihtiyar
Zaman University, Phnom Penh, Cambodia
e-mail: aihtiyar@zamanu.edu.kh

M. Öztürk
Burch University, Sarajevo, Bosnia and Herzegovina

S. Rezaei (ed.), *Quantitative Tourism Research in Asia*, Perspectives on Asian Tourism, https://doi.org/10.1007/978-981-13-2463-5_5

5.1 Introduction

Whether tourism sector/industry has a significant impact on economic growth is an important economic question for both policy makers and investors. Tourism influences on economic growth through flow of foreign currency into the domestic economy, which stimulates economic activities, advances gross domestic product, and generates and increases jobs, occupations, disposable income, and poverty reduction. Tourism industry has emerged as one of the leading service sectors in the global economy including Cambodia especially in recent 25 years, benefiting destinations and communities worldwide. Tourism plays an important role and contributes and influences on the whole economy in a country by direct, indirect, and induced ways.

Tourism is closely linked to sustainable development. Having recognized the importance of tourism to economic growth, most international organizations have begun to argue that tourism growth can influence, as well, the economic and sociocultural development of the society (Cárdenas-García et al. 2015). The fact of the matter is that tourism contributes to the three dimensions of sustainable development: economic, social, and environmental. Tourism has become a key factor for socioeconomic progress, sustainable economic growth and development, and poverty alleviation through the creation of jobs and enterprises, infrastructure development, and the export revenues earned. Gross domestic product (GDP) increases in a country as a consequence of the direct effects and implications of the tourism industry. On the other hand, tourism sector attracts a great amount of foreign direct investment, increases new innovation, advances the level of technology, and improves international economic and political relations.

Economic flows generated by tourism industry have become essential factors to realize economic development and to reduce poverty in many developing countries including Cambodia. At the same time, international tourism-driven economic flows have become the most important factors in international economic relations in almost every developing country.

Cambodia has a rich diversity of natural, historical, and cultural resources. Therefore, tourism sector which has had a steady growth is the third largest sector in Cambodia. Thus, over the course of the last 25 years, the rapid development of the tourism sector has become one of the main pillars of economic development and poverty reduction in Cambodia.

Tourism Development Strategic Plan 2012–2020 has started to be implemented by the government in order to make more contribution of the tourism sector in economic development and poverty reduction in Cambodia. This plan focuses on main strategic directions such as tourism product development, travel facilitation and transportation, regional and international connectivity, and tourism safety system.

This study will focus primarily on the following subtopics: definition of sustainable tourism development, theoretical approach about the economic impacts of tourism sector, and general information and data about the economic importance of tourism sector in the world.

Then, the case of Cambodia will be dealt with. In the later part of the chapter, the relationship between Cambodian tourism receipts and gross domestic product growth rate will be analyzed. According to the statistical outcomes of the study, some suggestions that should be taken into account in the government's strategy for the tourism sector will be offered.

5.2 Sustainable Tourism Development

Tourism is a major source of income for many countries, especially developing countries, as well as one of the fastest-growing service sectors in the world. Because it is a people-oriented sector, tourism also provides many jobs which have helped revitalize local economies. In other words, being a labor-intensive sector, tourism reduces poverty through its capacity to create decent jobs, especially in rural areas and among women and indigenous people. However, like other forms of development, tourism can also cause its share of problems, such as social dislocation, loss of cultural heritage, economic dependence, and ecological degradation. Learning about the impacts of tourism has led many people to seek more responsible holidays, including various forms of alternative or sustainable tourism such as "nature-based tourism," "ecotourism," and "cultural tourism" (Fien et al. 2010).

While interest in tourism as a tool for "development" started in the 1970s, the concentration on its role in alleviating poverty came out in the late 1990s, and since then, "pro-poor tourism" has quickly become a recognized, worthwhile, and more self-consciously moral approach to tourism, especially when practiced by community-based organizations under the "sustainable tourism" (Harrison 2015; Harrison and Schipani 2007).

Sustainable tourism is defined as "tourism that respects both local people and the traveler, cultural heritage and the environment" (Fien et al. 2010). A broader definition of sustainable tourism is as follows: "Tourism that takes full account of its current and future economic, social and environmental impacts, addressing the needs of visitors, the industry, the environment and host communities" (UNWTO 2005). Sustainable tourism is offered as an answer to both poverty and environmental concerns (Pleumarom 2012).

Sustainable tourism development meets the needs of the present tourists and host regions while protecting and increasing the opportunity for the future. Sustainable tourism development provides to management of all resources in such a way that economic, social, and aesthetic needs can be fulfilled while maintaining cultural integrity, essential ecological processes, biological diversity, and life-support systems (UNWTO 2002).

Sustainable tourism development requires that all relevant stakeholders share similar values and commitment to sustainable tourism. It also requires strong political leadership to build a consensus and provide wide and efficient participation of all stakeholders. Achieving sustainable tourism is a continuous process, and it requires constant monitoring of impacts. Thus, specific objectives should ensure to all parties a clear vision of what is to be accomplished and by whom. Measurable objectives should provide a clear basis for evaluating progress and guiding policy makers where to place efforts in the future (Athanasopoulou 2013).

According to the definition of sustainable tourism, all tourism activities of whatever motivation, such as holidays, business travel, conferences, adventure travel, and ecotourism, need to be sustainable. Sustainable tourism should thus make optimal use of environmental resources, respect host communities, and ensure viable, long-term economic operations so that benefits are equitably distributed among all stakeholders. Sustainable tourism drives inclusive economic growth and social development by promoting entrepreneurship, creating decent jobs, and eradicating poverty and can also encourage environmental protection, consideration of climate, cultural heritage preservation, and stronger peace and mutual understanding around the world while leaving no one behind (UNWTO 2017). Additionally, the importance of sustainable tourism as a tool for sustainable development (economic-social-environmental) is increasing more and more every day.

Like every form of development, tourism can have both positive and negative impacts. The aim of sustainable tourism is to maximize benefits such as job creation, foreign exchange revenues, and new infrastructure while safeguarding cultural heritage and living culture and minimizing negative environmental and social impacts (Fien et al. 2010).

As a conclusion, a sustainable approach to tourism means that neither the natural environment nor the sociocultural fabric of the host communities will be damaged by the tourist arrivals. On the contrary, the natural environment and the local communities should benefit from tourism, both economically and culturally. Sustainable tourism allows destinations and companies to minimize the negative effects and implications of tourism on the environment and on cultural heritage while maximizing its socioeconomic benefits. Sustainability implies that tourism resources and attractions should be utilized in such a way that their subsequent use by future generations is not compromised (UNWTO 2000, 2010).

5.3 The Economic Impacts of Tourism Sector

Tourism (including travel) is accepted as one of the most important and forefront sectors which has long been known that it can have impacts on overall economic activities. Tourism as an economic sector contributes to the economic activities with three different impacts: direct, indirect, and induced impacts. The sector has significant indirect and induced impacts as well as its direct economic impact. The total contribution of the sector includes its "wider impacts," in other words the indirect and induced impacts on the economy as well as direct impact. These three impacts are summarized in Table 5.1 by the WTTC (WTTC 2017a).

As shown in Table 5.1, the direct impact of the tourism sector to GDP reflects the "internal" spending on tourism and travel as well as government "individual" spending. Internal spending includes total spending within a particular country on tourism

Table 5.1 The economic impacts of tourism sector

Source: WTTC (2017a)

and travel by residents and nonresidents for business and leisure purposes. Individual spending includes the spending by government on tourism and travel services directly linked to visitors, such as cultural or recreational.

Shortly, the effects on earnings and employment in the sectors producing goods and services purchased by tourists are termed the "direct impact." On the other hand, these sectors purchase other goods and services as part of their production processes, and these, in turn, are produced by other sectors or imported. Thus, there is a supply chain of sectors that produce goods that ultimately satisfy consumption; the effects of tourism on an economy through this supply chain are called the "indirect impact" (Blake et al. 2008).

The "indirect" impact contains the GDP and jobs supported by (WTTC 2017a):

- Tourism investment spending: an important aspect of both current and future activities that includes investment activity such as the purchase of new aircraft and Roth construction of new hotels
- Government "collective" spending, which helps tourism activities in different ways as it is made on behalf of the "community at large," e.g., tourism marketing and promotion, aviation, administration, security services, resort area security services, resort area sanitation services, etc.
- Domestic purchases of goods and services by the tourism sector dealing directly with tourists: including purchases of food and cleaning services by hotels, of fuel and catering services by airlines, and IT (information technology) services by travel agents

As domestic residents earn money from tourism activities, and part of these additional earnings is spent on domestically produced goods and services, there is a further third round of effects termed "induced impact" (Blake et al. 2008). The "induced" impact measures the GDP and jobs supported by the spending of those who are directly or indirectly employed by the tourism sector.

In addition to direct, indirect, and induced impacts, according to some academic studies (such as Schubert et al. 2011; Chou 2013), tourism sector has positive impacts on the increase of long-run economic growth through different channels: An additional channel is that tourism is a significant foreign exchange earner which allows for payment of imported capital goods or the basic inputs used in the production process. The second channel is that tourism plays an important role in spurring investment in new infrastructure and competition between local companies and companies in other tourist countries. Third, tourism contributes to generating employment and increasing income. Fourth, tourism can cause positive exploitation of economies of scale in national companies. One more additional channel is that tourism is an important element in the diffusion of technical knowledge, stimulation of research and development, and the accumulation of human capital.

The faster growth of tourism induces an increase in households' disposable income and governments' revenues through multiplier effects, improvements in the balance of payments, and growth in the number of tourism-promoted government policies. For these reasons, the tourism and the development of tourism have usually been considered a positive contribution to economic growth (Oh 2005; Chou 2013).

According to Chao et al. (2006), an expansion of tourism increases the relative price of nontraded goods, improves the tertiary terms of trade, and yields a gain in revenue. But if this increase in the relative price of nontraded goods results in a lowering of demand for the capital used in the traded sector, subsequent deindustrialization in the traded goods sector may lower resident welfare. Chao et al.'s (2009) another study further points out that if the output effect is dominant, expansion of tourism raises employment and welfare; however, under realistic conditions, tourism may lower both labor employment and welfare due to rising costs.

On the other hand, according to Mazumder et al. (2013), tourism has an impact on a national economy at two levels: first at national or macro level and secondly at subnational or micro level. When tourism is considered to advance economic growth through foreign currency earnings, it contributes to an increase in state revenue at the first or macro level. At the second level or micro level, there will be an expansion in citizens' well-being in the form of income and employment generation.

Additionally, according to Balaguer and Cantavella-Jorda (2002), international tourism has an impact on economic growth at least in two different ways: The first way is enhancing efficiency through competition between local firms and the ones corresponding to other international tourist destinations. The second way is facilitating the exploitation of economies of scale in local firms.

5.4 The Case of Cambodia

During 25 years, Cambodia has achieved remarkable economic growth and development. Over the course of this period, Cambodia has experienced more than 7% gross domestic product (GDP) growth on average per year.

The country has seized growth opportunities from its tourism potential in addition to its rich natural resources and the evolving dynamics of the regional and global economy. Its unique cultural heritage has provided the country to be a tourism destination in Southeast Asia.

The tourism sector gradually improves due in part to newly established regional direct flights and a number of initiatives to boost tourist arrivals, year by year.

In this section of the chapter, the following points will be addressed: tourism infrastructure and potential, economic importance of tourism, and tourism development efforts in Cambodia.

5.4.1 Tourism Infrastructure and Potential in Cambodia

Cambodia is best known for the temples of Angkor in Siem Reap. Before the French protectorate, the Angkor monuments were regarded as a particularly religious site in Cambodia. After Cambodia gained independence from France in 1953, Angkor Wat has become the symbol of Cambodian identity.

Angkor Wat, which is one of the eight wonders of the world and also included in the UNESCO World Heritage site, is renowned internationally and the most important tourist attraction treasure for Cambodia. In addition to the temples of Angkor, there are additional but lesser-known temple sites mostly concentrated in the north and northwest of the country. There are also some historical and colonial sites in cities and towns like the capital Phnom Penh and Battambang in the northwest.

The southern coast boasts beaches and island leisure areas in Sihanoukville, and the southwest and northeast of the country still have vast areas of dense jungle, home to endangered flora and fauna. The Mekong River cuts through the country entering from Laos in the north, winding its way south and then east through Vietnam (RGC 2014a). Siem Reap and Phnom Penh, with relatively good urban services, account for about 60% of Cambodia's total accommodation stock and half of the country's international visitor arrivals (ADB 2017). Tourism sector in Siem Reap not only provides jobs for the locals but also attracts many Cambodians from other regions to come to work in tourism-related services and businesses (Chheang 2009). Consequently, Siem Reap, Phnom Penh, and Sihanoukville (and Kampot and Koh Kong) are the most accessible and developed centers from a tourism perspective.

Cambodian tourism has developed mainly on the basis of the comparative advantage of the Angkor Wat temples and the cost competitiveness of Cambodia in the traditional source markets in Europe and the USA. Despite the fact that Cambodia is endowed with countless historical, cultural, and natural tourist destinations, international tourist arrivals remain largely driven by the Angkor Archeological Park (Angkor Wat temples) located in the northeastern province of Siem Reap. Overreliance on the temples and price competitiveness, without quickly developing the competitive advantages of Cambodian tourism, lies at the heart of the problem facing the tourism sector. New destinations are needed in the country to solve the main problem facing the tourism sector. For example, ecotourism and coastal tourism have some potential in Cambodia for diversification of tourist attraction sites. In this context, the mountainous northeast provinces of Mondulkiri and Ratanakiri can be promoted for ecotourism. Ecotourism, especially community-based ecotourism, has a potential in Cambodia, particularly in terms of extending tourist flows into the northeast of the country and developing a less exploitative form of tourism where the Land Law and the rights of inhabitants are taken into account. However, ecotourism is likely to remain a small niche within Cambodia, and the development of coastal tourism provides an opportunity to diversify the tourism sector on a larger scale (UNDP 2009; Reimer and Walter 2013; ADB 2014; WB 2017).

Cambodia's tourism sector is overwhelmingly reliant on the leisure and holiday market. Leisure is the main purpose of visits, followed by business and visiting friends and relatives. As a component of tourism, leisure spending generates about 85% of total tourism spending, while business spending generates about 15% of total tourism spending. Cambodia's coastal resorts and casinos are also popular although they compete directly with similar offerings from elsewhere in the region. While Cambodia may present a lower-cost option, this is often at the expense of the quality of tourism services being offered, including in areas such as sanitation, infrastructure – including transport infrastructure – and customer service. Although tourism revenues are generated primarily around the Siem Reap-Phnom Penh-Sihanoukville triangle, fortunately, there are signs of early diversification toward other regions of the country. Regions identified as future tourist drawcards include the expansive waterways of Tonle Sap and the Mekong River, the northeast as an ecotourism base and the southern coastal zones (RGC 2014a; WTTC 2017b; ADB 2017).

On the other hand, in spite of the fact that Cambodia remains highly cost-competitive globally, it is surrounded by competitors that perform even better under this criterion. In the tourism sector strategic planning, in addition to diversify tourism attraction, it is a priority to consider that Cambodia is eroding its principal competitive advantage – cost (UNDP 2009).

The rapid growth in tourism sector has led to considerable interest in tourism as one of the main instruments for poverty alleviation especially in developing countries. Within this finding framework, it is seen that growth in tourism has been also accompanied by an apparent poverty reduction in Cambodia.

Tourism and associated service activities are heavily focused around the three major urban centers. As a matter of fact, although Cambodia's poverty has fallen most rapidly since the early 1990s, there are significant differences in poverty incidence between urban and rural areas. The majority of poverty is rural in the country. The improvement in urban living standards far exceeds the level in rural areas. As a result, the concentration of tourism and associated service activities in urban areas has a negative impact on equitable distribution of income. This also underscores the importance of diversifying Cambodia's tourism products and destinations beyond the three major urban centers to create a more resilient sector and provide more broad-based and geographically dispersed poverty reduction benefits (Hill and Menon 2013; RGC 2014a).

Additionally, in tourism sector's development strategies and policies, to deliver a lasting development effect with widespread poverty reduction benefits, it is important to take into account the need to target higher-spending visitors, including business visitors, high-end leisure tourists, and longer stays through more diverse product offerings.

Of course, air and road transport infrastructure is very important to the quality of tourism products in a country which can offer the diversity of experiences available to tourists. Especially, air transport is key to tourism growth in Cambodia. Because of this fact, upgrades to Siem Reap and Phnom Penh international airports have improved tourist experiences, and the introduction of e-visa arrangements has simplified arrival procedures. The government's commitment to an "open skies" policy since 1999 and deregulation of the airline industry allowed foreign carriers to operate in the country, resulting in competitive fares, additional flights, and increased tourist arrivals. The progress of air transportation services has significantly contributed to the development of tourism sector. While modernization in the major airports enhances the arrival/departure experience, increase in the number of regional low-cost airlines flying to Cambodian airports and the simplification of visa procedures have encouraged tourist arrival (RGC 2014a, b; ADB 2014). On the other hand, almost half of all tourists to Cambodia arrive by land. For this reason, developing the rural road network will be essential for boosting travel within and between provinces to diversify tourism beyond Angkor Wat and the temples near Siem Reap (ADB 2014).

Despite the many higher-education training programs available, the sector has a quite lack of "culture of professions." There is a need to strengthen curriculums to improve employment opportunities in tourism sector and reduce reliance on sourcing skilled labor force from offshore. Sector stakeholders report that the need for quality human resources far outstrips supply and the lack of well-trained workers is one of the main barriers to improving tourism sector's service quality and competitiveness. In addition, the industry needs better data on skill gaps and skill needs in order to plan and provide relevant training. These kinds of needs require ongoing public-private co-investment over the medium term in order to create a labor force that can sustainably service the sector (RGC 2014a; MoC 2016). The following issues need to be focused in the effort to provide the qualified labor force in the sector (ADB 2017):

(a) Expanding the number and geographic spread of quality public and private technical and vocational education and training facilities
(b) Increasing professional development for teachers
(c) Improving collaboration between government and private industry to develop and implement competency-based curriculum and certification frameworks
(d) Expanding training courses that are tailored to the time and resource constraints of learners

(e) Increasing enrollment, with due consideration for need-based scholarships to ensure inclusivity

Cambodia needs to invest strongly in modernizing its tourism infrastructure and creating the framework for delivering tourism services that meet future expectations from the sector, including in areas of sanitation, food hygiene, and customer service.

According to Gaughan et al. (2008), on one hand while international interest and tourism revenues flow into the region to support, develop, and exploit the Angkor World Heritage site, on the other hand, the Angkor basin continues to experience significant landscape changes. These changes should be considered in policies and development strategies that seek to balance sustainable local livelihoods with the benefits and growth of tourism in Cambodia.

Natural and cultural assets, including historic sites, cultural industries, national parks, and urban green space, make a destination attractive to visitors and livable for residents. Recognizing that protection of such assets is essential to maintaining the cultural values and ecosystem services that are important to many facets of their economies, Cambodia, the Lao PDR, Myanmar, and Vietnam have legally established extensive protected area systems in conjunction with other ASEAN countries.

Tourism can also help to improve the image and identity of Cambodia. It is argued that the Cambodian state considers tourism as the main contributor to the improvement of Cambodia's image and identity as well as to socioeconomic development (Chheang 2008).

5.4.2 The Economic Importance of Tourism in Cambodia

In fact, tourism has been strongly developed in Cambodia since the 1960s. However, civil war had seriously damaged tourism sector in the 1970s and 1980s. For this reason, the statistics for the period of 1970s and 1980s are not available due to records not conducted during the time (Chheang 2009).

After the cessation of conflict and the start of establishing political stability in 1993, the tourism as an economic sector has started to be an important and very dynamic engine of economic growth and generator of foreign exchange revenues in Cambodia. Especially, the Angkor basin of Cambodia, the site of the great Angkor temple complex, has experienced rapid development in tourism sector. As a result, today, tourism represents an important component of the economy in Cambodia.

Tourism, as one of the forefront service sectors, contributes in different channels to the development of the country's economy. It increases national income, creates employment opportunities, improves the country's foreign exchange revenues, attracts foreign direct investment (FDI) to the country, and encourages investment in related subsectors (such as hotel, restaurants, leisure services, and sporting facilities), transport, and finance sectors.

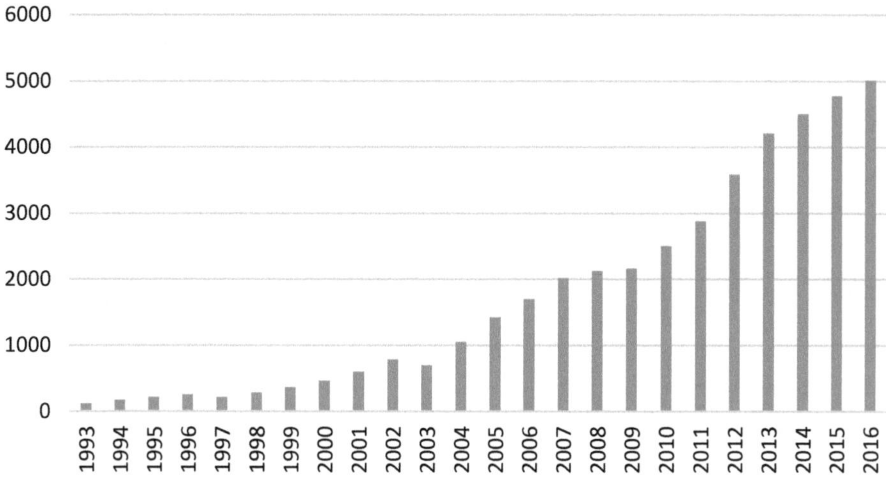

Graph 5.1 International tourist arrival number (thousand). (Source: MoT 2017)

Tourism is considered an export industry that mitigates against commercial imbalance and improves Cambodian competitiveness (RGC 2012).

Cambodia's international tourist arrivals have increased from a very low number in the mid-1990s to about 4.8 million arrivals in 2016 (see Graph 5.1). The number of tourist arrivals increased an average of 18% each year between 1994 and 2016. According to expectation of WTTC, in 2027, the country will attract about 8 million international tourist arrivals.

In addition to the increase in international tourist arrivals, as the domestic economy rapidly develops, there has also been an increase in domestic tourism. As a result of both domestic and international tourism, total tourism receipts are making a significant contribution to the national economy. Tourist spending on hotels, restaurants, and transport, using a strict definition, accounts for about 12% of the country's economy (see Graph 5.2). In Cambodia, the direct contribution of tourism (including travel) to GDP was $2.4 billion, 12.2% of total GDP in 2016. This contribution primarily reflects the economic activity generated by industries such as hotels, travel agents, airlines, and other passenger transportation services (excluding commuter services). It also includes the activities of the restaurant and leisure industries directly supported by tourists. The direct contribution of tourism to GDP is forecasted to rise by 6.5% for each year, from 2017 to 2027, to $5.0 billion, 11.7% of total GDP in 2027. However, taking into account the impact of tourist spending beyond the tightly defined "tourism" sector – for instance, in the retail and construction sectors – some suggest that tourism constitutes about 28% of the Cambodian economy. In Cambodia, the total contribution of tourism to GDP (including wider effects from investment, the supply chain, and induced income impacts) was $5.5 billion, 28.3% of total GDP in 2016. The direct contribution of tourism to GDP is forecasted to rise by 9.9% each year, from 2017 to 2027, to $11.5 billion, 27.3% of total GDP in 2027. The role of tourism as a pioneer sector, better able than other

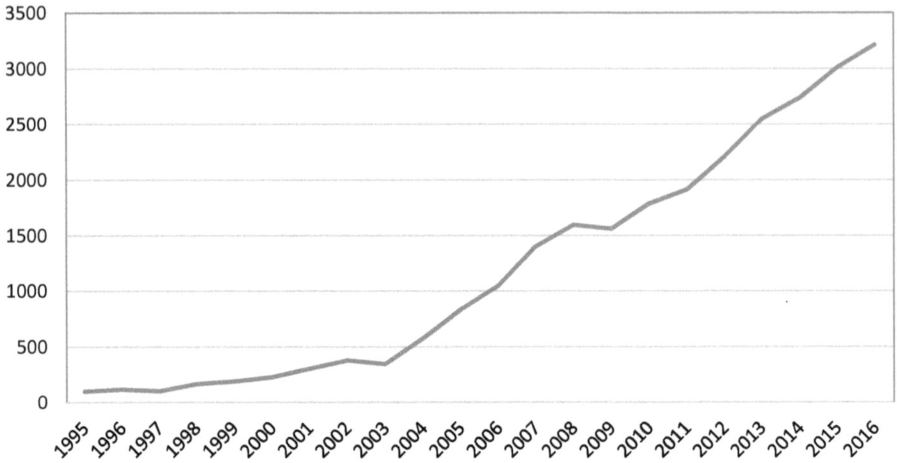

Graph 5.2 Tourism receipts (million US$). (Source: MoT 2017)

sectors to cope with the challenges posed by economies in transition, has been demonstrated in Cambodia. Tourism receipts also have important social benefits for Cambodia and help preserve historical, cultural, and environmental assets (UNDP 2009; RGC 2014a; WTTC 2017b).

Overall, tourism is an important net contributor to foreign exchange revenues and national income for the country. However, growth in tourism revenue is less than the growth in international arrivals. This reflects the decline in average daily expenditure per tourist as well as the decline in visitors' average length of stay. These trends are, at least in part, due to the changing mix of international visitor profiles and the rising portion of visitors from GMS countries. GMS countries are Cambodia, China, Lao PDR, Myanmar, Thailand, and Vietnam. Especially, growth in number of low-spending tourists from China, Korea, and Vietnam means a proportional decrease in tourist revenues. In addition, the economic slowdown in China negatively affects the tourism sector in Cambodia. Thus, the most important challenge for Cambodia in tourism sector is to attract visitors that will spend more and stay longer. On the other hand, this fact indicates that while it is important for Cambodia to be an attractive destination for visitors from within its own region, offering new world-class tourism products is also important to reverse the trends in average length of stay and average expenditure (RGC 2014a; MoC 2016).

Additionally, stronger revenue growth requires a concerted effort to target higher-spending business travelers as well as encourage leisure tourists to stay longer. This requires improvements in the tourism products being offered – especially in relation to the quality of tourism infrastructure – as well as a diversification beyond the traditional Siem Reap-Phnom Penh-Sihanoukville visitor triangle. It is more than adding destinations; it is also about developing service infrastructure: sporting facilities such as golf or tennis, hiking facilities, kayaking, scuba diving, restaurant business, circuits to visit pagodas (a very rich, largely unknown, and unadvertised resource), ecotourism, etc. (RGC 2014a).

Apart from the different reasons (see Sun 2015), the current potential of tourism sector is one of the most important sectors that attracts especially foreign investors to invest in Cambodia. Investment in tourism sector has accounted for nearly one-third of the FDI flows to Cambodia since the mid-1990s. Overall, total investment in the tourism sector was 15.5% of total investment ($0.6 billion) in 2016. According to WTTC expectations, it should rise by 5.1% in 2017 and rise by 6.4% per year over the next 10 years to $1.2 billion in 2027, 14.1% of total investment.

At the same time, tourism sector creates considerable direct and indirect employment. The sector accounts for more than 11% of GDP and contributes significant employment in the country. In Cambodia, tourism (including travel) directly supported 988,000 jobs (11.4% of total employment) in 2016. This includes employment by hotels, travel agents, airlines, and other passenger transportation services excluding commuter services. It also includes the activities of the restaurant and leisure industries directly supported by tourists. This is expected to rise by 9.1% in 2017 and rise by 4.8% pa to 1720,000 jobs (16.8% of total employment) in 2027, because tourism is a labor-intensive service sector activity with strong subsector linkages to a large number of other service industries, such as accommodation, restaurant, banking, and transport services. Moreover, in Cambodia, the total contribution of tourism to employment, including jobs indirectly supported by the industry (in other words, including wider effects from investment, the supply chain, and induced income impacts), was 25.9% of total employment (2252,500 jobs) in 2016. This is expected to rise by 6.7% in 2017 to 2.403,500 jobs and rise by 3.5% per year to 3389,000 jobs in 2027 (33.0% of total employment) (WTTC 2012, 2017b).

The high percentage of international tourist spending as a percentage of GDP does not mean that tourism displaces other viable economic activities; rather, tourism provides a source of foreign exchange revenues for the country (Goodwin 2008).

Visitor exports are a key component of the direct contribution of tourism sector. The ratio of visitor exports to total exports also increases year by year in Cambodia (see Graph 5.3). Visitor exports generated $3356.1 million, 26.5% of total exports in 2016. This is forecasted to grow by 11.8% in 2017 and grow by 6.2% per year, from 2017 to 2027, to $6868.5 million in 2027, 26.4% of total exports (WTTC 2017b).

In addition to rapid growth of tourist arrivals, in the tourism sector, Cambodia has moved quickly from dependency on distant US and European source markets to a much more balanced spread of markets, with well over half of the international market coming from ASEAN member states. Of course, this diversity of source markets reduces the vulnerability of the tourism sector to external shocks emanating from any one country (UNDP 2009). The sector is benefiting from closer regional ties and the rise of a middle class in Asia with money to spend on leisure in Cambodia. About 77% of visitors to Cambodia originate from other Asian and Pacific countries (ADB 2014) (Table 5.2).

The linkages between tourism and local communities were not only weak but are also much weaker than was observed across a wide range of other low-income countries in Southeast Asia and sub-Saharan Africa. For these facts, ecotourism and community-based tourism projects were developed by the GTZ, UNDP, International Finance Corporation (IFC), and the Netherlands Development Organization in the northeastern

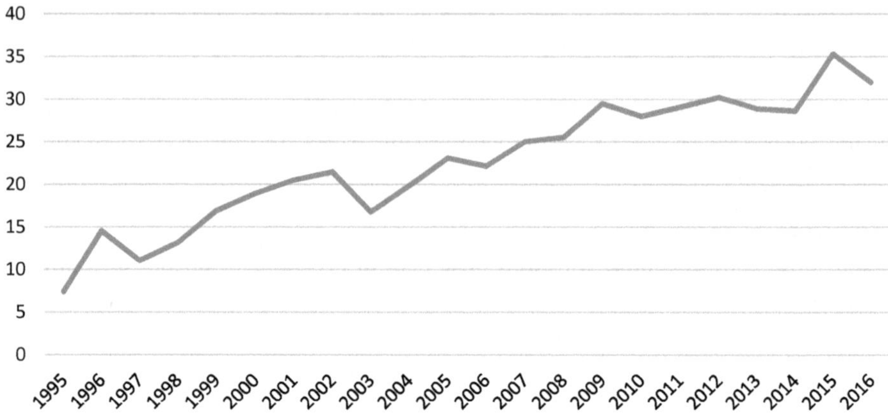

Graph 5.3 International tourism receipts (% of total exports). (Source: MoT 2017)

Table 5.2 Estimates and forecast about travel and tourism (Cambodia)

Cambodia	2016 USDbn (1)	2016 % of total	2017 Growth (2)	USDbn (1)	2027 % of total	Growth (3)
Direct contribution to GDP	2.4	12.2	10.7	5.0	11.7	6.5
Total contribution to GDP	5.5	28.3	9.9	11.5	27.3	6.6
Direct contribution to employment (4)	988	11.4	9.1	1720	16.8	4.8
Total contribution to employment (4)	2252	25.9	6.7	3389	33.0	3.5
Visitor exports	3.4	24.6	11.8	6.9	24.5	6.2
Domestic spending	0.8	3.9	5.6	1.6	3.8	7.4
Leisure spending	3.5	10.4	11.1	7.1	9.8	6.3
Business spending	0.6	1.9	8.2	1.4	1.9	7.4
Capital investment	0.6	15.5	5.1	1.2	14.1	6.4

Source: WTTC (2017b)
(**1**) 2016 constant prices and exchange rates, (**2**) 2017 real growth adjusted for inflation (%), (**3**) 2017–2027 annualized real growth adjusted for inflation (%), and (**4**) 000 jobs

provinces, the Greater Mekong subregion, and Kratie. Smaller NGOs developed small pro-poor projects in the Cardamom Mountains region and elsewhere (UNDP 2009). Tourism's potential role as a mechanism to transfer resources from largely affluent international tourists to local communities has not being yet realized in Cambodia.

5.4.3 *Tourism Competitiveness Index and Cambodia*

The World Economic Forum (WEF) has been actively engaged in an investigation of national competitiveness to better understand the drivers of growth and prosperity. The objective has been to provide benchmarking tools that enable

Table 5.3 The travel and tourism competitiveness index (Cambodia)

Year	Rank/number of countries	Score
2007	96/124	3.64
2008	112/130	3.32
2009	108/133	3.43
2011	109/139	3.4
2013	106/140	3.56
2015	105/141	3.24
2017	101/136	3.32

Source: WEF's Reports (2007–2017)

countries to identify key obstacles to competitiveness and to provide a platform for dialogue among government, business, and civil society to discuss the best ways of removing them. The World Economic Forum (WEF) has, since 2007, engaged leaders in travel and tourism to carry out an in-depth analysis of the travel and tourism competitiveness of about 140 countries across the world. The main objective of the Travel and Tourism Competitiveness Index (TTCI) is to help explore the factors driving travel and tourism competitiveness worldwide, thus providing a basis for implementing policies on a country-by-country basis (WEF 2007; Dwyer et al. 2011).

The WEF ranks selected nations according to the Travel and Tourism Competitiveness Index (TTCI), which scores from 1 to 6 the performance of a given country in each specific subindex. The TTCI measures "the set of factors and policies that enable the sustainable development of the travel and tourism sector, which in turn, contributes to the development and competitiveness of a country." This index considers countries on a scale of 1 to 7, based on natural and cultural resources, the local environment, and infrastructure, among other factors (WEF 2007, 2017).

According to the first WEF's "The Travel and Tourism Competitiveness Index" (WEF 2007), Cambodia was ranked 96 of 124 countries in 2007. In 2017, according to WEF's Index (WEF 2017), Cambodia was ranked 101 of 136 countries (see Table 5.3).

It achieved a high score in the prioritization of tourism, price competitiveness, and affinity for travel and tourism but received low scores for health and hygiene, tourism infrastructure, information and communications technology, and cultural resources. Consequently, developing infrastructure for better health and hygiene, particularly water and sanitation, will improve the health of the population and the country's attractiveness as a tourist destination (ADB 2014).

As a result, considering the index data, it can be said that Cambodia has a very low competitive power in the tourism sector.

5.4.4 Tourism Development Efforts in Cambodia

Tourism is one of the top five national and socioeconomic development priorities of Cambodia (Chheang 2009). The factors that have influenced the development of the tourism sector in recent years include (MoEF 2016; MoT 2017; WB 2015, 2017):

- Multilateral/regional tourism cooperation with Asian countries.
- Efforts to attract additional arrivals by establishing more direct flights with Asian countries and introducing new initiatives such as the "China Ready" initiative and joint tour packages: China's structural shift from manufacturing to service is likely to boost demand for tourism, which in return could have positive spillover for Cambodia. To take advantage of this opportunity, Cambodia's Ministry of Tourism has launched the "China Ready" policy that targets to attract about two million tourists from China by 2020.
- Bilateral collaboration with Thailand under "Two Kingdoms, One Destination."
- Strengthening the role and organizational structure of the Ministry of Tourism.
- The Ministry of Tourism's leadership in the management and development of the tourism sector, including human resource development.
- Collaboration between public and private sectors: The Ministry of Tourism has been tasked to collaborate with other public agencies and the private sector to implement the Law on Tourism and the Tourism Development Strategic Plan.
- The preparation of a Tourism Marketing Strategy 2015–2020 by the authorities.

After the Tourism Strategic Development Plan 2009–2013, a new strategic plan was prepared. The Tourism Strategic Development Plan covering the 2012–2020 period was prepared by the government of the country in 2012. The Cambodian government's vision for the tourism sector is set out in the plan to support the sector's efficiency and set up the tourism sector as a source of sustainable growth. This plan contains a wide range of short-term and long-term measures to help address these challenges facing the tourism sector, such as a clear need to diversify product offerings by opening up new destinations with supporting infrastructure and services that meet international standards.

The launch of Cambodia's Tourism Development Strategic Plan 2012–2020 was an important period of time for the country in identifying the key challenges currently facing the tourism sector and laying down a strategy to ensure that tourism continues to support economic development and poverty reduction (RGC 2014a).

Strategic directions of Cambodia's Tourism Development Strategic Plan 2012–2022 can be summarized as follows (ADB 2014, 2017):

(a) Enhancing tourism product development and product quality
(b) Improving marketing and promotion
(c) Improving travel facilitation, transportation, and regional and international connectivity
(d) Improving tourism safety and management of negative environmental, social, cultural, and economic impacts
(e) Strengthening legal systems and management mechanisms
(f) Developing human resources

Moreover, as a result of a study (Chheang 2009), related to the role and with respect to the state in tourism development policies in Cambodia, nine factors were determined: security and safety for tourists; infrastructure and tourism facility development; collaboration between stakeholders, including private sector, NGOs, and government; cultural heritage preservation; environmental protection; human resource development; tourism product promotion; simplification of travel procedures; and regional cooperation.

Furthermore, as an umbrella institution, the Cambodia Tourism Federation (CTF) has been established in 2015. The creation of the CTF to federate a number of associations specific to the sector is opening opportunities for private sector investors and operators involved in tourism to develop and dialogue around shared positions (MoC 2016).

The Ministry of Environment has created a green buffer zone to prevent encroachment on the protected areas through developing agro-industry projects and ecotourism projects (RGC 2014b).

After all, it can be seen that the Cambodian government set tourism as one of the key sectors for invigorating Cambodia's long-term economic growth. International assessments also recognize the Cambodian government's considerable efforts to prioritize the tourism sector in the country.

A mechanism for private sector participation at the national and subregional level through the development of the tourism marketing board has been established in Cambodia. The Cambodia Community-Based Ecotourism Network (CCBEN) was established in 2002 to promote and support community-based ecotourism network (CBET) for the conservation of natural and cultural resources and for equitably raising the standard of living and quality of life of local communities. The CCBEN is a network of organizations, travel agencies, educational institutions, and communities in Cambodia which are involved in CBET. It operates to be a partner with the government and endeavors to develop pro-poor CBET in Cambodia, as a way of reducing poverty and protecting social, cultural, and environmental resources. Besides, the UNWTO is assisting Cambodia in formulating a long-term plan for CBET (Khanal and Babar 2007).

5.4.5 ASEAN, Cambodia, and Tourism Sector

Cambodia is a member of the Association of Southeast Asian Nations (ASEAN), which was established on 8 August 1967. In addition to Cambodia, the member countries of ASEAN are Brunei Darussalam, Indonesia, Lao PDR, Malaysia, Myanmar, the Philippines, Singapore, Thailand, and Vietnam.

Tourism provides opportunities for ASEAN countries to overcome their geographic limitations (such as Singapore) as other neighboring countries are endowed with adequate resources to essentially deem the local demand of leisure. This regional cooperation also helps minimize risky competition between destination areas and empower tourism development to reach to a greater height. Likewise, some of the ASEAN countries have comparative advantage in well-organized tourism infrastructure (e.g., Singapore) and resources (e.g., Malaysia) from which other countries can be benefited (Mazumder et al. 2013).

ASEAN is the best performing region in the world in terms of growth in international visitors.

ASEAN countries are experiencing remarkable regional economic cooperation among themselves. After the ASEAN Tourism Strategic Plan 2011–2015, a new tourism strategic plan covering the years 2016–2025 was prepared in 2015. With the new tourism strategic plan, it is aimed to contribute more to the "comprehensive," "green," and "knowledge-based" economic growth goal of ASEAN countries.

According to the ASEAN Tourism Strategic Plan 2016–2025, there is a need to consolidate the gains already made and take a more strategic approach to addressing the single destination marketing, quality standards, human resource development, connectivity, investment, community participation, safety and security, and natural and cultural heritage conservation challenges facing the development of ASEAN as a competitive, sustainable, and more socioeconomically inclusive and integrated tourism destination. In this context, the vision for ASEAN tourism is to be a quality tourism destination offering a unique, diverse ASEAN experience and to be committed to responsible, sustainable, inclusive, and balanced tourism development, so as to contribute significantly to the socioeconomic well-being of ASEAN people (Yonn 2017; ASEAN 2015).

Natural and cultural assets, including historic sites, national parks, and urban green space, make a destination attractive to tourists and livable for residents. Recognizing that protection of such assets is crucial to maintaining the cultural values and ecosystem services that are important to many aspects of their economies, Cambodia has legally established extensive protected area systems in conjunction with other ASEAN countries (ADB 2017).

As a member of ASEAN, Cambodia is ideally placed to draw on the strong growth in international tourism in ASEAN countries. The principal competitors to Cambodia's tourism sector in ASEAN countries are Thailand and Vietnam. Compared to Cambodia, these countries attract larger number of international tourists who stay longer and spend more. Indeed, Thailand was the favorite tourist destination among ASEAN countries (see ASEAN 2016). However, these countries are also complementary markets as visitors incorporate Cambodia into a broader GMS (the Greater Mekong subregion) itinerary.

Asia continues to be the major source of tourists, and the top five Asian countries (Vietnam, China, Thailand, South Korea, and Lao PDR) account for about 60% of the market share. Especially, Vietnam, China, and Thailand are the three main countries of the tourism market share. Thus, more international tourists arrive in Cambodia via land border crossings compared to international airports. This expresses the practice of tourists combining Cambodia with a wider regional visit as well as the growing importance of Vietnamese, Laotian, and Thai tourists in the Cambodian market. Although Asia continues to be the main tourism market for Cambodia, the increase in the number of tourist arrivals from Western Europe and North America is noteworthy.

5.5 The Relationship Between Tourism Revenues and GDP in Cambodia

5.5.1 Review of the Literature

In starting the review of literature, it must first be stated that there are many studies on the relationship between tourism sector and economic growth/development. Considering the academic studies on the relationship between tourism and economic growth, it is found that this topic has attracted a lot of interest among researchers. These studies can be grouped into two main categories: single-country studies and country-group studies. On the other hand, the empirical results of these studies belonging to both categories are mostly mixed and divergent. Some studies support that tourism sector leads to economic growth (the tourism-led economic growth hypothesis), while other studies maintain that the economic growth stimulates tourism sector (the economic-driven tourism-growth hypothesis or the growth-led tourism hypothesis). At the same time, there are several studies which suggest either a bidirectional relationship between tourism and economic growth or no relationship at all.

For example, Lanza et al. (2003), Narayan (2004), Oh (2005), and Katircioglu (2009) find evidence supporting the economic-driven tourism-growth hypothesis.

The tourism-led economic growth hypothesis is supported in studies by, for instance, Lee and Chang (2008), Fayissa et al. (2008), Blake et al. (2008), Kreishan (2010), Schubert et al. (2011), Arslanturk et al. (2011), Ekanayake and Long (2012), Aslan (2014), and Cárdenas-García et al. (2015).

The reciprocal hypothesis is supported by, for example, Shan and Wilson (2001), Kim et al. (2006), Ridderstaat et al. (2014), and Tang and Tan (2015).

Although according to relatively few studies, such as Webster and Ivanov (2014) and Antonakakis et al. (2015), there is no significant impact on tourism's contribution to economic growth.

Furthermore, there are some studies, such as Sharif et al. (2017) and Wu et al. (2017), which find different results for different countries.

The results of some of these studies will be mentioned in more details.

Schubert et al. (2011) investigate the impacts on economic growth of a small tourism-driven economy caused by an increase in the growth rate of international tourism demand. The study uses annual data of Antigua and Barbuda for the period 1970–2008. Results of the study indicate that an increase in the growth of tourism demand leads to transitional dynamics with gradually increasing economic growth and increasing terms of trade.

Arslanturk et al. (2011) examined the causal link between tourism receipts and GDP in Turkey from 1963 to 2006. It used the rolling-window and time-varying coefficient estimation methods to analyze the Granger causality based on vector error correction model (VECM) in this study. Results of the study indicate that there is no Granger causality between the series, while the findings from the time-varying coefficient model based on the state-space model and rolling-window technique show that GDP has no predictive power for tourism receipts. However, tourism receipts have a positive-predictive content for GDP following the early 1980s.

Ekanayake and Long (2012) explore the relationships between tourism development and economic growth in developing countries using the developed heterogeneous panel co-integration technique using the annual data for the 1995–2009 period. According to the results of the study, though the elasticity of tourism revenue with respect to real GDP is not statistically significant for all regions, its positive sign shows that tourism revenue makes a positive contribution to economic growth in developing countries.

Chou (2013) examines causal relationships between tourism spending and economic growth in ten transition countries for the period 1988–2011. Panel causality analysis, which accounts for dependency and heterogeneity across countries, is used herein. The empirical results of Chou's study support the evidence on the direction of causality and are consistent with the neutrality hypothesis for three of these ten transition countries (i.e., Bulgaria, Romania, and Slovenia). The growth hypothesis holds for Cyprus, Latvia, and Slovakia, while reverse relationships were found for the Czech Republic and Poland. The feedback hypothesis also holds for Estonia and Hungary.

Li et al. (2013) investigated the relationship between tourism receipts and economic growth in Malaysia. The primary objective of this study is to determine the long-run triangular relationship for economic growth, tourism receipts, and other variables (physical capital, education, health, exports, and government tourism expenditure) in Malaysia for the period 1974–2010 based on Granger causality in VECM. According to the empirical results of the study associated with Granger causality among economic growth, tourism and exports within the neoclassical framework are inconsistent in Malaysia.

Ridderstaat et al. (2014) explore the long-run relationship between tourism development and economic growth in a small island destination. The study employs an econometric methodology consisting of unit root testing, co-integration analysis, vector error correction modeling, and Granger causality testing. Results confirm the reciprocal hypothesis.

Webster and Ivanov (2014) investigate the impact of a destination's competitiveness upon tourism's contribution to economic growth using a cross section with 131 countries. Destination competitiveness is measured with the World Economic Forum's Travel and Tourism Competitiveness Index, while tourism's contribution to economic growth is measured with the growth decomposition methodology. According to the results of the study, destination competitiveness has no statistically significant impact on tourism's contribution to economic growth.

Another study by Tang and Tan (2015) attempts to further verify the validity of the tourism-led growth hypothesis in Malaysia using a multivariate model derived from the Solow growth theory. It employed annual data from 1975 to 2011 in this study. Tang and Tan found that economic growth, tourism, and other determinants are co-integrated. Specifically, tourism has a positive impact on Malaysia's economic growth both in the short run and in the long run.

Antonakakis et al. (2015) examine the time-varying relationship between tourism and economic growth in Europe. According to their study, the tourism-economic growth relationship is not stable over time in terms of both its magnitude and direction. Rather, it is very responsive to major economic events.

Another study by Cárdenas-García et al. (2015) attempts to determine whether the economic growth experienced in some countries as a result of the expansion of the tourism activity over the last two decades influences an increase in the level of economic development. For this purpose, a sample of 144 countries has been used. Results of the study verify that this relationship happens, especially in more developed countries, which calls into question the conception of tourism as a driving force of economic development for the least developed countries and even in developing countries.

Du et al. (2016) explored if tourism development is an additional determinant of income in the presence of the standard income determinants (such as capital accumulation) or if the effects of tourism development on economic growth work through the standard income determinants, instead. Empirically, Du, Lew, and Ng developed a tourism-growth model that is an extension of Solow (1956) and estimate their model with a cross section of 109 countries. Their findings indicate that investments in tourism in and of itself appear to be insufficient for economic growth. Instead, tourism's contribution to the long-term growth of an economy comes through its role as an integral part of a broader development strategy that is more generally focused on standard income determinants.

A recent study by Sharif et al. (2017) investigated the relationship between tourism development and economic growth in the USA for the 1996–2015 period. Three innovative techniques that are continuous wavelet, wavelet coherence power spectrum, and wavelet-based Granger causality that consider the decomposition of time series at different time frequencies are utilized in this study. According to the results of the study, there is a significant long-run relationship that occurs between tourism development and economic growth in the USA. Additionally, the results of the study show that there is a unidirectional causal influence of economic growth on tourism development in the short run, whereas in the long run, the opposite causal relationship is evident in the USA.

Another recent study by Wu et al. (2017) explores the causal relationship between international tourism receipts and economic growth in China's 11 eastern provinces, accounting for both dependency and heterogeneity across provinces. The empirical results support evidence for the growth hypothesis in the province such as Hebei. A reverse relationship supports evidence on the conservation hypothesis for the provinces, such as Liaoning, Zhejiang, Fujian, Shandong, and Guangdong. A neutrality hypothesis supported 5 of these 11 major provinces (i.e., Beijing, Tianjin, Shanghai, Jiangsu, and Hainan).

Table 5.4 Variables used in the model

Variables	Code of variable	Type
GDP	lnGDP	Endogenous
Tourism	lnTOURISM	Endogenous
Export	lnEXPORT	Endogenous

5.5.2 *Variables and the Methodology*

There are three variables, tourism receipts, exports, and gross domestic product, used in this study. All variables are annual time series and obtained from the World Bank databank. Since the series are available from 1993 to 2016, the study is limited with this period (Table 5.4).

Logarithmic values of the variables are included in the analysis. First of all, stationarity of the variables is tested with augmented Dickey-Fuller (ADF) (Dickey and Fuller 1979, 1981; Phillips and Perron 1988). After then Johansen co-integration test is performed to identify whether there is a long-term relationship between variables. In the third step, vector error correction mechanism (VECM), Wald test, and Granger causality tests are performed to identify the existence of short-run relationship between variables. Finally, residual tests, serial correlation LM test, heteroscedasticity ARCH test, and histogram normality test are performed to check if the model is stable and desirable. E-views 8 statistical program is used for data analysis.

5.5.3 *Unit Root Test*

Variables used in regression analysis can be time series but they should be stationary. Time series which is nonstationary or has unit root may lead to spurious regression that has meaningless and biased results.

Most of the time series are not stationary, not normally distributed around the fixed mean or the characteristics of the stochastic proceedings varies in time. When a series is stationary, its mean, variance, and covariance don’t change in time.

Elimination of trend and seasonal effects, getting the logarithmic values and differencing are the ways that the series become stationary. However, any operation performed on the series causes loss of information and weakens the relations between the series. Co-integration models were developed to use the series in original form without sticking in spurious regression trouble.

For co-integration tests, all variables are put in the model in their level I (0), and also they should be stationary at the same level. For that, the variables were checked via augmented Dickey-Fuller test through the following regression equations:

$$\Delta X_t = a + \alpha X_{t-1} + \beta \sum_{i=1}^{m} \Delta X_{t-i} + e_t \quad (5.1)$$

Table 5.5 Stationary of the variables

Code of variables	Without trend					With trend				
	τ	%1	%5	%10	Prob	τ	%1	%5	%10	Prob
lnGDP	−3.39	−3.77	−3.01	−2.65	0.023	−3.39	−4.44	−3.63	−3.26	0.078
lnTOURISM	−5.05	−3.77	−3.01	−2.64	0.006	−4.89	−4.50	−3.66	−3.27	0.005
lnEXPORT	−6.22	−3.77	−3.01	−2.64	0.000	−6.44	−4.44	−3.63	−3.26	0.000

Number 1 in codes of variable shows that the first-level difference of that series is taken. * symbolizes level of the series as %1 and ** as %5

$$\Delta X_t = a + bt + \alpha X_{t-1} + \beta \sum_{i=1}^{m} \Delta X_{t-i} + e_t \tag{5.2}$$

First Eq. (5.1) represents the stationarity without a trend, and the second one (5.2) represents with a trend. Null hypotheses for these equations are H_0, $\alpha = 0$, and H_0, $b = 0$, which means X_t series is not stationary. If H_0 is rejected, then alternative hypothesis is accepted; that means X_t series is stationary. According to the results of ADF unit root test, the series are not stationary in their level I (0), but the first difference of each is stationary I (1).

Table 5.5 shows the test results of the variables on both levels and first difference with two stages: for intercept and for intercept and trend. The null hypothesis of nonstationary is not rejected for all variables in their level at the 1% level of significance. But it is rejected for all variables in the first difference at the level 1% and 5% level of significance.

5.5.4 *VAR Lag Order Selection*

Variables are included in the Johansen co-integration model with their lagged values. The best available method to select optimum lag order is VAR model.

VAR model uses the criterions such as the likelihood ratio (LR), final prediction error (FPE), Akaike information criterion (AIC), Hannan-Quinn information criterion (HQ), and Schwarz information criterion (SC) to determine the optimum lag order. As it is shown in Table 5.6, the optimum lag level for the analysis is 2 according to all criteria.

5.5.5 *Long-Term Causality Between the Gross Domestic Product and the Tourism Receipts: Co-integration Test*

Co-integration models identify the long-term relationship between the variables and display how the short-term disequilibrium turns over to the long-term equilibrium. The advantage of co-integration models is that the variables are included in the

Table 5.6 VAR lag order selection

Lag	LogL	LR	FPE	AIC	SC	HQ
1	64.46286	NA	1.03e-06	−5.282177	−4.834525	−5.185025
2	82.88752	26.32094[a]	4.37e-07[a]	−6.179764[a]	−5.284459[a]	−5.985460[a]
3	88.57799	6.503391	6.80e-07	−5.864570	−4.521613	−5.573114

[a]Indicates lag order selected by the criterion

model at their levels; hereby the series don't suffer from the loss of information since they are not subject to the elimination process from the unit root.

To identify the relations between the series x_t and y_t, an equation can be established as shown below:

$$y_t = \alpha x_t + u_t \tag{5.3}$$

If the error term of the equation u_t which is shown below is stationary, x_t and y_t are said to be co-integrated:

$$y_t - \alpha x_t = u_t \tag{5.4}$$

Johansen co-integration method is a kind of dynamic co-integration estimator that lagged values of the variables which are included in the model. It also presents asymptotically efficient estimates of the co-integrating vectors (the β's) and of the adjustment parameters (the α's). It bases on the maximum likelihood estimator and makes all co-integrating vectors possible to be estimated when there are more than two variables. The model estimates at most **n − 1** co-integrating vectors for **n** non-stationary variables.

Johansen co-integration method is explained with the AR(k) model as shown below:

$$x_t = [\mu + \Pi] x_{t-1} + \dots\dots\dots\dots + \Pi_k x_{t-k} + \varepsilon_t \tag{5.5}$$

where ε is the error term and μ, Π_1 ... Π_k parameters are restricted so they are estimated by the help of vector autoregressive model:

$$\Delta x_t = \mu + \Gamma_{1[\Delta x]_{t-1}} + \ldots + \Gamma_{k-1} [x]_{t-k+1} + [\Pi x]_{t-1}\, \varepsilon_t \tag{5.6}$$

The series in the model have unit root so the first difference of each variable is taken with the formula (5.6) to make them stationary:

$$\Gamma_i = -(i - \Pi_1 - \cdots \Pi_i) i = 1, \ldots k - 1 \,\text{ve}\, \Pi = -(\mathrm{I} - \Pi_1 - \cdots \dots\dots \Pi_k)$$

With this process coefficient matrix Π is searched weather it has information about the relations between variables in the data vector. There are three conditions:

Table 5.7 Unrestricted co-integration rank test (trace)

Hypothesized no. of CE(s)	Eigenvalue	Trace statistic	0.05 Critical value	Prob.[b]
None[a] ($r = 0$)	0.781833	55.77124	42.91525	0.0016
At most 1 ($r \leq 1$)	0.543206	23.79884	25.87211	0.0886
At most 2 ($r \leq 2$)	0.295139	7.344863	12.51798	0.3099

Trace test indicates two co-integrating eq(s) at the 0.05 level
[a]Denotes rejection of the hypothesis at the 0.05 level
[b]MacKinnon et al. (1999) p-values

Table 5.8 Unrestricted co-integration rank test (maximum Eigenvalue)

Hypothesized no. of CE(s)	Eigenvalue	Max-Eigen statistic	0.05 Critical value	Prob.[b]
None[a] ($R = 0$)	0.781833	31.97240	25.82321	0.0068
At most 1 ($r \leq 1$)	0.543206	16.45398	19.38704	0.1269
At most 2 ($r \leq 2$)	0.295139	7.344863	12.51798	0.3099

Max-Eigenvalue test indicates two co-integrating eq(s) at the 0.05 level
[a]Denotes rejection of the hypothesis at the 0.05 level
[b]MacKinnon et al. (1999) *p*-values

Rank (Π) = 0. Coefficient matrix (Π) is zero, and Eq. (5.5) is appropriate with the traditional time series differential vector.

Rank (Π) = p. Coefficient vector (Π) rank is a whole and X vector process is stationary.

0 < Rank (Π) = $r < p$. Coefficient matrix (Π) is multiplication of two matrices formed as $\alpha\beta$ and p^*r dimensions; that means there is a long-term relation between the variables.

All variables are on the same order so Johansen co-integration test can be applied to find whether the variables are integrated or not. If the variables are integrated, then there is at least unilateral causality between the variables (Granger 1969). Causality is checked with standard Granger test (1969). Also VECM model is performed to identify the speed of adjustment from the short-term disequilibrium into the long-term equilibrium (Granger 1988).

Johansen co-integration method uses the trace and the maximum Eigenvalue tests to determine how many co-integrating vectors are there between the variables. Results of these two tests are shown in Tables 5.7 and 5.8.

The value of the trace statistic in the first line which is based on the assumption that there is no co-integration vector (r=0) is bigger than the critical value at the 0.16% level of significance. So we can reject the null hypothesis which means there is no co-integration association among the variables. But the trace statistic in the second line is less than the critical value at the 8.9% level of significance so we can't reject the null hypothesis that there is at most one co-integration equation ($r \leq 1$).

Maximum Eigenvalue statistic confirms the outcomes of the trace statistic: that there is one co-integration equation between the variables and they have long-run positive relationship. Max-Eigen statistic is bigger than the critical value at the 0.0068 level of significance, so we can reject the null hypothesis that there is no co-

Table 5.9 Normalized vector

	lnGDP	lnTOURISM	lnEXPORT
Coefficient	1.000000	−0.425228	1.738583
Std Error		(0.11636)	(0.17361)

integration equation between the variables. But in the second line, max-Eigen statistic is less than the critical value at the 12% significance level so the null hypothesis cannot be rejected.

According to the normalized co-integrating coefficients shown in Table 5.9, the coefficient of the tourism receipts is 0.425. These coefficients are the parameter estimations of the long-term elasticities. So 42.5% of every increase in the tourism receipts reflects to the GDP growth in the long run.

5.5.6 Vector Error Correction Mechanism

According to the co-integration theory and the empirical studies, though there is a long-run association between variables, there may be deviations from the equilibrium in the short run. The vector error correction model is used to investigate how the short-run disequilibrium evolves to the long-run equilibrium.

The results of the VECM are shown in Table 5.10. Granger suggests the causality analysis via the VECM model if the series are co-integrated. The VECM model of this study is shown below:

$$\Delta \ln \mathrm{GDP}_t = \propto_0 + \sum_{i=1}^{n} \propto_{1i} \Delta \ln \mathrm{GDP}_{t-i} + \sum_{i=1}^{n} \propto_{2i} \Delta \ln \mathrm{TOURISM}_{t-1} + \sum_{i=1}^{n} \propto_{3i} \Delta \ln \mathrm{EXPORT}_{t-1} + \gamma \mathrm{ECM}_{t-1} + \varepsilon_t$$

where γ coefficient of ECM_{t-1} is the error correction term and the speed of adjustment of the co-integration model. Δ ln GDP_{t-i}, Δ ln $\mathrm{TOURISM}_{t-1}$, and Δ ln EXPORT_{t-1} are the lagged values of the variables that reflect the effects of short-term deviations. ECM_{t-1} which is a lagged value of the error terms is obtained from the co-integration equation. $\propto_{1i}$, $\propto_{2i}$, and $\propto_{3i}$ are the short-term parameters that show the direct effects on the dependent variable.

According to the VECM model, if the error term is negative in sign, then there is a long-run causality between variables. In this model, γ is negative in sign and significant (probability value is less than 5%), so there is a long-run causality running from export revenue and tourism receipts to GDP. The coefficient ECM_{t-1} estimated for error correction is −0,55 which is between zero and minus one in appropriate with the theory ($-1 < -0{,}55 < 0$). This result shows that the long-run

Table 5.10 Vector error correction mechanism

Dependent variable, Δ $\ln GDP_t$				
Independent variable	Coefficient	Std. error	t-statistic	Prob.
ECM_{t-1}	−0.551023	0.252725	−2.180327	0.0482
Δ $\ln GDP_{t-1}$	1.150459	0.352781	3.261113	0.0062
Δ $\ln GDP_{t-2}$	0.511570	0.394764	1.295888	0.2176
Δ $\ln TOURISM_{t-1}$	−0.016657	0.112889	−0.147556	0.8850
Δ $\ln TOURISM_{t-2}$	−0.207084	0.078922	−2.623894	0.0210
Δ $\ln EXPORT_{t-1}$	0.285100	0.239283	1.191476	0.2548
Δ $\ln EXPORT_{t-2}$	0.067059	0.109789	0.610804	0.5519
ε_t	−0.071665	0.075693	−0.946786	0.3610
R-squared	0.622465	Mean dependent variable		0.082372
Adjusted R-squared	0.419177	S.D. dependent variable		0.066804
SE of regression	0.050913	Akaike info criterion		−2.835077
Sum squared residuals	0.033697	Schwarz criterion		−2.437164
Log likelihood	37.76831	Hannan-Quinn criterion		−2.748720
F-statistic	3.061987	Durbin-Watson stat		2.144029
Prob (F-statistic)	0.038861			

relationship between the variables is not influenced by the short-run disequilibrium. The formula shown below is used to calculate the speed of adjustment:

$$1/\gamma = 1/0,55 = 1,87$$

This result shows that the long-term co-integration equilibrium is provided in less than two terms (1.87).

5.5.7 *Short-Term Causality Between GDP and the Tourism Receipts*

Short-term causality is searched applying the Wald test and checking how significant the parameters are. Wald test results are shown in Table 5.11.

α_{11} and α_{12} are the coefficients of $\Delta\ln GDP_{t-1}$ and $\Delta\ln GDP_{t-2}$. The probability of chi-square for null hypothesis $[\alpha_{11} = \alpha_{12} = 0]$ is 0.4% which is less than 5%. So $\Delta\ln GDP_{t-1}$ and $\Delta\ln GDP_{t-2}$ have short-term association with lnGDP.

α_{21} and α_{22} are the coefficients of $\Delta\ln TOURISM_{t-1}$ and $\Delta\ln TOURISM_{t-2}$, and the probability of chi-square for null hypothesis $[\alpha_{21} = \alpha_{22} = 0]$ is 4.8% which is less than 5%. So the tourism receipts have short-term association with GDP.

Table 5.11 Wald test

Null hypothesis	Statistic	Value	df	Probability
$\alpha_{11} = \alpha_{12} = 0$	F-statistic	5.474854	(2, 13)	0.0188
	Chi-square	10.94971	2	0.0042
$\alpha_{21} = \alpha_{22} = 0$	F-statistic	3.858295	(2, 13)	0.0484
	Chi-square	7.716590	2	0.0211
$\alpha_{31} = \alpha_{32} = 0$	F-statistic	0.778267	(2, 13)	0.4795
	Chi-square	1.556535	2	0.4592

α_{31} and α_{32} are the coefficients of ΔlnEXPORT$_{t-1}$ and ΔlnEXPORT$_{t-2}$, and the probability of chi-square for null hypothesis [$\alpha_{31} = \alpha_{32} = 0$] is 48% which is more than 5%. So there is no short-term association between tourism receipts and GDP.

5.5.8 *Model Stability Tests*

How significant the outcomes of the model depend on how effective the model is. For that the residuals of the model should not be serially correlated and shouldn't be under ARCH effect, but it should be normally distributed. So the residuals of the model will be checked with histogram normality, heteroscedasticity, and Breusch-Godfrey serial correlation LM tests.

5.5.8.1 Histogram Normality Test

Histogram normality test was performed to check how the residual of the model is distributed (Graph 5.4).

Null hypothesis is that the residual of the model is normally distributed. Jarque-Bera statistic and the corresponding probability value confirm the null hypothesis at the 90% level of significance.

5.5.8.2 Heteroscedasticity Test: ARCH

The second concern for model efficiency is that the model shouldn't be under ARCH effect. If the squared residual errors of the model exhibit autocorrelation, then ARCH effects are present.

Heteroscedasticity ARCH test outcomes in Table 5.12 verify that the model is not under ARCH effect. The probability value of the chi-square which is 57% approves the null hypothesis that there is no ARCH effect in the model.

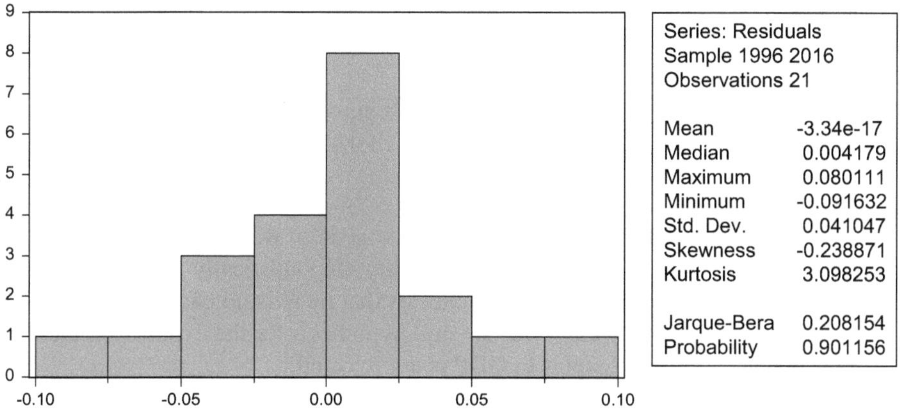

Graph 5.4 Histogram normality test

Table 5.12 Heteroscedasticity test: ARCH

F-statistic	0.498610	Prob. F(2,16)	0.6165
Obs*R-squared	1.114722	Prob. chi-square(2)	0.5727

Table 5.13 Breusch-Godfrey serial correlation LM test

F-statistic	1.580034	Prob. F(2,11)	0.2493
Obs*R-squared	4.686519	Prob. chi-square(2)	0.0960

5.5.8.3 Breusch-Godfrey Serial Correlation LM Test

The third concern for the model efficiency is that there shouldn't be serial correlation on the residuals of the model.

Breusch-Godfrey serial correlation LM test results shown in Table 5.13 verify that there is no serial correlation on the residuals of the model. Chi-square probability is significant at the level of 9,6% which is more than 5%, and it approves the null hypothesis that there is no serial correlation on the residuals of the model.

5.5.9 VEC Granger Causality/Block Exogeneity Wald Tests

Following the identification of long-term relationship between variables, the causality and the direction of it are determined. Therefore, VEC Granger Causality/Block Exogeneity Wald Tests were performed to identify causality from tourism receipts and export revenues to GDP.

Table 5.14 VEC Granger Causality/Block Exogeneity Wald Tests

	Variables	Chi-sq	df	Prob.
Dependent variable	Δ $\ln GDP_t$			
Excluded	Δ $\ln TOURISM_t$	7.716590	2	0.0211
Excluded	Δ $\ln EXPORT_t$	1.556535	2	0.4592

According to the VEC Granger Causality/Exogeneity Wald Test results shown in Table 5.14, chi-square value and the probability are statistically significant at the level of 2% significance so the null hypothesis that there is no causality from tourism receipts to GDP is rejected. But the null hypothesis for the export revenue that there is no causality from export to GDP is not rejected.

5.6 Discussion and Policy Recommendations

In order for the contribution of the tourism industry to the economic growth to be sustainable and in higher level, the following points should be given importance. The improvement and development of the tourism industry in Cambodia need to involve a variety of elements.

The role of private, public, NGO, media, and international institutions and organizations should be coordinated together to promote tourism, because the harmony among the perceptions, values, and attitudes of all stakeholders related to tourism is decisive in reaching the objectives related to the tourism sector. For example, tourism is one of the fastest-growing economic sectors in Cambodia, but local community-based products that can provide economic independency to the poor and are recognized by the tourism industry remain largely unsupported as poverty reduction strategies and rationales for developed education. For this and similar reasons, there is an important necessity to increase sensitivity on the coordination of institutions and organizations related to tourism industry.

Furthermore, in the tourism sector's development strategies and policies, greater attention needs to be paid to the fact that more parts of the tourist spending reach the poorer sections of the population. Generating net benefits for the poorer sections of the population is called pro-poor tourism. Of course, it is important that tourism sector not be pursued independently as a sure-fire pro-poor strategy; instead all sectors should be directed toward pro-poor purposes. Reducing regulatory frameworks at a policy level for a single sector contradicts the systems approach which sustainability demands. It should also be noted that pro-poor tourism cannot succeed without a successful development of the entire tourism destination.

On the other hand, temporarily, the government's support to the tourism industry is vital. Cambodian government should focus more on providing adequate and necessary infrastructure to the tourism areas, because preparing the transportation infrastructure and also the ease traffic are the main components of fluency in the tourism activities. Especially, it is quite difficult for the tourism to increase its market share without the good and fast transport service.

The capability and talents of management of the tourism services need to improve. Because the development of sector depends on the increase in the number of visitors, the tourism is an industry that emphasizes hospitality. For this reason, those who manage tourism sector should be well educated and have high managerial talents. Of course, effective and well-managed tourism activities will greatly satisfy the tourists. Their satisfaction will be seen from the intensity of repetitive visits and their verbal marketing.

Compared to some other ASEAN countries, the average length of stay of visitors to Cambodia is short. Therefore, the Cambodian government should also spend more effort to increase the average length of stay of visitors to Cambodia. Of course, in addition to reducing costs and increasing the quality of the tourism product, the extension of the average length of stay of visitors to Cambodia is also dependent on diversification in the tourism sector.

The development and improvement of tourism industry need to prioritize the elements of local culture and the active involvement of the local community. The local community should participate in maintaining and improving the tourism areas.

Emphasis on the focal attraction of the Angkor Wat complex has a negative effect on the full consideration of both the economic and social dimensions of tourism industry. Due to the problems experienced in regional development, still Siem Reap and Angkor Wat and a few other places receive most tourists. This problem requires restructuring the tourism promotion strategy with a wider perspective.

Cambodia should spend more effort for diversification of tourist attraction sites beyond the key attraction of Angkor Archeological Park in Siem Reap and business tourism in Phnom Penh. To these ends, interagency coordination, as well as collaboration with the private sector to develop and access new attraction sites and to improve currently underdeveloped sites, is essential. Considering the construction expansion, helping other emerging urban areas to become more tourist-friendly destinations with improved tourism infrastructure, facilities, and accessibility would help support diversification in the tourism sector. Additionally, the tourism sector can also be improved gradually through newly established direct flights and a number of initiatives and promotional efforts to diversify tourism attraction beyond the Angkor Wat complex to boost tourist arrivals (UNDP 2009; Reimer and Walter 2013; ADB 2014; WB 2017).

In contrast to the high expectations, because of a lack of human capital and tourism experience in local communities and similar reasons, contribution of ecotourism to the tourism industry has not yet enough realized. Therefore, in order to realize the expected contribution from ecotourism to the tourism industry, especially and primarily the human capital of the sector and tourism experience in local communities need to be developed.

In Cambodia, tourism strategies and policies must also have an inclusive, responsible, and sustainable dimension, because sustainable tourism would provide a rationale for heritage protection, promote wider environmental protection initiatives, and contribute to the social welfare.

5.7 Conclusion

It can be seen that tourism is one of the fastest-growing service sectors in the world, considering the contribution of tourism to the world's GDP and the increase in the number of international tourist arrivals. Tourism plays an increasingly important role in the socioeconomic development of countries year after year. Being a major source of income and employment for many countries, tourism is one of the most important assistants in addressing problems caused by poverty.

Tourism has been a strong contributor to Cambodia's economic development over the past 25 years, and recent developments toward attracting more and more international tourist arrivals have increased the share of the sector in the country's economy.

According to the results of this study, it can be concluded that the hypothesis of tourism-led growth is proven for Cambodia in the long run. Johansen co-integration test verifies that GDP, tourism receipts, and export revenues have long-term association. Speed of adjustment from short-term disequilibrium to the long-term equilibrium is 55% for every term in the model. The VEC Granger Causality/Block Exogeneity Wald Tests show that the causality from tourism receipts to GDP is statistically significant. But the causality from export revenues to GDP is not significant. Changing foreign trade policies and the weak performance of the commodity export in Cambodia may have negative effect on the determination of relations between export and the GDP. And also this study shows that the tourism policies and the tourism revenues are more effective on the GDP than the export policies and the export revenues.

The results of the overall study suggest that the Cambodian government should focus on economic policies to promote tourism as a potential source of economic growth, taking into account that the tourism sector has an important potential for contributing to economic growth. Of course, maximizing the benefit of sector depends upon the quality of the product and the tourism services and upon the national and international markets. Especially, quality infrastructure, qualified human capital, and progressive development strategies and policies will help reduce the risks to sustainable and inclusive tourism growth in Cambodia. Besides, the negative economic, social, and environmental impacts of tourism should be minimized which is necessary to maximize the net benefit of the sector.

Acknowledgment The research was supported by Zaman University in Cambodia, Research Office, via Exploratory Research Grant Scheme of Zaman University, grant no. ERGS/3/2017/ZURO/09/17.

References

ADB (Asian Development Bank). (2014). *Cambodia: diversifying beyond garments and tourism*. Economics and Research Department, Asian Development Bank. Philippines.

ADB (Asian Development Bank). (2017). *Tourism sector assessment, strategy, and road map for Cambodia, Lao People's Democratic Republic, Myanmar, and Vietnam (2016–2018)*. Economics and Research Department, Asian Development Bank. Philippines.

Antonakakis, N., Dragouni, M., & Filis, G. (2015). Tourism and growth: The times they are a-changing. *Annals of Tourism Research, 50*, 165–169.

Arslanturk, Y., Balcilar, M., & Ozdemir, Z. A. (2011). Time-varying linkages between tourism receipts and economic growth in a small open economy. *Economic Modelling, 28*(1), 664–671.

ASEAN. (2015). *ASEAN tourism strategic plan 2016–2025*. Jakarta: The Association of Southeast Asian Nations.

ASEAN. (2016). *ASEAN economic community, chartbook*. Jakarta: The Association of Southeast Asian Nations.

Aslan, A. (2014). Tourism development and economic growth in the Mediterranean countries: Evidence from panel Granger causality tests. *Current Issues in Tourism, 17*(4), 363–372.

Athanasopoulou, A. (2013). *Tourism as a driver of economic growth and development in the EU-27 and ASEAN regions*. Singapore: EU Center.

Balaguer, J., & Cantavella-Jorda, M. (2002). Tourism as a long-run economic growth factor: The Spanish case. *Applied Economics, 34*(7), 877–884.

Blake, A., Arbache, J. S., Sinclair, M. T., & Teles, V. (2008). Tourism and poverty relief. *Annals of Tourism Research, 35*(1), 107–126.

Cárdenas-García, P. J., Sánchez-Rivero, M., & Pulido-Fernández, J. I. (2015). Does tourism growth influence economic development? *Journal of Travel Research, 54*(2), 206–221.

Chao, C. C., Hazari, B. R., Laffargue, J. P., Sgro, P. M., & Yu, E. S. (2006). Tourism, Dutch disease and welfare in an open dynamic economy. *The Japanese Economic Review., 57*(4), 501–515.

Chao, C. C., Hazari, B. R., Laffargue, J. P., & Yu, E. S. (2009). A dynamic model of tourism, employment and welfare: The case of Hong Kong. *Pacific Economic Review, 14*(2), 232–245.

Chheang, V. (2008). The political economy of tourism in Cambodia. *Asia Pacific Journal of Tourism Research., 13*(3), 281–297.

Chheang, V. (2009). State and tourism planning: A case study of Cambodia. *Tourismos: An International Multidisciplinary Journal of Tourism., 4*(1), 63–82.

Chou, M. C. (2013). Does tourism development promote economic growth in transition countries? A panel data analysis. *Economic Modelling, 33*, 226–232.

Dickey, D. A., & Fuller, W. A. (1979). Distribution of the estimators for autoregressive time series with a unit root. *Journal of the American Statistical Association, 74*(366a), 427–431.

Dickey, D. A., & Fuller, W. A. (1981). Likelihood ratio statistics for autoregressive time series with a unit root. *Econometrica: Journal of the Econometric Society, 1*, 1057–1072.

Du, D., Lew, A. A., & Ng, P. T. (2016). Tourism and economic growth. *Journal of Travel Research, 55*(4), 454–464. https://doi.org/10.1177/0047287514563167.

Dwyer, L., Forsyth, P., & Dwyer, W. (2011). The travel and tourism competitiveness index as a tool for economic development and poverty reduction. In L. Moutinho (Ed.), *Strategic management in tourism* (pp. 33–52). Wallingford: CAB International.

Ekanayake, E. M., & Long, A. E. (2012). Tourism development and economic growth in developing countries. *The International Journal of Business and Finance Research., 6*(1), 51–63.

Fayissa, B., Nsiah, C., & Tadasse, B. (2008). Impact of tourism on economic growth and development in Africa. *Tourism Economics, 14*(4), 807–818.

Fien, J., Calder, M., & White, C. (2010). *Sustainable tourism (module 16). Teaching and learning for a sustainable future*. UNESCO.

Gaughan, A. E., Binford, M. W., & Southworth, J. (2008). Tourism, forest conversion, and land transformations in the Angkor basin, Cambodia. *Applied Geography, 29*(2), 212–223. https://doi.org/10.1016/j.apgeog.2008.09.007.

Goodwin, H. (2008). Tourism, local economic development, and poverty reduction. *Applied Research in Economic Development., 5*(3), 55–64.

Granger, C. W. (1969). Investigating causal relations by econometric models and cross-spectral methods. *Econometrica: Journal of the Econometric Society, 1*, 424–438.

Granger, C. W. (1988). Some recent development in a concept of causality. *Journal of Econometrics, 39*(1–2), 199–211.

Harrison, D. (2015). Development theory and tourism in developing countries: What has theory ever done for us? *International Journal of Asia-Pacific Studies:IJAPS., 11*(1), 53–82.

Harrison, D., & Schipani, S. (2007). Lao tourism and poverty alleviation: Community-based tourism and the private sector. *Current Issues in Tourism, 10*(2–3), 194–230. https://doi.org/10.2167/cit310.0.

Hill, H., & Menon, J. (2013, January) *Cambodia: rapid growth with institutional constraints* (ADB Economics Working Paper Series. No.331).

Katircioglu, S. T. (2009). Revisiting the tourism-led-growth hypothesis for Turkey using the bounds test and Johansen approach for cointegration. *Tourism Management, 30*(1), 17–20.

Khanal, B. R, & Babar, J. T. (2007). *Community based ecotourism for sustainable tourism development in the Mekong region*. Policy Brief. Hanoi: Hanoi Resource Centre.

Kim, H. J., Chen, M. H., & Jang, S. C. (2006). Tourism expansion and economic development: The case of Taiwan. *Tourism Management, 27*(5), 925–933.

Kreishan, F. M. (2010). Tourism and economic growth: The case of Jordan. *European Journal of Social Sciences, 15*(2), 63–68.

Lanza, A., Temple, P., & Urga, G. (2003). The implications of tourism specialisation in the long run: An econometric analysis for 13 OECD economies. *Tourism Management, 24*(3), 315–321.

Lee, C. C., & Chang, C. P. (2008). Tourism development and economic growth: A closer look at panels. *Tourism Management, 29*(1), 180–192.

Li, C. C., Mahmood, R., Abdullah, H., & Chuan, O. S. (2013). Economic growth, tourism and selected macroeconomic variables: A triangular causal relationship in Malaysia. *Margin: The Journal of Applied Economic Research., 7*(2), 185–206. https://doi.org/10.1177/0973801013483503.

MacKinnon, J. G., Haug, A. A., & Michelis, L. (1999). Numerical distribution functions of likelihood ratio tests for cointegration. *Journal of Applied Econometrics, 14*(5), 563–577.

Mazumder, M. N., Sultana, M. A., & Al-Mamun, A. (2013). Regional tourism development in Southeast Asia. *Transnational Corporations Review., 5*(2), 60–76.

MoC (Ministry of Commerce). (2016). *2016–2020 Medium term plan for the implementation of Cambodia's trade SWAp*. Phnom Penh.

MoEF (Ministry of Economy and Finance). (2016 August). *Cambodia macroeconomic monitor: Mid-year assessment 2016*. Phnom Penh.

MoT (Ministry of Tourism). (2017, January). *Tourism statistics report*. Phnom Penh: Ministry of Tourism, Statistics and Tourism Information Department.

Narayan, P. K. (2004). Economic impact of tourism on Fiji's economy: empirical evidence from the computable general equilibrium model. *Tourism Economics, 10*(4), 419–433.

Oh, C. O. (2005). The contribution of tourism development to economic growth in the Korean economy. *Tourism Management, 26*(1), 39–44.

Phillips, P. C., & Perron, P. (1988). Testing for a unit root in time series regression. *Biometrika, 75*(2), 335–346.

Pleumarom, A. (2012). *The politics of tourism, poverty reduction and sustainable development*. Penang: Third World Network (TWN).

Reimer, J. K., & Walter, P. (2013). How do you know it when you see it? Community-based ecotourism in the Cardamom Mountains of southwestern Cambodia. *Tourism Management, 34*, 122–132.

RGC (Royal Government of Cambodia). (2012). *Tourism development strategic plan 2012–2020*. Phnom Penh.

RGC (Royal Government of Cambodia). (2014a, January). *Cambodia trade integration strategy 2014–2018*. Phnom Penh.

RGC (Royal Government of Cambodia). (2014b). *National strategic development plan 2014–2018*. Phnom Penh.

Ridderstaat, J., Croes, R., & Nijkamp, P. (2014). Tourism and long-run economic growth in Aruba. *International Journal of Tourism Research, 16*(5), 472–487. https://doi.org/10.1002/jtr.1941.

Schubert, S. F., Brida, J. G., & Risso, W. A. (2011). The impacts of international tourism demand on economic growth of small economies dependent on tourism. *Tourism Management, 32*(2), 377–385.

Shan, J., & Wilson, K. (2001). Causality between trade and tourism: Empirical evidence from China. *Applied Economics Letters, 8*(4), 279–283.

Sharif, A., Saha, S., & Loganathan, N. (2017, December 25). *Does tourism sustain the economic growth? A wavelet based evidence from United States*. Tourism Analysis: An Interdisciplinary Journal. ISSN 1083-5423.

Solow, R. M. (1956). A contribution to the theory of economic growth. *The Quarterly Journal of Economics, 70*(1), 65–94.

Sun, C. (2015). *Investing in Cambodia*. http://www.moc.gov.kh/Portals/0/Docs/OfficialDocs/H-20151028085854498.pdf. Accessed 2 Sept. 2017.

Tang, C. F., & Tan, E. C. (2015). Does tourism effectively stimulate Malaysia's economic growth? *Tourism Management, 46*, 158–163.

UNDP (United Nations Development Programme). (2009). *Cambodia country competitiveness: Driving economic growth and poverty reduction* (Discussion Paper No.7). United Nations Development Programme. Phnom Penh. Available from: https://www.researchgate.net/publication/303699040_Cambodia_Country_Competitiveness_Driving_Economic_Growth_and_Poverty_Reduction. Accessed 19 Sept 2017.

UNWTO (World Tourism Organization). (2000). *Report of multistakeholder working group on tourism*. World Tourism Organization.

UNWTO (World Tourism Organization). (2002, February 11–13). *Asia Pacific ministerial conference on sustainable development in ecotourism: Final report*. Male: World Tourism Organization.

UNWTO (World Tourism Organization). (2005). *Making tourism more sustainable – A guide for policy makers*. Madrid: World Tourism Organization and United Nations Environment Programme-2005.

UNWTO (World Tourism Organization). (2010). *Tourism and the millennium development goals*. Madrid. http://icr.unwto.org/publication/tourism-mdgs. Accessed 21 Sept. 2017.

UNWTO (World Tourism Organization). (2017). *Measuring sustainable tourism: A call for action report of the 6th international conference on tourism statistics*. Manila. 21–23 June 2017. Philippines.

WB (World Bank Group). (2015, October). *Cambodia economic update: Adapting to stay competitive*. World Bank Group.

WB (World Bank Group). (2017, April). *Cambodia economic update: Staying competitive through improving productivity*. World Bank Group.

Webster, C., & Ivanov, S. (2014). Transforming competitiveness into economic benefits: Does tourism stimulate economic growth in more competitive destinations? *Tourism Management, 40*, 137–140. https://doi.org/10.1016/j.tourman.2013.06.003.

WEF (World Economic Forum). (2007–2017). *The travel and tourism competitiveness report*. Geneva: World Economic Forum.

WTTC (World Travel and Tourism Council). (2012). *Travel and tourism economic impact 2012 – Cambodia*. London.

WTTC (World Travel and Tourism Council). (2017a, March). *The economic impact of travel and tourism – world*. London.

WTTC (World Travel and Tourism Council). (2017b). *Travel and tourism economic impact 2017 – Cambodia*, London.

Wu, T. P., Wu, H. C., Liu, S. B., & Hsueh, S. J. (2017). The relationship between international tourism activities and economic growth: Evidence from China's economy. *Tourism Planning & Development., 13*, 1–7 https://doi.org/10.1080/21568316.2017.1324809.

Yonn, R. (2017). The effect of Cambodia economy on ASEAN economic moving forward. *Journal of Management, Economics, and Industrial Organization, 1*(2), 1–16.

Part II
Current Trends and Applications

Chapter 6
Regression Analysis with Dummy Variables: Innovation and Firm Performance in the Tourism Industry

Jung Wan Lee and Parahny Manorungrueangrat

Abstract The chapter demonstrates an application of correlation and regression analysis in tourism research. The chapter provides the concepts of product moment correlation, partial and part correlation, regression, and regression with dummy variables. The chapter provides, in particular, an application of regression analysis with dummy variables to estimate the economic effects of innovation on firm financial performance in the tourism industry of an economy in Asia. The paper examines signs and significance of the relationship between technological innovation and firm financial performance in the tourism industry of Thailand. The study employs multiple regression analysis with dummy variables, correlation analysis (i.e., the Pearson correlation), and chi-square tests (i.e., the Pearson chi-square test), and the associated statistics are explained. The dataset includes a total of 188 data points from 46 listed firms in tourism in the Stock Exchange of Thailand over a period of five fiscal years, 2011–2015.

Keywords Regression analysis · Correlation analysis · Chi-square test · Cross-sectional data · Dummy variable · Technological innovation · Firm performance · Tourism research · Thailand

6.1 Introduction of Methodology

Chapter Objectives

- To introduce the concepts of product moment correlation, partial correlation, and part correlation.
- To introduce the concept of regression analysis. The regression is a powerful technique for analyzing the association between a metric-dependent variable and one or more independent variables. However, beyond explaining the nature and

J. W. Lee (✉) · P. Manorungrueangrat
Administrative Sciences Department, Metropolitan College, Boston University, Boston, MA, USA
e-mail: jwlee119@bu.edu; parahny@bu.edu

S. Rezaei (ed.), *Quantitative Tourism Research in Asia*, Perspectives on Asian Tourism, https://doi.org/10.1007/978-981-13-2463-5_6

degree of association between the variables, regression does not imply or assume causality.

- To explain the nature and methods of multiple regression analysis and describe the general model, estimation of parameters, standardized regression coefficient, significance testing, residual analysis, and the meaning of partial regression coefficients.
- To describe specialized techniques used in multiple regression analysis, particularly regression with dummy variables and analysis of variance and covariance with regression.

6.1.1 Product Moment Correlation and Partial Correlation

The product moment correlation is a statistic used to determine whether a linear relationship exists between two metric variables. From a sample of n observations, X and Y, the product correlation coefficient r is given as:

$$r = \frac{\mathrm{cov}_{xy}}{S_x S_y} \qquad -1 < r < 1$$

where Cov_{xy} denotes the covariance between X and Y, and S_x and S_y are the standard deviations of X and Y. Also known as the correlation coefficient, r is used as an index to determine if a linear relationship exists between X and Y. If the variables X and Y are metric and their distributions have some shape, r is an estimate of R, the product moment correlation of the population. The product correlation coefficient cannot reveal the presence of nonlinear relationships, if any, between the variables. It can only indicate the degree to which variation in one variable is related to the variation in another, if the relationship is linear.

The partial correlation is the correlation between variables X and Y after controlling for the effect of one or more additional variables (in this case Z) which is denoted by $r_{xy.z}$ and is given by:

$$r_{xyz} = \frac{r_{xy} - (r_{xz})(r_{yz})}{\sqrt{1 - r_{xz}^2}\sqrt{1 - r_{yz}^2}}$$

where r_{ij} represents the correlation coefficient between i and j.

Partial correlation coefficients have an order associated with them equal to the number of variables being controlled. For example, r has order 0. This statistic can be useful in detecting spurious relationships between variables. A partial correlation coefficient is a useful statistic that measures the association between two variables after adjusting or controlling for the effects of one or more additional variables. The order of the partial correlation coefficient indicates the number of variables being adjusted or controlled for.

The coefficient represents the correlation between Y and X when the linear effects of the other independent variables have been removed from X (but not from Y). It is expressed as:

$$r_{y(x,z)} = \frac{r_{xy} - r_{yz} r_{xz}}{\sqrt{1 - r_{xz}^2}}$$

However, the part correlation coefficient is generally viewed as less important than the partial correlation coefficient since the partial conveys the *absolute* contribution of an independent variable to the model fit, whereas the part conveys the *relative* improvement to the model attributable to an independent variable. The product moment, as well as the partial and part correlation coefficients, provides a conceptual foundation for bivariate as well as multiple regression. For further information, the concepts of product moment correlation and partial correlation are well documented in the book (Malhotra 2010).

6.1.2 Multiple Regression

The multiple regression involves a single dependent variable and two or more independent variables. The assumptions made in bivariate regression are true for multiple regression also. The general form of the regression model is:

$$Y = \beta_0 + \beta_1 X_1 + \beta_2 X_2 + \beta_3 X_3 + \ldots\ldots + \beta_k X_k + e$$

which is estimated by:

$$\hat{Y} = a + b_1 X_1 + b_2 X_2 + b_3 X_3 + \ldots\ldots + b_k X_k$$

where the b_is represent the partial regression coefficients.

The b_is represent the partial regression coefficients (i.e., b_1 represents the expected change in Y when X_1 is changed by one unit and the X_2 through X_k are held constant). It can also be interpreted as the simple regression coefficient b for the regression of Y on the residuals of X_1 when the effect of X_2 through X_k has been removed from X_1. The beta coefficients are the partial regression coefficients when all the variables (Y, X_1, $X_2 \ldots X_k$) have been standardized to mean 0 and variance 1 before estimating the regression equation. The intercept, a, and the bs are estimated by solving a system of simultaneous equations derived by differentiating and equating the partial derivatives to zero.

For measuring the strength of association in multiple regression and its interpretation, the total variation is decomposed as:

$$SS_y = SS_{\text{reg}} + SS_{\text{res}}$$

where

$$SS_y = \sum_{j=1}^{n} \left(Y_j - \bar{Y}\right)^2 \; SS_{\text{reg}} = \sum_{j=1}^{n} \left(\hat{Y}_j - \bar{Y}\right)^2 \; SS_{\text{res}} = \sum_{i=1}^{n} \left(Y_i - \hat{Y}_i\right)^2$$

The strength of association is measured as the square of the multiple correlation coefficient:

$$R^2 = \frac{SS_{\text{reg}}}{SS_y}$$

R^2 is large when the correlations between the independent variables are low. It cannot be less than the highest simple r^2 of any individual independent variables with the dependent variable. To overcome the diminishing returns found in R^2 with the addition of more independent variables to the model, R^2 is adjusted for the number of independent variables and sample size as:

$$\text{Adjusted } R^2 = R^2 - \frac{k\left(1 - R^2\right)}{n - k - 1}$$

where k is the number of independent variables and n is the sample size.

The significance testing involves testing the statistical significance of the overall regression equation as well as specific partial regression coefficients. The null hypothesis for the overall test is:

$$H_0 : R_p^2 = 0$$

The overall test can be conducted by using the statistic:

$$F = \frac{SS_{\text{reg}} / k}{SS_{\text{res}} / \left(n - k - 1\right)} = \frac{R^2 / k}{\left(1 - R^2\right) / \left(n - k - 1\right)}$$

This statistic has an F distribution with k and $(n - k - 1)$ degrees of freedom. If the overall null hypothesis is rejected, additional tests are carried out to determine which specific coefficients β_i have non-zero values. Testing for the significance of theβ_is is done by using t-tests or partial F-tests. In the partial F-test, significance of a specific coefficient β_i is done by using an incremental F-statistic:

$$F = \frac{SS_x / 1}{SS_{\text{res}} / (n-k-1)}$$

which has an F distribution with 1 and $(n - k - 1)$ degrees of freedom.

For further information, the principles of regression analysis, the meaning of partial regression coefficients, and the associated statistics are well documented in the book (Malhotra 2010).

6.1.3 Regression with Dummy Variables

The dummy variables as categorical variables are used as independent variables. This method is also called variable respecification. A nominal variable with k-categories is respecified into $(k - 1)$ dummy variables (with values 0 or 1) and the regression equation becomes:

$$\hat{Y}_i = a + \sum_{i=1}^{k-1} b_i D_i$$

For the kth category, $D_i = D_2 = \ldots D_{k-1} = 0$ and $\hat{Y}_i = a$ is the regression equation. This serves as a reference. For category i, $D_i = 1$ and all other $D_i = 0$. The regression becomes $\hat{Y}_i = a + b_i$. Here b_i represents the difference between the predicted $\hat{Y}_i$ for category i compared to category k. The dummy variable regression bears an equivalence with one-way ANOVA. Also n-way analysis of variance and analysis of covariance can be performed using regression with dummy variables.

6.2 Literature Review and Hypotheses

6.2.1 A New Agenda in Tourism Research

Innovation refers to the capacity to adopt and use new technologies in the production and management process. In the tourism industry in particular, the definition of technological innovation is subject to a number of peculiarities due to the specific characteristics of the industry. Despite this, information technology is often represented in tourism. In the past, information technology was mainly used to provide a platform for information sharing, communications, operational excellence in the management process, and business transactions. Now it has evolved to provide new business opportunities, customer relationship management, and direct marketing tools. What are the best examples of technological innovation, i.e., use of information technology, in tourism? Since technological innovation can be implemented in so many different ways, this question is pretty broad. Despite this, it can include the use of home automation system in a room, online reservation platforms, an online customer relationship management system, and the use of social media and Internet websites for their business. For tourism business, i.e., hotel, tourism agencies, and transportation company, information technology provides a new integrated communication channel for the business to enable and empower customers to interact with the company before, during, and after their experience of tourism services. For this purpose, the use of social media and Internet websites is absolutely necessary to keep customers coming back. For example, Hyatt Hotels launched its Twitter Concierge Service in 2009, providing answers to specific customer questions, making room and dinner reservations, and finding nearby attractions. In doing so, tourism firms are actively involving themselves in interaction with customers so that they are able to keep their customers happy and build loyalty. By offering better services on the front end and increasing efficiency on the back end, tourism firms can create the best experience possible for their customers.

In recent years, there has been a growing interest in the importance of technological innovation and its effect on the growth and profitability of firms in the tourism industry. We can reasonably assume that technological innovation can lead to the growth of tourism firms and increase profitability. In order to provide empirical evidence for the proposition, this study aims to investigate the economic spillover effects of technological innovation on firm financial performance in the tourism industry of Thailand. It would be interesting to see if technological innovation of tourism firms in Thailand is positively reflected in the profitability of the firm and then the greater the capacity to harvest the profits from technological innovation, the greater the financial benefits to promote technological innovation.

6.2.2 Overview of Innovation in the Tourism Industry of Thailand

Thailand has been one of the world's top tourism destinations since a long time ago as it is famous for its beautiful landscape and amazing sceneries ranging from mountains to beaches, historic landmarks, and cultural sites. Additionally, the cost of travel in Thailand is affordable, ranked 20th in global cheapest tourism destinations (Wade 2016). According to a report (Department of Tourism, Thailand 2016), the overall visitor number has been increasing over the recent 5 years, 2011–2015, even though it dropped to about 7% in 2014 due to the political unrest in Thailand. Thailand is still ranked the 4th world's top tourism destination by CNN (Neild 2016). In this light, Thailand tourism is not limited only to leisure, pleasure, recreation, and entertainment but also extends to healthcare service known as medical tourism. Thailand aims to serve first-class quality of healthcare service with affordable price. The Ministry of Public Health has implemented a medical hub policy to position Thailand as an international medical tourism center. As a result, Thailand has become one of the world's largest medical tourism markets (Thailand Board of Investment 2014).

In the tourism industry of Thailand, a number of tourism firms listed in the Stock Exchange of Thailand have applied technological innovation in order to increase their financial performance and customer satisfaction as well. The use of Internet websites plays an important role to drive technological innovation in the tourism industry of Thailand. Most of tourism firms exploit the benefits from *Web 2.0*. Only 3 out of 46 tourism firms do not provide their corporate websites. The other 43 tourism firms provide their websites with much information, including room rates, promotions, news, hot deals, spotlights, activities, and nearby attractions. The Internet website allows tourism firms to lead sales, launch promotions, and add tourism business portfolio. Tourism customers tend to book a room in advance through a hotel website because they can receive a better promotion on online booking (Charoenphon 2014).

In the medical tourism industry in Thailand, top hospitals also provide substantial information on their websites. The technological innovation in their business can be seen in the form of interactions between firms and customers and cooperation between firms and firms. Bumrungrad International Hospital is one of best practices of innovation-driven firm in Thailand. It is ranked 9th in the top ten hospitals in the world (Bumrungrad Holdings 2014). Bumrungrad has exploited the use of Web 2.0 providing six languages on its website, which is used as an effective communication tool. Prospective medical tourists can look up on the website to find information about medical treatment, doctors, medical tourism promotions and medical tour packages, as well as estimated cost of medical tourism and make an appointment. Bumrungrad claims to be one of the most trusted hospitals that post actual costs online through its *Real Cost* program (Patient Beyond Boarder 2015). Furthermore, Bumrungrad not only provides its own alternative accommodations such as Bumrungrad residences and suites, but it also cooperates with other hotels to accommodate medical tourists. Thus, the flow of communication is lifted by the use of Web 2.0. The cooperation between firms has allowed the firm to provide convenience and serve the best experience possible to customers. Consequently, this results in increasing overall financial performance in their business.

In the transportation industry in Thailand, many of the transportation firms also apply technological innovation to reduce operation costs, increase customer satisfaction, and adopt environmentally friendly solutions. Transportation firms introduce electronic service (*E-service*) through websites and mobile devices. For example, Air Asia allows customers to preorder their preference meals in advance through their website. E-boarding that displays on mobile devices and self-check-in are applied as environmentally friendly solutions (AAV PLC 2015). Furthermore, in 2015 Air Asia introduced Air Asia mobile application, which the firm claims that its app was one of the world's leading low-cost airline apps. The app provides convenience to customers to enable them to easily manage their flight schedule on mobile devices (The Star 2015). In addition, BTS is another example of firms using technological innovation to increase their financial performance. BTS introduced an online payment system, so-called Rabbit Card, which has been created since 2012. By partnering with AEON Thana Sinsap (Thailand) PLC, a credit and financial service firm in Thailand, BTS was able to provide convenient financial services at retail outlets and tourism sites through *Rabbit Card*. Consequently, it generates a stream of income through its network of stakeholders (BTS Group Holdings 2015). Aforementioned examples of applying technological innovation to improve the service quality and its profitability in the tourism industry also are wildly seen in other firms listed in the Stock Exchange of Thailand.

To some extent, the tourism industry of Thailand relies on social media to communicate with international visitors. Since the country has been facing political unrest, which affects the number of international visitors, the Tourism Authority of Thailand promoted the safety for international visitors through bloggers. The credibility of each blogger, who has millions of followers, could encourage his or her followers to be confident in the worry-free situation in Thailand. As bloggers are self-publishers and today storytellers, international visitors would quickly receive the message, which they believe is reliable through those bloggers (Hospitality Net 2014). In sum, those tourism firms in Thailand that employ technological innovation into their business are more likely to survive in the rapidly changing market and be able to increase their financial performance.

6.2.3 Innovation and Firm Performance

Technological innovation and technological proficiency can also affect economic performance of tourism firms. Knowledge spillover occurs when one company can use another's knowledge without or with compensation but at less than the market value. Market spillovers occur when competitive forces in the company offer a new product at a lower price. Some studies report that there was a positive correlation between technological innovation and firm performance in particular with stock price (e.g., Cooper and Kleinschmidt 2007; Hanel and St-Pierre 2002; Nobelius 2004). They report that while technological innovation may result in high uncertainty and unclear rates of return, companies that succeed in commercializing new technology tend to have higher market shares and premium pricing that lead to a competitive edge. Moreover, spillover effects of technological innovation positively drive new product and service developments in the firm. At the same time, competitors can also gain knowledge and help reduce its costs. The creation of new knowledge can bring significant economic growth to firms if the knowledge spreads to other departments (Decelle 2004; Scheidegger 2004). For example, developing a superior design or reducing overall costs for an entry into a new marketplace allows the firm to offer a greater value to customers and improve its competitive advantage (Hill and Jones 2009). Some research in this field reports that firms investing a large portion of their revenues in technological innovation tend to yield higher levels of firm performance than those that are investing lesser (e.g., Artz et al. 2010; Yeh et al. 2010). They report that firm growth rates are positively correlated with technological innovation capability.

However, the ability to measure the spillover effects of technological innovation on firm performance is rather limited. For example, techniques such as discounted cash flow models and internal rates of return are much difficult to measure and have several shortcomings. A study examines the spillover effects of technological innovation through cash flow and reports that the marginal rate of return exceeds the marginal cost of capital in Japanese companies (Mahlich and Roediger-Schluga 2006). An issue in the tourism industry, more than any others, is the potentially longer time horizon to determine returns. In addition, innovation costs are more difficult to measure as salaries and wages take up a lot of the expenditures. Varila and Sievänen (2005) looked at performance estimations of innovation projects; their analysis includes capital budgeting requirements and expected returns. They suggest an expected value analysis that allows for probability calculations and sensitivity analysis.

Some researchers discuss how high uncertainty of project assumptions and long lag times make simple models inadequate (e.g., Loch and Tapper 2002). A study finds that it has a lag period before commercialization while technology transfers can have a more rapid effect on productivity growth while it raises rates of innovation (Cameron et al. 2005). The common issues with risks and rewards induce managers to make a simple short-term decision instead of decisions toward high-potential, high-risk, and long-term projects. In fact, technological innovation strategies must identify trade-offs and priorities among conflicting goals. Performance measurements must consider the alignment of the project with overall corporate strategies as well as technological proficiency for key personnel and progression along with a learning curve.

According to Yeh et al. (2010), there is a level exceeding that the increased technological innovation does not yield proportional rewards. Return on equity, return on assets, and net profit growth were used as proxies for firm performance and technological innovation classified as the threshold variables. A study reports there is a strong correlation between technological innovation and sales growth: the study further reports the effect of innovation on several aspects of financial performance such as profit margins, return on assets, growth, and labor productivity measured by sales per employee (Morbey and Reithner 1990). The study reports there was a positive relationship between technological innovation and firm productivity. This indicates that profit margin would be governed by firm productivity and modified by technological innovation. Another study finds that technological innovation is positively related to firm performance (Mudambi and Swift 2013). In addition, it is said that top management has positive and direct impact on the level of technological innovation and affects firm performance as well (Gentry and Shen 2013; Kor 2006). For example, top management is likely to cut technological innovation expenditures when they position under pressure after missing their forecasts.

Considering technological innovation and economic output, technological innovation has been frequently used to increase economic outputs. A study reports that a great capacity of a firm to adopt both internal and external technologies is required in order to implement technological innovation that contributes a positive effect on the growth rate of the turnover (Haned 2010). That having said, if a firm finds their investment in technological innovation is not profitable, then the firm easily cut their investment in technological innovation (Rosenberg 2004). The firm, consequently, would not be fast enough to benefit from economies of scale. It is also said that private firms consider high costs and risks of innovation before they undertake technological innovation that is expected to be profitable for the firm (Weiermair 2004). In this light, the positive correlation between economic performance and technological innovation capability over time will be strong, as innovation output is significantly higher than the internal R&D cost (Haned 2010). However, for those firms in other business sectors between high technology and low technology, the cost and risk of innovation might be different: high or low. For example, the service sector is arranged in low technology, and it would not require a large investment for technological innovation (Thirtle et al. 2003). On the other hand, it is said that the tourism industry relies on the Internet, and it becomes one of the largest online businesses (Kim 2004; Rosenberg 2004). It is said that the increasing number of Internet users allows tourism firms to provide online tourism services.

6.2.4 Hypotheses

Tourism firms in particular are becoming increasingly aware of the critical role of technological innovation on their business performance. However only a few incorporated a long-term growth strategy in their business. While there is much existing literature about the relationship between technological innovation and firm performance in the manufacturing industry, studies that examine the relationship in the tourism industry in particular appear to be extremely limited. Hence, to fill the research gap, the authors propose the following hypotheses:

Hypothesis 1: Technological innovation is likely associated with firm performance in accounting-based measures in particular.

Hypothesis 2: Technological innovation is likely associated with firm performance in market-based measures in particular.

Hypothesis 3: Technological innovation is likely associated with organizational characteristics and the size of a firm in particular.

6.3 Data and Research Methods

6.3.1 Measurement Items and Data

This section describes data and outlines the methodology used in the selection of indicators and the normalization of data. The measurement item and descriptions of data and sources are described in Table 6.1.

The dataset includes a total of 188 data points (after excluding the data missing financial information) from 46 listed firms (12 firms in tourism, 16 firms in hospitality, and 18 firms in transportation) in the Stock Exchange of Thailand over a period

Table 6.1 Measurement items and descriptions of data

Measurement items	Descriptions of data and sources
Innovation	Determining how technological innovation is measured is critical but difficult. In some cases, researchers use corporate disclosure statements, company reports, and corporate news. However, there is no way to empirically determine whether the technological innovation information provided by companies is overreported or not. Therefore to some extent, the measurements of technological innovation are open to questions about subject bias. As an alternative, the authors propose the technological innovation capabilities of a firm such as electronic commerce capability and information technology capability. Corporate websites are used here as a proxy for the technological innovation capabilities that a firm possesses. A firm employs technological innovation by adopting information technology (i.e., using Internet, corporate websites, and social media) in a given year which gives a value of 1 and otherwise 0
Accounting-based measures	The short-term effect of technological innovation on firm performance is measured here by accounting-based measures (i.e., return on assets, return on sales, and return on the shareholders' equity). If a firm has performed well in technological innovation that is conductive to superior performance, it should have strong bottom-line benefits. To minimize the impact of accounting variations on accounting-based measures across firms, the authors use the data published in the Stock Exchange of Thailand
Market-based measures	The long-term effects of technological innovation on firm performance are measured by market-based measures (i.e., Tobin's Q and market value), which reflect the market's perception for earnings in the future. If a firm has performed well in technological innovation, then it should yield high investor expectations for future profits. Tobin's Q is a measure of the economic rents from long-term assets and the long-term growth expectations of a firm. If Tobin's Q is above 1, it means that the firm is worth more than the cost of its asset, which indicates that the firm is overvalued. Market value is a measure of the market capitalization of a listed company, which refers to the current price for a market-traded security and represents what investors believe the firm is worth
Control variables	Considering some factors that may influence the linkage between technological innovation and firm performance, the inclusion of some organizational characteristic variables is necessary. Although there would be some controversial issues in the selection of these organizational characteristics, there is a widely accepted control variable in literature. This variable includes firm size, among others. Firm size is viewed as one of the most validated variables in determining firm performance in terms of its effect on creating superior firm performance. A larger firm can afford the firm several economic benefits due to its ability to leverage their resources, even in comparison with other size firms. Economies of scale, low cost of sales, and the standardization of operations are a few of the benefits larger firms enjoy. Thus, larger firms with significant market power are likely to give a high priority to technological innovation and reap more profitability gains in the market (Mukhopadhyay and AmirKhalkhali 2010; Tsai and Wang 2005). However, experts appear to be split on the overall effect of firm size on firm financial performance. For example, small- and medium-sized high-tech firms in Japan tend to yield superior financial performance from their technological capabilities in comparison with large firms (Yasuda 2005). Therefore, it is expected that firm size is likely associated with firm performance. Per firm size, a dummy with a value of 1 indicates that the firm has been listed in the Stock Exchange of Thailand Index Series – Large Cap and otherwise 0, which indicates small- and medium-sized firms. The Large Cap refers to the 30 most representative stocks listed on the Stock Exchange of Thailand

Table 6.2 Descriptive statistics of the sample (2011–2015)

Descriptive	Tourism	Hospitality	Transportation	Total	
Statistics	Mean	Mean	Mean	Mean	Std. deviation
Technological innovation[a]	0.83	0.93	0.91	0.89	0.31
Return on assets[b]	2.05	13.25	2.70	6.42	11.54
Return on equity[b]	2.60	14.99	−2.21	5.73	18.36
Return on sales[b]	−10.49	11.12	−8.32	−1.78	57.63
Risk ratio[b]	0.88	0.59	1.52	0.97	1.15
Tobin's Q	1.30	2.34	0.80	1.53	1.52
Market value[c]	226	942	1232	803	2044

[a]If a company uses any type of information technology tools (i.e., Internet website) and innovation programs, it gives 1, otherwise 0
[b]Percent (%)
[c]Unit is million US dollars (an average exchange rate given during the period of 2011–2015 was 31 Thai Baht per dollar)

of five fiscal years, 2011–2015. Table 6.2 provides descriptive statistics of the sample.

6.3.2 Empirical Model

To investigate the effect of technological innovation on firm performance, a regression technique is employed. Multiple regression equations are as follows:

$$\sum_{1}^{k} Y_{ij} = \beta_{0j} + \sum_{1}^{q} \beta_{1j} \left(\sum_{1}^{q} X_{ij} \right) + \varepsilon_{ij} \tag{6.1}$$

$$\sum_{1}^{k} Y_{ij} = \beta_{0j} + \sum_{1}^{q} \beta_{1j} \left(\sum_{1}^{q} X_{ij} \right) + \beta_{2j} \left(\mathrm{CONT}_{ij} \right) + \varepsilon_{ij} \tag{6.2}$$

where Y_{ij} refers to the dependent variable, subscript i refers to the individual case, subscript j refers to the group, and k refers to the measurements of firm performance, including return on assets, return on the shareholders' equity, return on sales, Tobin's Q, and market value.

X_{ij} refers to the regressors; q refers to the explanatory variables, including technological innovation, revenue, and cost of sales.

CONT_{ij} refers to the regressor of control variables, including firm size and leverage ratio.

β_{0j} refers to the regression intercept.

β_{1j} and β_{2j} refer to the slope coefficients of the relationship between the regressors and the dependent variable.

ε_{ij} refers to the error term.

6.4 Empirical Results

Table 6.3 presents the correlations between the explanatory variables and the firm performance measures. Table 6.3 indicates that technological innovation has a positive correlation with accounting-based measures: return on assets ($p < 0.01$) and return on sales ($p < 0.05$) in particular. Table 6.3 indicates that technological innovation has no correlation with market-based measures: neither Tobin's Q nor market value.

A multiple regression is employed on the stacked data of 188 data points from 46 companies for the fiscal years 2011–2015 to test the hypotheses. Table 6.3 describes the results of the regression analysis with dummy variables.

In testing hypothesis 1 that technological innovation is likely associated with firm performance in the short term, i.e., accounting-based measures, Table 6.4 indicates that technological innovation is positively associated with accounting-based performance, in particular return on assets ($p < 0.01$) and return on sales ($p < 0.01$). The results support that technological innovation has a positive impact on short-term performance measures.

In testing hypothesis 2 that technological innovation is likely associated with firm performance in the long term, i.e., market-based measures, Table 6.4 indicates that technological innovation has no impact on long-term performance measures, neither Tobin's Q nor market value ($p > 0.1$).

Table 6.5 indicates the results of chi-square tests for the correlation between technological innovation and firm size. In testing hypothesis 3 that technological innovation is likely associated with the size of a firm, Table 6.5 indicates that technological innovation is not related to firm size ($p > 0.1$), which indicates that technological innovation can occur quite independent of firm size.

Table 6.3 Results of correlation coefficients between the variables

	Firm size	Innovation	ROA	ROE	ROS	Tobin's Q	Market value
Innovation	0.135						
ROA	0.211**	0.192**					
ROE	0.229**	0.097	0.893**				
ROS	0.141	0.184*	0.724**	0.792**			
Tobin's Q	0.198**	0.068	0.318**	0.306**	0.038		
Market value	0.704**	0.123	0.190**	0.179*	0.143	0.233**	
Risk ratio	−0.003	0.138	−0.265**	−0.451**	−0.276**	−0.228**	−0.024

Correlation is significant at the 0.05 level, two-tailed (*, $p < 0.05$ and **, $p < 0.01$)
ROA refers to return on assets; ROE refers to return on the shareholders' equity; and ROS refers to return on sales

Table 6.4 Results of multiple regression analysis with dummy variables

	Accounting-based measures					
Variables	Return on assets		Return on equity		Return on sales	
Technological innovation	0.535 [2.346]*	0.636 [2.864]**	0.249 [1.101]	0.401 [1.926]	0.530 [2.236]*	0.666 [2.982]**
Revenue	3.326 [3.875]**	3.263 [2.641]**	3.594 [4.220]**	2.861 [2.492]*	1.672 [1.876]	1.213 [0.976]
Cost of sales	−3.428 [3.988]**	−3.175 [2.570]**	−3.791 [4.443]**	−2.764 [2.409]*	−1.674 [1.874]	−0.954 [0.768]
Firm size		−0.079 [0.247]		0.098 [0.331]		0.021 [0.066]
Risk ratio		−0.322 [3.837]**		−0.507 [6.494]**		−0.446 [5.286]**
Constant	−0.426	−0.507	−0.166	−0.323	−0.454	−0.583
R-square	0.121	0.188	0.134	0.301	0.051	0.180
Adjusted R-square	0.106	0.165	0.121	0.281	0.035	0.156
F-statistic	8.234**	8.257**	9.350**	15.308**	3.244*	7.816**

	Market-based measures			
Variables	Tobin's Q		Market value	
Innovation	0.186 [0.788]	0.238 [1.018]	0.036 [0.381]	0.044 [0.513]
Revenue	2.212 [2.495]*	0.576 [0.443]	6.376 [17.932]**	4.211 [8.793]**
Cost of sales	−2.354 [2.650]**	−0.603 [0.464]	−6.202 [17.412]**	−3.999 [8.358]**
Firm size		0.506 [1.502]		0.730 [5.887]**
Risk ratio		−0.218 [2.467]*		−0.095 [2.940]**
Constant	−0.131	−0.263	−0.009	−0.139
R-square	0.057	0.099	0.668	0.732
Adjusted R-square	0.042	0.074	0.662	0.724
F-statistic	3.677*	3.925**	120.865**	97.458**

Note: The numeric values of coefficients are standardized coefficients. The numeric values in square brackets [] are t-statistics. Coefficients are significant at the 0.05 significance level (*, $p < 0.05$, and **, $p < 0.01$)

Table 6.5 Results of chi-square tests for the correlation between innovation and firm size

			Value	Std. Error	Approx. T[a]	Significance
Nominal by nominal	Phi		0.106			0.346
	Cramer's V		0.106			0.346
	Contingency coefficient		0.106			0.346
Ordinal by ordinal	Kendall's tau-b		0.098	0.066	1.465	0.143
	Gamma		0.250	0.164	1.465	0.143
	Somers' d	Innovation	0.167	0.111	1.465	0.143
		Firm size	0.058	0.040	1.465	0.143
Chi-square tests		Pearson chi-square	2.124			0.346
		Likelihood ratio	2.235			0.327

[a]Using the asymptotic standard error assuming the null hypothesis

6.5 Discussion and Implication for Managerial Actions

The study investigated the spillover effect of technological innovation on firm performance both in the short term and in the long term in the tourism industry of Thailand. The results of the study support that greater technological innovation results in superior financial performance in the short term, i.e., accounting-based measures. Greater technological innovation tends to lead to a reduction in costs, which results in superior financial performance in the short term. In this way technological innovation provides firms large opportunities and helps build up a competitive advantage in the market. Therefore it is essential to look more closely at those aspects that build up technological innovation capabilities and a greater diffusion of information technology. For example, the best practices in technological innovation in the tourism industry of Thailand include waste management, fresh water conservation, and energy efficiency. The best practices in technological innovation in the hospitality and healthcare industry of Thailand include robot doctor, consult online, and online innovation exhibition. The best practices in technological innovation in the transportation industry of Thailand include electronic boarding pass, self-check-in, fly thru, pre-order meal, membership card, flight simulation, and travel green competition to reduce operational costs. The findings of the study suggest that the adoption and implementation of technological innovation affect directly to cut operating costs, resulting in better short-term profitability ratios, in particular return on assets and return on sales.

The study found no conclusive evidence of the direct effect of technological innovation on firm long-term performance, i.e., market-based measures. Some plausible explanations include the following: the spillover effect of technological innovation on firm long-term performance is likely diversified or redirected to, and the time lag involved in realizing the long-term performance makes it more difficult to measure the direct effect of technological innovation on firm long-term performance. In other words, technological innovation directly affects to cut operating costs, thus resulting in better accounting-based bottom-line performance. However, it does not guarantee that the best practices in technological innovation would yield long-term profitability and better market value.

The study found that large firms generally allocate more resources to technological innovation than small- and medium-sized firms (mean = 0.89, standard deviation = 0.39), while the size of a firm does not seem to justify the firm performance associated with their technological innovation (correlation coefficient is 0.135, with $p > 0.05$). The findings suggest that large firms tend to yield higher market value (regression coefficient is 0.730, with $p < 0.01$) than the other. The finding also suggests that small- and medium-sized firms generally have less favorable expectations of investors for their market value than large firms, even though technological innovation likely occurs quite independent of firm size.

In sum, the findings of the study suggest that technological innovation be played an important role in improving firm short-term profitability. To adapt successfully to today's dynamic business environment, tourism management leadership generally has a very important role in determining which type of technological innovation firms can develop and how firms implement necessary policies in conjunction with technological innovation. It is clear for tourism business to develop such technological innovation capabilities that allow them to move more nimbly in today's dynamic business environment; the odds are better that they will ultimately succeed in the market.

6.6 Conclusion

The findings of the study show that technological innovation has a strong and positive impact on firm short-term performance, i.e., accounting-based measures, in particular, return on sales and return on assets in the tourism industry of Thailand. The study concludes that technological innovation has played an important role in improving firm short-term profitability in the tourism industry. The results support that a company that incorporates technological innovation and develops technological innovation capabilities is likely well positioned for success in the market. Continuous technological innovation and greater technological innovation are crucial to superior firm performance and growth.

The results, however, should be treated with caution because the results may be biased from the selection of samples used in this study. Further tests using a large number of samples with more data points can uncover different findings about the spillover effects of technological innovation on firm performance. Another direction for future study is the use of different measurement of technological innovation.

References

AAV PLC. (2015). *Annual report 2015*. Available from: http://www.aavplc.com/index.html?menu=ar&lang=en. Accessed 6 Apr 2016.

Artz, K. W., Norman, P. M., Hatfield, D. E., & Cardinal, L. B. (2010). A longitudinal study of the impact of R&D, patent, and product innovation on firm performance. *Journal of Product Innovation Management, 27*, 725–740.

BTS Group Holdings. (2015). *Annual report 2014/15*. Available from: http://bts.listedcompany.com/misc/ar/20150622-bts-ar-20142015-en-06.pdf. Accessed 6 Apr 2016.

Bumrungrad Holdings. (2014). *Top 10 best hospitals in the world*. Available from: https://www.bumrungrad.com/BIHFiles/db/dbeec672-199e-4a3b-a1f5-c17efdad4a2e.pdf. Accessed 6 Apr 2016.

Cameron, G., Proudman, J., & Redding, S. (2005). Technological convergence, R&D, trade and productivity growth. *European Economic Review, 49*, 775–897.

Charoenphon, V. (2014). *The future of Thailand's tourism industry*. Available from: https://www.scbeic.com/en/detail/product/458. Accessed 6 Apr 2016.

Cooper, R. G., & Kleinschmidt, E. J. (2007). Winning businesses in product development: The critical success factors. *Research-Technology Management, 50*(3), 52–66.

Decelle, X. (2004). *A conceptual and dynamic approach to innovation in tourism* (OECD Working Paper). Available from: https://www.oecd.org/cfe/tourism/34267921.pdf. Accessed 6 Apr 2016.

Department of Tourism, Thailand. (2016). *Visitor statistics 2015, 2014, 2013, 2012, 2011*. Available from: http://www.tourism.go.th/home/listcontent/11/221/276. Accessed 6 Apr 2016.

Gentry, R. J., & Shen, W. (2013). The impacts of performance relative to analyst forecasts and analyst coverage on firm R&D intensity. *Strategic Management Journal, 34*(1), 121–130.

Haned, N. (2010 September). *An evolutionary view of innovation and economic performance: A firm level analysis*. Presented at European Conference on Innovation and Entrepreneurship. Reading: Academic Conferences International Limited.

Hanel, P., & St-Pierre, A. (2002). Effects of R&D spillovers on the profitability of firms. *Review of Industrial Organization, 20*(4), 305–322.

Hill, C. W., & Jones, G. R. (2009). *Strategic management, an integrated approach* (9th ed.). Nashville: South-Western College Publishing.

Hospitality Net. (2014). *Thailand's tourism industry turned to social media and travel blogs to send the message that the country is safe for visitors*. Available from: http://www.hospitalitynet.org/news/4066125.html. Accessed 6 Apr 2016.

Kim, C. (2004). *E-tourism: An innovation approach for the small and medium-sized tourism enterprises (SMTES) in Korea* (OECD Working Paper). Available from: https://www.oecd.org/cfe/tourism/34268048.pdf. Accessed 6 Apr 2016.

Kor, Y. Y. (2006). Direct and interaction effects of top management team and board compositions on R&D investment strategy. *Strategic Management Journal, 27*(11), 11081–11099.

Loch, C. H., & Tapper, V. A. S. (2002). Implementing a strategy-driven performance measurement system for an applied research group. *Journal of Product Innovation Management, 19*, 185–198.

Mahlich, J., & Roediger-Schluga, T. (2006). The determinants of pharmaceutical R&D expenditures: Evidence from Japan. *Review of Industrial Organization, 28*(2), 145–164.

Malhotra, N. K. (2010). *Marketing research: An applied orientation* (6th ed.). Upper Saddle River: Pearson.

Morbey, G. K., & Reithner, R. M. (1990). How R&D affects sales growth, productivity and profitability. *Research-Technology Management, 33*(3), 11–14.

Mudambi, R., & Swift, T. (2013). Knowing when to leap: Transitioning between exploitative and explorative R&D. *Strategic Management Journal, 35*(1), 126–145.

Mukhopadhyay, A., & AmirKhalkhali, S. (2010). Profitability performance and firm size-growth relationship. *Journal of Business & Economics Research, 8*(9), 121–126.

Neild, B. (2016). Irresistible cities: World's 25 top tourism destinations. *CNN News*. Available from: http://www.cnn.com/2016/01/28/travel/most-visited-cities-euromonitor-2016/. Accessed 8 Apr 2016.

Nobelius, D. (2004). Towards the sixth generation of R&D management. *International Journal of Project Management, 22*, 369–375.

Patient Beyond Boarder. (2015). *Bumrungrad international hospital*. Available from: http://www.patientsbeyondborders.com/hospital/bumrungrad-international-hospital. Accessed 8 Apr 2016.

Rosenberg, N. (2004). *Innovation and economic growth* OECD (Working Paper). Available from: https://www.oecd.org/cfe/tourism/34267902.pdf. Accessed 8 Apr 2016.

Scheidegger, E. (2004). *Can the state promote innovation in tourism? Should it? The example of Switzerland* (OECD Working Paper). Available from: https://www.oecd.org/cfe/tourism/34268075.pdf. Accessed 8 Apr 2016.

Thailand Board of Investment. (2014). *Thailand, medical hub of Asia*. Available from: http://www.boi.go.th/upload/content/BOI-brochure2014-Medical-20140709_20272.pdf. Accessed 8 Apr 2016.

The Star. (2015). *AirAsia launches all-new mobile app*. Available from: http://www.thestar.com.my/business/business-news/2015/11/19/airasia-launches-allnew-mobile-app/. Accessed 8 Apr 2016.

Thirtle, C., Lin, L., & Piesse, J. (2003). *The impact of research-led agricultural productivity growth on poverty reduction in Africa, Asia and Latin America*. Available from: http://ecsocman.hse.ru/data/609/661/1219/118.pdf. Accessed 8 Apr 2016.

Tsai, K. H., & Wang, J. C. (2005). Does R&D performance decline with firm size? A re-examination in terms of elasticity. *Research Policy, 34*, 966–976.

Varila, M., & Sievänen, M. (2005, June 26–29). *Profitability estimation of pioneering R&D project*. Proceedings of 2005 AACE international transactions (pp. ES191–ES198), AACE international's 49th annual meeting. New Orleans.

Wade, R. (2016). *136 world cities ranked cheapest to most expensive: Backpacker index for 2016*. Available from: http://www.priceoftravel.com/world-cities-by-price-backpacker-index/. Accessed 8 Apr 2016.

Weiermair, K. (2004). *Product improvement or innovation: What is the key to success in tourism?* (OECD Working Paper). Available from: https://www.oecd.org/cfe/tourism/34267947.pdf. Accessed 8 Apr 2016.

Yasuda, T. (2005). Firm growth, size, age and behavior in Japanese manufacturing. *Small Business Economics, 24*(1), 1–15.

Yeh, M. L., Chu, H. P., Sher, P. J., & Chiu, Y. C. (2010). R&D intensity, firm performance and the identification of the threshold: Fresh evidence from the panel threshold regression model. *Applied Economics, 42*, 389–401.

Chapter 7
Quantitative Dimension of India's Tourism Trajectory in the Post-reform Era

K. R. Pillai

Abstract Tourism occupies a prominent place in the development strategy of the country, due to its vast potential and multiplier effect through spillovers and linkages. Tourism multiplier is an all-encompassing indicator to measure its economic essence owing to its impressive inter-sectoral linkages. India could leverage the travel-centric traits of the global population by creating a conducive destination attribute through enabling policy and advisory services. Both due to the splendid destination attributes and facilitating policy milieu, India claims a sizeable slice in the global visitation pie. The 'Incredible India' call has conferred ample mileage in its tourism campaign in the international fora. The various initiatives taken by the government to promote tourism as a growth engine have amply supported the phenomenal growth in visitation and its positive spillover in the country. In this backdrop, the proposed study traces the longitudinal trends of various tourism-related indicators, relevant to India, corroborating policy levels. The scope of the study is limited to foreign tourists' arrival in the country and its associated quantitative dimensions. The relevant data for the research will be elicited from Centre for Monitoring of Indian Economy's online repository through institutional access. The database comprises of foreign tourist arrival in gender and selected nationality dimensions, FOREX earnings from tourism, profitability and industry returns positions. The major analytical tools were CAGR, correlation, t-test, one-way ANOVA and MANOVA. The results of analysis confirm impressive strides in the variables taken for analysis.

Keywords Tourism · India · Economic reform · Tourism impact · Inbound tourism

K. R. Pillai (✉)
School of Management, Manipal Academy of Higher Education, Manipal, India
e-mail: Kr.pillai@manipal.edu

S. Rezaei (ed.), *Quantitative Tourism Research in Asia*, Perspectives on Asian Tourism, https://doi.org/10.1007/978-981-13-2463-5_7

7.1 Introduction and Background

Driven by the exuberance of pro-market reform experiences in the country, the Indian industry, in general, and tourism, in particular, have displayed astounding growth over the years. Tourism has already been established that high demand growth and product innovation have net job-creating effect in productive sectors (Pianta 2004). The tourism-led growth hypothesis has got much appreciation among academic researchers (Eadington and Redman 1991; Bezbaruah 2005; Li et al. 2013; Tugcu 2014; Cárdenas-García et al. 2015; Pillai 2017; Shahzad et al. 2017). Li et al. (2013) outline its potential to be a source of gains for public and private sectors through multiplier effects and can result in business expansion.

Among the service sector industries, tourism has a niche in itself due to its economic significance. Spreading over a wide range of sectors and activities, tourism has now become a prioritised development vector (Shahzad et al. 2017; Mathieson and Wall 1982; Edgell 1990). It has experienced an explosive growth from strenuous travel, bored sight-seeing and the rustic one-manned inn-keeping to luxurious travel, sumptuous sight-seeing and elegant multi-skilled and multimillion corporate hotel business, palatial travel facilities, panoramic landscapes and diversified entertainments and recreations. It implies that the growth of tourism industry is not only vertical but horizontal also. The growth of trade and commerce, man's inherent lust to travel and the development of transport and communication networks necessitated the refinement and reinforcement of the hospitality business. Along with the growth of the industry, it generates a multitude of direct, indirect and induced employment opportunities in a large number of interrelated sectors.

The significance of tourism, both from demand and from supply angles, has been rising over the years. In the economics parlance, the role of tourism is assessed in terms of its contribution towards employment, income and foreign exchange. The service sector economy is fast becoming significant in recent times. The main characteristic of the development of service economy is the removal of borderlines within production and consumption activities (Singh 2000). By generating employment and earning income, tourism plays a pivotal role in promoting service sector economy.

Tourism is an economic activity of immense global importance today. Ryan (2010) argues that tourism is both cause and consequence of global nature of business and travel networks of current era. As a development vector (Baladacchino 2013), it has already become the most important civil industry in the world (Lindberg and Hawkins 1999). Travel and tourism is one of the fastest growing industries in the world and accounts for 10.6% of the world's employment (Ahluwalia 2002). It has carved a niche for itself as a very effective instrument for generating employment, earning foreign exchange and thereby facilitating overall development. With its immense potentiality of labour absorption, the economic significance of tourism industry is especially unique.

India's tourism sector obtained a fresh lease of life with the embarkation of the country on the path of economic reforms during the early 1990s. The reforms intended to liberate its industry from unnecessary bureaucratic shackles and, thereby, to make its economy internationally competitive. There observed an explicit revival of industrial sector in the post reform scenario. Tourism industry has its preponderance owing to ease of travel and other elements of structural reforms. As per the World Travel and Tourism Competitiveness Index 2017, India has improved its rank by 25 points over the last 3 years (Crotti and Misrahi 2017). In this backdrop, the proposed study traces the longitudinal trends of various tourism-related indicators, relevant to India, lending due thoughts on significant macroeconomic transitions. The scope of the study is limited to foreign tourists' arrival in the country and its associated quantitative dimensions.

7.2 Economic Significance of Tourism

Tourism occupies a prominent place in the development strategy of the country, due to its vast potential and multiplier effect through spillovers and linkages. Tourism multiplier is an all-encompassing indicator to measure its economic essence owing to its impressive inter-sectoral linkages. As a cross-border service delivery, tourism has, unarguably, become a first-choice option in the development strategy of nation states across the world. Revenue generated from international tourist arrivals is a fresh infusion of resource for developmental activities in the destination region, through interindustry linkages. Hence, the entire process stimulates further economic activities in the host country and triggers further economic expansion (Li et al. 2013; Shahzad et al. 2017). Tugcu (Tugcu 2014) argues that as a major component of the overall economy, tourism occupies an important role in the development strategy of an economy. This notion holds an uncontestable hegemony in India's socio-economic landscape, which is reiterated by its resource potential and developmental needs. Development stimulates economic growth. Increases in tourist arrivals and/or tourism receipts subsequently lead to an increase in economic growth, suggesting that investments in the tourism industry will induce economic growth by increasing the income of the current workforce and creating new jobs within and beyond the tourism industry (Tugcu 2014; Shahzad et al. 2017).

Tourism, unarguably, qualifies for being considered for people-centric growth strategy (Pillai 2017). The potential of tourism to contribute to growth process can be identified in many respects. Firstly, tourism motivates the development in infrastructural facilities and thereby creates external economies. Secondly, it creates a multitude of employment, directly and indirectly, as the industry is highly personal oriented and comprises a cross-section of economic activities. Thirdly, foreign tourists' expenditure improves the balance of payment position of the host country.

Tourism being a global industry, an effort to boost tourism is an export-oriented strategy of growth (Gray 1974). Fourthly, the industry is not a single entity but a joint set of various activities that move in tandem and, therefore, reinforces inter-sectoral linkages. Mckee (1988) endorses tourism as an industry with strong backward linkage.

The economic significance of the industry can be better understood by tourism multiplier. The expenditure of tourists in a destination creates new employment, income and outputs in the region which, in turn, produce further impetus to macro-economic indicators. The ultimate multiplier effect is created through direct, indirect and induced impacts of tourist spending (Mathieson and Wall 1982). Primary effects are direct and immediate, and they involve the effect of actual tourist expenditure on all activities engaged in the destination. This causes an increase in employment and output to other sectors and areas to cater to the needs of the visitors and, thereby, fetch income to far and wide. Prevalence of inter-sectoral linkages among the different sectors in the destination/economy is essential to create total impacts in the overall economic landscape.

The multiplier effects of tourism are widespread, and the benefits are shared by an incredibly large number of service providers like lodging, food and beverages, handicrafts, local transport, guides and shopping (Bezbaruah 2005). In the total range of services, many of the service providers are not visible at all. When tourists move across borders, tourism becomes an export of a nation or region, though no commodities are being physically exported. Its return becomes a source of income and jobs for people involved directly, in providing tourism services, and indirectly support other sectors of the economy as the newly created income is spent in the purchase of other goods and services produced in the region (Eadington and Redman 1991). The total impact of these primary and subsequent flows of income in all sectors in the region is technically called multiplier effect. The size of the tourist multiplier is regarded as a significant measure of the economic benefit of the visitor expenditure as it reflects the interrelations among the economic system (Hall and Page 2014). Hence, tourism can ideally be branded as a catalyst for economic development through its multiplier effects on other sectors of the economy. The WTTC's (2017) latest report estimates the total contribution of tourism to the global GDP is 10.2% and that of global employment is 9.6% in the year 2016.

7.3 India's Resolve to Tourism Industry

The service sector is gaining increasing importance in Indian Economy, and tourism has its own niche in service industry. Having noted the immense potential of tourism industry as a catalyst of economic development, by bestowing impetus to other industries through forward as well as backward linkages, let us examine India's resolve to the development to fuel its economy. Identifying the significance of tourism, as a major engine of growth, the Committee of Special Group on Targeting Ten Million Employment Opportunities per year over the Tenth Plan Period points out the necessity of an appropriate tourism policy (Gupta 2002). The Committee is of the opinion to harness tourism's direct and multiplier effects for employment and poverty eradication in an environmentally sustainable manner. The following strategies have also been put forward to achieve the objectives of the National Tourism Policy.

1. Positioning and monitoring tourism development as a national priority activity
2. Enhancing and maintaining the competitiveness of India as a tourism destination
3. Improving India's existing tourism products and expanding these to meet new market requirements
4. Creating world-class infrastructure
5. Developing sustained and effective marketing plans and programmes (Gupta 2002) (p436)

Hannam and Diekmann (2010) remark that the government has taken initiatives to redevelop and rebrand many of its destination and begun to invest in tourism infrastructure on a proactive basis. The country has acknowledged the primacy of infrastructure development to facilitate the growth of tourism sector. Development of infrastructure connected to tourism destination is a top priority of India as per the latest national policy guidelines on tourism (GoI 2017). The policy milieu also envisages to promote tourism in a sustainable manner by active participation and partnership of various stakeholders and to project the country as a 365-day tourism destination (GoI 2017). Leveraging its comparative advantages in some specific sectors, certain niche tourism products like cruise, adventure, medical and wellness, golf, polo, etc. have been identified to overcome the seasonality nature of tourism in the country (GoI 2017).

7.4 Review of Relevant Literature

Ngee (2011) examines the perspectives of various stakeholders to position India as a global healthcare and medical tourism destination, using grounded theory framework. The study has underlined the importance of tourism as a medical hub in positioning efforts and role of public-private partnerships in exploring emerging avenues in healthcare tourism. Li et al. (2013) empirically establish the triangular causality among economic growth, tourism and exports, keeping government expenditure on tourism, physical capital, education and health as control variables, whereas Oh (2005), some years behind, could not find any empirical evidence for tourism-led economic growth in Korean economy.

Drawing upon the data from 144 countries, Cardenas-Gracia et al. (2015) concede that tourism as a driving force of economic development is more relevant in developed countries than of developing or least developed countries.

In their critical analysis on tourism in India, Hannam and Diekmann (2010) vividly outline the contemporary geography, political, economics, sociology, culture and environment in the country that can be impetus and deterrent to tourism promotion. The authors contempt that the country has always maintained a rather uneasy and ambivalent relationship with tourism.

7.5 Research Design

The research followed a longitudinal design and quantitative approach. The study captured the chronological trends in India's tourism industry since introduction of structural reforms in the country. No study, previously, has addressed that the current study attempted to take up. The relevant data for the research were obtained from Centre for Monitoring of Indian Economy's (CMIE) online repository through institutional access. The database comprised of foreign tourist arrival, FOREX earnings from tourism, total income, long-term investments, gross fixed assets, net worth, total capital and total and operating expenses, among others (refer table in annexure for details).

Selected indicators were presented in line graphs to capture their chronological transition. The CAGR (compound annual growth rate) of crucial indicators was calculated offering realistic growth metrics of such indicators. The analytical tools used to treat the data statistically were correlation, t-test, one-way ANOVA and MANOVA (multivariate analysis of variance). The MANOVA was loaded on selected indicators to examine the equality of vectors of means on multiple dependent variables across groups (Hair et al. 2011; Hinton and McMurray 2014). In this analysis, inflow of foreign tourists was taken as independent variable. The indicators such as total income from tourism, FOREX earning, long-term investments, gross fixed assets, total capital and profit after tax were considered as dependent variables, conceding the impact of tourist inflow on all these variables. The entire period, relevant for the study, was grouped into three to facilitate the examination of vector means' equality (Hair et al. 2011; Hinton and McMurray 2014).

7.6 Results and Analysis

This longitudinal study, which spans across two decades, has examined the trends in tourism in India to capture the quantitative dimension of tourism in the country. The major variables taken for analysis include annual data on arrival of foreign

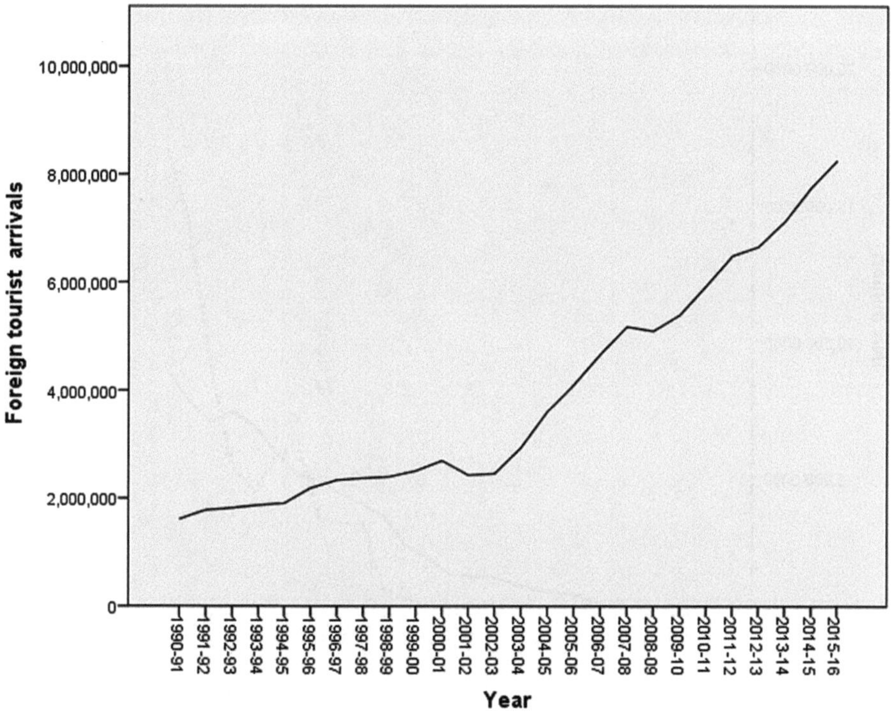

Fig. 7.1 Trends in annual arrival of foreign tourists. (Source: Current Research)

tourists, long-term investments, gross fixed capital, net worth, profit after tax, total expense, operating expense, total capital, total income and FOREX earning. Let us examine the chronological trends in key variables before we move on to statistical analysis. Annual visitation of foreign tourists, as the key variable in this study, will be examined first, as depicted in Fig. 7.1.

The trends on foreign tourists' arrival display overall growing pattern in the period of reference. However, the visitation has been treading on a steep growth path after a short snag during the period between the years 2000 and 2003, as is evidenced in the figure. The global cues of economic downfall can be witnessed from the transition of the line graph during the years 2007–2009. The CAGR estimated for visitation over the study duration is 6.67%.

Examining the long-term investment trends on infrastructure in tourism can be an explicit evidence of the inclination of the agencies/authorities concerned to facilitate future growth of tourism with long-term vision. Figure 7.2 depicts the longitudinal trends in long-term investments and gross fixed assets in the national tourism sector during the period of reference. Both the indicators have registered sectoral

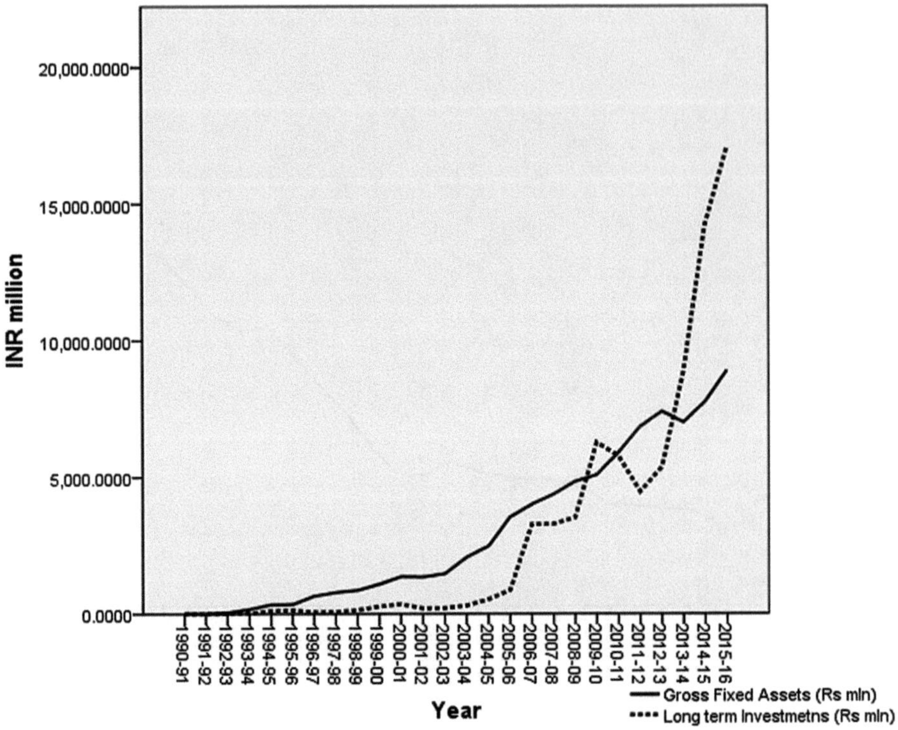

Fig. 7.2 Long-term investments and gross fixed assets. (Source: Current Research)

upward trends. However, long-term investment has recorded an impressive exponential growth, especially the period after the year 2005. Its growth has been steady and upward since the year 2012. The growth trends have been corroborated with the respective CAGR values. The long-term investments and gross fixed assets have registered 61.9% and 28.0% of CAGR, respectively, during the time.

Another result from the study is the net worth of Indian tourism industry. A steady rise in net worth is an uncontestable signal of good financial health. The study has taken profit after tax as well to understand the trends in effective profit of the industry over the years. Both are represented in Fig. 7.3.

The financial health of India's tourism industry is awfully impressive. The growth of net worth has registered an exponential pattern since the year 2008. Profit after tax has recorded an unsteady staggering growth over the years. The estimated CAGR for net worth and profit after tax are 40.21% and 33.54%, respectively.

Bivariate correlation was loaded on such variables as foreign tourists' arrival, long-term investment, FOREX earning, total income, gross fixed assets, net worth, profit after tax, total expenses and operating expenses of tourism industry. The analytical results are presented in Table 7.1.

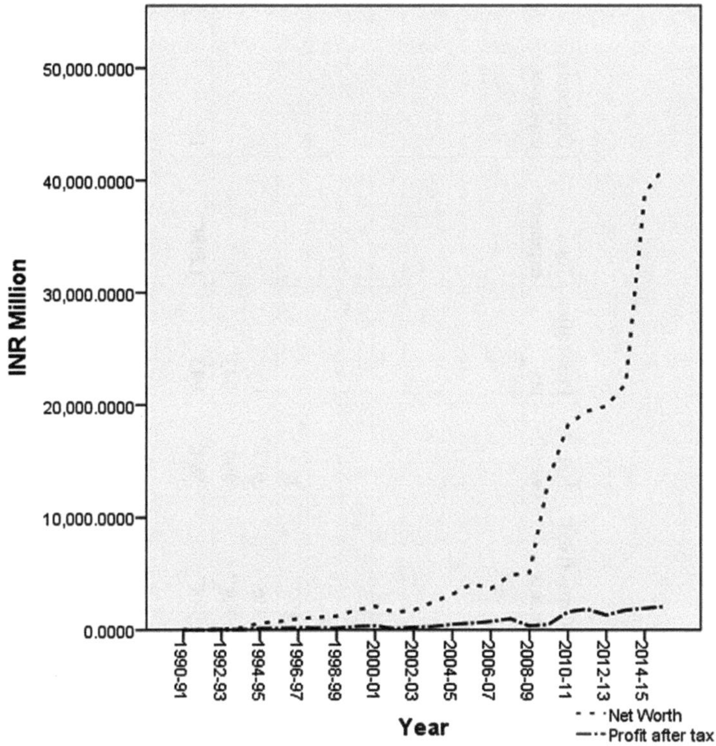

Fig. 7.3 Net worth and profit after tax. (Source: Current Research)

An examination of bivariate correlation coefficients indicates excellent and superior interdependencies among the variables of interest. All correlation results are statistically significant, manifesting robust theoretical relations among the macroeconomic variables of importance. Two-sample t-test was loaded on selected indicators to examine the mean variations during the periods prior and post global crisis. This tool is suitable to compare means of two vectors of any metric variables (Hair et al. 2011; Hinton and McMurray 2014). The results are presented in Table 7.2.

The results obtained from independent sample t-test offer ample empirical evidence to infer significant differences in the average values of all variables od relevance during the reference period. These trends were evidenced in the trend analysis in the beginning of this section.

The study has loaded one-way ANOVA to capture the dynamics of selected macroeconomic indicators in a more detailed way of shortening the duration of coverage of time. Indian economy has undergone series of policy-related changes in the macro level, which have repercussions on the country's tourism industry. Furthermore, the country experienced a major economic transition in the aftermath

Table 7.1 Correlations among selected variables

	Foreign tourist arrivals	Long-term investments	FOREX earning	Total income	Gross fixed assets	Net worth	Profit after tax	Total expenses	Operating expenses
Foreign tourist arrivals	1								
Long-term investments	.902**	1							
FOREX earning	.946**	.830**	1						
Total income	.988**	.872**	.930**	1					
Gross fixed assets	**.995****	.881**	.945**	.989**	1				
Net worth	.909**	.971**	.811**	.906**	.897**	1			
Profit after tax	.945**	.869**	.861**	.954**	.936**	.917**	1		
Total expenses	.987**	.867**	.932**	**.999****	.989**	.899**	.943**	1	
Operating expenses	.986**	.861**	.933**	**.999****	.988**	.894**	.943**	**1.000****	1

Source: Current research
Note: ** Correlations are significant at 0.000 level

Table 7.2 Mean comparisons (prior and after 2008): independent sample test

		Levene's test for equality of variances		t-test for equality of means						
									95% confidence interval of the difference	
		F	Sig.	t	df	Sig. (2-tailed)	Mean difference	Std. error difference	Lower	Upper
Foreign tourists' arrivals	Equal variances	1.063	.313	−10.089	23	.000	−4017485.118	398214.612	−4841254.806	−3193715.430
Total tax from tourism	Unequal variances	5.003	.035	−8.415	9.255	.000	−807.1052941	95.9075034	−1023.1558197	−591.0547685
Long-term investments	Unequal variances	27.893	.000	−4.476	7.167	.003	−7796.0970588	1741.7059091	−11895.2467660	−3696.9473516
Gross fixed assets	Equal variances	.268	.610	−10.208	23	.000	−5502.6463235	539.0676244	−6617.7926673	−4387.4999798
Net worth	Unequal variances	18.739	.000	−4.811	7.072	.002	−20735.4367647	4309.5623027	−30905.0466000	−10565.8269294
Total capital	Equal variances	1.948	.176	−11.935	23	.000	−2124.1654412	177.9753566	−2492.3355170	−1755.9953653
FOREX earning	Equal variances	.098	.758	−13.658	23	.000	−3193.2392647	233.8019739	−3676.8954973	−2709.5830321
Total income	Unequal variances	4.415	.047	−9.823	9.480	.000	−12779.15397	1300.91438	−15699.45077	−9858.85717
Profit after tax	Unequal variances	15.142	.001	−4.957	7.702	.001	−1176.2055882	237.2933906	−1727.1072537	−625.3039228
Total expenses	Equal variances	3.559	.072	−12.404	23	.000	−11602.6658824	935.4196345	−13537.7288281	−9667.6029366
Operating expenses	Equal variances	2.447	.131	−12.548	23	.000	−9430.97956	751.58174	−10985.74485	−7876.21426

Source: Current research

Note: Irrelevant rows were removed for ease of comparison and reference. The second column indicates the retained rows

Table 7.3 ANOVA results of selected variables

Indicators	F	Sig.
Foreign tourists' arrival	58.790	.000
Total tax from tourism	59.842	.000
Long-term investments	20.751	.000
Gross fixed assets	75.379	.000
Net worth	25.924	.000
Total capital	52.309	.000
FOREX earning	45.169	.000
Total income	78.779	.000
Profit after tax	24.194	.000
Total expenses	89.670	.000
Operating expenses	91.779	.000

Source: Current research

Table 7.4 Multivariate tests[a]

Effect		Value	F	Hypothesis df	Error df	Sig.
Intercept	Pillai's trace	.853	16.415[b]	6.000	17.000	.000
	Wilks' lambda	.147	16.415[b]	6.000	17.000	.000
	Hotelling's trace	5.793	16.415[b]	6.000	17.000	.000
	Roy's largest root	5.793	16.415[b]	6.000	17.000	.000
VISIT	Pillai's trace	.978	124.272[b]	6.000	17.000	.000
	Wilks' lambda	.022	124.272[b]	6.000	17.000	.000
	Hotelling's trace	43.861	124.272[b]	6.000	17.000	.000
	Roy's largest root	43.861	124.272[b]	6.000	17.000	.000
Group	Pillai's trace	1.177	4.286	12.000	36.000	.000
	Wilks' lambda	.156	4.341[b]	12.000	34.000	.000
	Hotelling's trace	3.279	4.373	12.000	32.000	.000
	Roy's largest root	2.386	7.158[c]	6.000	18.000	.001

Source: Current research
[a]Design: Intercept + VISIT + Group
[b]Exact statistic
[c]The statistic is an upper bound on F that yields a lower bound on the significance level

of global economic meltdown. Hence, vector means comparison test using ANOVA was thought to be more insightful. The entire span of time was grouped into three (1991–1999; 2000–2008; and 2009–2016). This statistical tool (ANOVA) compares the group mean across the data set. Table 7.3 indicates the analytical results obtained from NAOVA

The output of ANOVA has clearly evidenced statistically significant difference of mean across groups for all relevant macroeconomic variables taken for analysis. This inference is explicit from the statistically significant "p" values associated with the respective "F" values. Hence, ANOVA results support the statistically significant differences in the vector means of the relevant indicators across the grouped periods.

Table 7.5 Levene's test of equality of error variances[a]

	F	df1	df2	Sig.
Long-term investments	6.791	2	23	.005
Gross fixed assets	4.903	2	23	.017
Total capital	7.006	2	23	.004
FOREX earning	1.741	2	23	.198
Total income	18.618	2	23	.000
Profit after tax	7.821	2	23	.003

Source: Current research
Tests the null hypothesis that the error variance of the dependent variable is equal across groups
[a]Design: Intercept + VISIT + Group

The research also examined the impact of inflows of foreign tourists on such macroeconomic variables as total income, long-term investments, gross fixed assets, total capital, FOREX earning and profit after tax. MANOVA was loaded on the relevant indicators to examine simultaneous causation (Hair et al. 2011) of variation on multiple outcome variables. The relevant results are presented in Tables 7.4, 7.5, and 7.6.

Box's M test result, $F (42; 1524) = 3.743$; $p < 0.05$, indicates that the test fails to violate the assumption of equality of covariances of the dependent variables across independent groups (Hair et al. 2011). The multivariate test approves the overall statistically significant effect of independent variables on all dependent variables. It is evidenced from the empirically significant Wilks' Lambda test statistic (refer Table 7.4). Levene's test for equality of error variances indicates that the test fails to violate the homogeneity assumption of variances for each dependent variable, except FOREX earnings (refer Table 7.6). The tests' between-subject effects obtained from MANOVA confirm statistically significant effect of visitation on all relevant dependent variables that can be attributed by the independent variable.

7.7 Discussion

We have observed impressive growth in all macroeconomic variables taken for study. The increased flow of tourists to the country attests to the fact of meeting visitor satisfaction. The 'Incredible India' call has conferred ample mileage in its tourism campaign in the international fora. The various initiatives taken by the government to promote tourism as a growth engine have amply supported the phenomenal growth in visitation and its positive spillover in the country. Ryan (2010) concedes that tourists are increasingly used to exercising individual choice and preferences with suppliers of services in an era of ever-expanding customization. The results of the current study validate the observation of Hannam and Diekmann (2010) with regard to the outward looking and more proactive engagement of India government with the global tourism economy of late.

Table 7.6 Tests of between-subject effects

Source	Dependent variable	Type III sum of squares	df	Mean square	F	Sig.
Corrected model	Long-term investments	430340509.938[a]	3	143446836.646	40.450	.000
	Gross fixed assets	201775423.271[b]	3	67258474.424	1240.105	.000
	Total capital	27697028.032[c]	3	9232342.677	79.721	.000
	FOREX earning	61676280.699[d]	3	20558760.233	65.839	.000
	Total income	1027541913.027[e]	3	342513971.009	443.685	.000
	Profit after tax	10300238.032[f]	3	3433412.677	69.070	.000
Intercept	Long-term investments	45553805.107	1	45553805.107	12.846	.002
	Gross fixed assets	2861311.935	1	2861311.935	52.757	.000
	Total capital	217762.296	1	217762.296	1.880	.184
	FOREX earning	435510.465	1	435510.465	1.395	.250
	Total income	13253913.222	1	13253913.222	17.169	.000
	Profit after tax	782571.698	1	782571.698	15.743	.001
VISIT	Long-term investments	103250575.842	1	103250575.842	29.116	.000
	Gross fixed assets	25673387.204	1	25673387.204	473.363	.000
	Total capital	2903138.062	1	2903138.062	25.069	.000
	FOREX earning	7040649.727	1	7040649.727	22.548	.000
	Total income	116071027.762	1	116071027.762	150.356	.000
	Profit after tax	2577350.807	1	2577350.807	51.849	.000
Group	Long-term investments	16515831.662	2	8257915.831	2.329	.121
	Gross fixed assets	662274.619	2	331137.309	6.105	.008
	Total capital	209149.514	2	104574.757	0.903	.420
	FOREX earning	317213.895	2	158606.947	0.508	.609
	Total income	7417653.024	2	3708826.512	4.804	.019
	Profit after tax	123456.522	2	61728.261	1.242	.308

Source: Current research; Note: Unimportant rows were removed
[a]R Squared = .847 (Adjusted R Squared = .826)
[b]R Squared = .994 (Adjusted R Squared = .993)
[c]R Squared = .916 (Adjusted R Squared = .904)
[d]R Squared = .900 (Adjusted R Squared = .886)
[e]R Squared = .984 (Adjusted R Squared = .982)
[f]R Squared = .904 (Adjusted R Squared = .891)

The impressive contribution of tourism to Indian economy has theoretical underpinnings. This calls for the prominence to be given to tourism in the country. One of the prominent theories in development economics (Hirschman 1958) propagates that since the investible capital is scarce in developing countries, preference should be given to that sequence of projects that generate maximum induced investment.

Steady growth witnessed in the long-term investment in tourism is an encouraging factor. Previous research has noted that significantly higher amount of investment in India's tourism sector is an evidence of the country's preparation to respond to the growing number of tourist inflow (Pillai 2017). The destination image plays crucial role in making a destination popular. The economic importance of tourism should be discerned not in terms of the physical crowding in the destination, rather in terms of dynamics of its spending. The concept of destination image is a multisensory component (Pearce 2010), which represents the amalgam of beliefs, mental representations and emotional attachment to the destination (Ryan 2010; Ditcher 1985; Baloglu and McCleary 1999).

7.8 Conclusion

The results of the study have ample evidence to infer that tourism can be a niche industry for India. The country could leverage the travel-centric traits of the global population by creating a conducive destination attribute through enabling policy and advisory services. Both due to the splendid destination attributes and facilitating policy milieu, India claims a sizeable slice in the global visitation pie. The country's resolve to tourism sector, acknowledging its potential as an engine of growth, has been manifested through specific policy frameworks. Conducive policy regimes have rendered sufficient impetus to the longitudinal growth of tourism. The 'Incredible India' call has conferred ample mileage in its tourism campaign in the international fora. In addition, targeted marketing campaigns have amply supported the phenomenal growth in visitation and its positive spillover in the country. India's advantage in low-cost high-quality healthcare service facilities offers ample scope for tapping this niche market internationally.

This longitudinal examination has given ample evidences to conclude that tourism has been an integral part of globalization initiatives in the country. This inference has been corroborated by the robust CAGR in the arrival of foreign tourists and its consequent macroeconomic variables. The exponentially growing net worth shows the solvency position of the industry, indicating considerably rise in the financial robustness of the industry.

The study is beset with certain limitations as well. The current investigation has not considered the implications of gender and nationality in tourist inflow. Outbound tourism was not in the scope of this study, which seems to have compromised the coverage of tourism industry. Investment in tourism sector and destination-specific studies deserve detailed investigation.

Appendix: Selected Indicators (in Rs. Million, Unless Otherwise Mentioned)

Year	Foreign tourist arrivals (Nos)	Total tax from tourism	Long-term investments	Gross fixed assets	Net worth	Total capital	FOREX earning	Total income	Total expenses	Operating expenses	Profit after tax
1990–1991	1,613,681	0.3	0.1	18.6	8.8	3.5	145.9	53.3	51.8	47.5	1.5
1991–1992	1,781,892	0.3	0.1	18.6	8.8	3.5	145.9	53.3	51.8	47.5	1.5
1992–1993	1,820,239	0.2	17.9	45.9	51.2	7.5	0.4	114	54.2	35.8	59.8
1993–1994	1,871,262	10.7	24.3	172.8	182.8	49.2	516.7	335.2	301.9	255.8	32.5
1994–1995	1,907,322	100.9	114.4	343.2	622.8	120.6	665.8	753.5	609.6	445.9	143.8
1995–1996	2,190,334	109.4	134.3	349	748.3	170.2	160.2	724.3	571.6	394.6	152.6
1996–1997	2,334,438	137	66	651.6	1005.30	202.6	427	1117.60	923.1	700.3	194.5
1997–1998	2,370,985	96.2	78.7	786.4	1128.80	292.7	553.3	1345.10	1157.80	949.2	187.3
1998–1999	2,397,457	78.9	140.3	859.5	1228.60	302.4	546.9	1900.20	1737.50	1439.50	162.7
1999–2000	2,505,446	125.7	272.5	1084.80	1761.20	327.6	432.8	2256.00	1941.20	1586.50	314.8
2000–2001	2,694,287	164.4	358.6	1360.30	2109.50	510.7	991.7	2691.70	2328.20	1830.60	363.5
2001–2002	2,428,383	110.87	203.7	1346.30	1575.60	490	569.9	2521.88	2400.21	1949.11	121.67
2002–2003	2,453,622	156.1	210.1	1463.70	1754.60	491.6	861.8	2651.90	2396.90	1882.60	255
2003–2004	2,933,061	191.9	305.7	2058.90	2477.50	551.6	786.7	2672.90	2376.90	1870.30	296
2004–2005	3,603,165	313.7	528.3	2473.00	3158.60	811.2	1032.80	4756.60	4247.90	3528.70	511.6
2005–2006	4,100,283	449.9	870.4	3542.20	4095.20	925.3	1886.20	5666.60	5062.80	4105.30	604.8
2006–2007	4,667,396	493.24	3280.00	4007.80	3657.00	1185.30	1798.74	7166.93	6415.29	5292.87	751.24
2007–2008	5,174,726	535.4	3291.00	4374.00	4853.20	2402.50	4246.30	9124.40	8120.80	6648.90	1003.20
2008–2009	5,092,721	653.5	3522.50	4842.30	5081.00	2668.70	4056.60	9516.70	9170.90	7357.40	345.8
2009–2010	5,387,565	625.3	6260.40	5078.20	13,375.10	2341.80	3376.10	10,142.00	9638.40	7995.70	503.6
2010–2011	5,928,714	852.8	5774.30	5897.80	18,240.90	2279.20	3752.42	13,964.26	12,354.82	10,383.40	1609.44
2011–2012	6,491,828	908.1	4467.50	6839.40	19,431.90	2133.00	3946.00	16,362.90	14,519.40	12,025.00	1843.50
2012–2013	6,657,514	920	5365.20	7393.50	19,916.80	2039.80	3270.40	16,792.60	15,477.10	12,330.90	1315.60
2013–2014	7,123,086	1282.30	8853.50	6994.00	21,898.20	2060.60	3350.10	16,850.20	15,088.50	12,234.70	1761.70
2014–2015	7,756,209	1200.90	14,211.90	7761.50	38,888.00	2865.90	4191.26	18,296.00	16,383.06	13,213.48	1912.94
2015–2016	8,254,363	1209.10	17,021.90	8900.40	41,086.70	3637.50	5025.50	17,617.00	15,544.30	12,313.60	2072.70

Source: Economy Indicator, Centre for Monitoring Indian Economy

References

Ahluwalia, M. S. (2002). Report of the task force on employment opportunities. In R. Kapila & U. Kapila (Eds.), *Planning commission reports on labour and employment*. New Delhi: Academic Foundation.

Baladacchino, G. (2013). *Global tourism and informal labour relations: The small-scale syndrome at work*. London: Routledge.

Baloglu, S., & McCleary, K. W. (1999, October 1). A model of destination image formation. *Annals of Tourism Research, 26*(4), 868–897.

Bezbaruah, M. P. (2005, Jun). Prospects for tourism. *Seminar, 550*, 32–35.

Cárdenas-García, P. J., Sánchez-Rivero, M., & Pulido-Fernández, J. I. (2015, March). Does tourism growth influence economic development? *Journal of Travel Research, 54*(2), 206–221.

Crotti, R., & Misrahi, T. (2017). *The Travel & Tourism competitiveness report 2017. Paving the Way for a More Sustainable and Inclusive Future*. Geneva: World Economic Forum.

Ditcher, E. (1985). What is an image? *Journal of Consumer Marketing, 2*, 39–52.

Eadington, W. R., & Redman, M. (1991, January 1). Economics and tourism. *Annals of Tourism Research, 18*(1), 41–56.

Edgell, L. D. (1990). *International tourism policy*. New York: Van Nostrand Reinhold.

GoI. E-book of the Ministry of Tourism, Government of India; 2017 [Accessed 5 July 2017]. Available from http://tourism.gov.in/sites/default/files/E-book-of-the-Ministry-of-Tourism-2016-17-as-on-May-22-2017.pdf

Gray, H. P. (1974, September). Towards an economic analysis of tourism policy. *Social and economic studies, 1*, 386–397.

Gupta, S. P. (2002). Report of the special group on targeting ten million employment opportunities per year over the tenth plan period. In R. Kapila & U. Kapila (Eds.), *Planning commission reports on labour and employment*. New Delhi: Academic Foundation.

Hair, J. F., Black, W. C., Babin, B. J., Anderson, R. E., & Tatham, R. L. (2011). *Multivariate data analysis*. Prentice hall: Upper Saddle River.

Hall, C. M., & Page, S. J. (2014). *The geography of tourism and recreation: Environment, place and space*. Abingdon: Routledge.

Hannam, K., & Diekmann, A. (2010). *Tourism development in India: A critical introduction*. Abingdon: Routledge.

Hinton, P. R., McMurray, I., & Brownlow, C. (2014). *SPSS explained*. Abingdon: Routledge.

Hirschman, A. (1958). *Strategy of economic development*. New Haven: Yale University Press.

Li, C. C., Mahmood, R., Abdullah, H., & Chuan, O. S. (2013, May). Economic growth, tourism and selected macroeconomic variables: A triangular causal relationship in Malaysia. *Margin: The Journal of Applied Economic Research., 7*(2), 185–206.

Lindberg, K., & Hawkins, D. E. (Eds.). (1999). *Ecotourism; A guide for planners and managers* (Vol. 1). Dehra Dun: Natraj Publishers.

Mathieson, A., & Wall, G. (1982). *Tourism, economic, physical and social impacts*. Harlow: Longman.

McKee, D. L. (1988). *Growth, development, and the service economy in the third world*. New York: Praeger Publishers.

Ngee, K. N. (2011). Towards a 6Cs positioning typology of medical tourism destination in India. *ASEAN Journal on Hospitality and Tourism., 10*(1), 27–38.

Oh, C. O. (2005, Febuary 1) The contribution of tourism development to economic growth in the Korean economy. *Tourism Management* 26(1):39–44.

Pearce, P. L. (2010). *Tourist behaviour: Themes and conceptual schemes*. New Delhi: Viva Books.

Pianta, M. (2004). The impact of innovation on jobs, skills and wages. *Economia e Lavoro., 1*, 7–26.

Pillai, K. R. (2017, July 3). Tourism in South Asia: An economic leverage to India. *Asia Pacific Journal of Tourism Research., 22*(7), 709–719.

Ryan, C. (2010). *Recreational tourism: Demand and impacts*. New Delhi: Viva Books.

Shahzad, S. J., Shahbaz, M., Ferrer, R., & Kumar, R. R. (2017, June 1). Tourism-led growth hypothesis in the top ten tourist destinations: New evidence using the quantile-on-quantile approach. *Tourism Management, 60*, 223–232.

Singh, S. (2000). Employment in restaurant and hotel sector in India. *The Indian Journal of Labour Economics., 43*(4), 963–976.

Tugcu, C. T. (2014, June 1). Tourism and economic growth nexus revisited: A panel causality analysis for the case of the Mediterranean region. *Tourism Management, 42*, 207–212.

WTTC. Travel & Tourism: Economic Impact 2017. World Travel and Tourism Council. 2017 [Accessed 10 May 2017]. Available from https://www.wttc.org/-/media/files/reports/economic- impact.2017/world2017.pdf

Chapter 8
A Quantitative Analysis of Intercultural Communication and Personality: The Case of Coffee Shop Consumers in Cambodia

Ali Ihtiyar, Osman Nuri Aras, and Mustafa Öztürk

Abstracts The article aims to better understand the impact of intercultural communication and personality on customer satisfaction and word of mouth and intention to revisit and to pay more in coffee stores in Cambodia. In this endeavour, the study illustrates the unique context of intercultural communication to highlight several improvements and to encourage the advancement of intercultural communication in the hospitality and tourism literature in Cambodia. Based on the established theories, the study assesses the role of customers' personality on intercultural communication competence and perceived cultural distance and its impact on interrole congruence and interaction comfort empirically. The research assists in strengthening communication strategies, which is required intercultural communication adjustments in multicultural hospitality business environment. The intercultural communication is expected to improve the tourism industry competitiveness when it positively influences interrole congruence and interaction comfort among service encounters in the industry. To initiate the research, data was gathered by questionnaires within selected coffee stores in Phnom Penh, Cambodia. The measurement of the constructs and their interrelationships were examined based on partial least square structural equation modelling (PLS-SEM). The proposed framework provides partly statistically significant relationships among the constructs. Furthermore, the study reveals additional insights into some managerial and theoretical solutions for addressing the intercultural communication of service encounters in the Cambodian hospitality industry. These contributions will postulate an impetus for future research in various service settings.

Keywords Personality · Intercultural communication · Customer satisfaction · PLS-SEM · Coffee shops and Cambodia

---

A. Ihtiyar (✉)
Azman Hashim International Business School, Universiti Teknologi Malaysia, Johor, Malaysia

O. N. Aras
Zaman University, Phnom Penh, Cambodia

M. Öztürk
Burch University, Sarajevo, Bosnia and Herzegovina

S. Rezaei (ed.), *Quantitative Tourism Research in Asia*, Perspectives on Asian Tourism, https://doi.org/10.1007/978-981-13-2463-5_8

8.1 Introduction

Understanding consumption patterns of customers is a topic not far from mind when trying to understand the key features of our age. It is popular interest for marketer due to its economic, financial, operational roles and its impact on the process of pre-/post-purchasing part. It is also significant indicator for individuals because it is a way of marking social status.

Concentration on consumption patterns of customers is also an impressive theme, and it is not relatively clear precisely where sociological, psychological and managerial studies should focus. It involves many disciplines with various layers. One of the crucial difficulties for studies of consumption has been the difficulty of interdisciplinary connection and of finding an explanation, which might be appropriate for the different disciplines. For years, sociology defines consumption as involving the choice, buying, having, use, repair and disposal of any products and/or services. This definition links and involves.a lot of ground and makes studies of many different kinds relevant to understanding the phenomenon of consumption patterns of customers, including public provision of waste services, big data analysis, formal economic exchange, household management, retailing, real estate management and so on. Among them, one of the key points for understanding of customer's consumption patterns is culture.

Culture was renominated as a fundamental part of everyday life, which comprised larger aesthetic modules than earlier as a consequence of commercial uses of aesthetic design (Jameson 1998, Haug and Bock 1986) which is required procedures to the understanding and interpreting of cultural artefacts. Culture influences the priorities customers place on certain activities, consuming features and perceptions of services and/or products (Polsa et al. 2013). Therefore, understanding the cultural differences of culturally diverse customers is an essential outline for understanding customer behaviour in an intercultural marketing context. Understanding themes and concepts based on a multicultural perspective has gained the attention of marketing researchers as a means to explain the attractive features, price perceptions and service satisfaction among customers (Huang et al. 2016; Ihtiyar and Ahmad 2015; Polsa et al. 2013).

The significance of service industries has increased expressively within the past decades, particularly among industrialised countries. During the period of development, new challenges for service providers have arisen due, in part, to the varying demands of culturally diverse customers. Thus, service providers should create an efficient, innovative, competitive and steady marketing orientation for national and/or international markets (Ihtiyar and Shannon 2016). Furthermore, driven by the recent growth of internationalisation and the mobilisation of services, a number of studies have tested and investigated relationships between culture and customer satisfaction (Kuo et al. 2009), purchase behaviour (Souiden and Pons 2009), loyalty (Asiah Omar and Musa 2011) and interaction of service encounters (Sharma et al. 2012). The rationale of these studies within the context of culture is understanding, analysing and improving the comprehensive solutions for service settings in the societies (Matos and Leis 2013). Furthermore, the cognitive inferences of the numerous implementations in practical and/or theoretical fields present culture as a significant and remarkable determinant in the marketing context (Wang and Mattila 2010; Souiden and Pons 2009), leading scholars or marketers not to disregard the influence of intercultural issues on customer behaviour.

Cultural indicators in a business discipline are becoming the most popular subject among Asian societies, particularly Pacific-Asia (Awang-Rozaimie et al. 2011). Pacific-Asia is one of the distinctive samples of a multicultural atmosphere that involves favourable intercultural communication competence and understanding in sustaining a harmonious cross-cultural relationship (Awang-Rozaimie et al. 2011). Among the Asian countries, Cambodia is one of the well-known destinations for travellers, and it is a country with a multicultural or multiracial peculiarity with a greater need for intercultural understanding (Tham et al. 2017; Chang 2015). For instance, according to statistics, which is obtained from Tourism Statistics Report (2017), there is significant increase on number of visitors from 2012 (3,584,307) to 2016 (5,011,712), and the rise percentage was almost 40%. Regarding the diversified traveller profile of Cambodia, the visitors come from ASEAN countries to Oceania from Northern Europe to North America and from Africa to Middle East; please see the further statistics at Table 8.1 (Tourism Statistics Report 2017).

Due to the international mobilisation and cultural discrepancies, there is a clear deficiency of behaviour homogeneity where the nature of Cambodia's tourism market

Table 8.1 International tourist arrivals to Cambodia by country of residence 2016 and 2017

Regions	2016			2017			Share	Change
Country of residence		Purposes of visit			Total	Female	(%)	(%)
		Holiday	Business	Others			2017*	2017*/16
ASEAN (Southeast Asia)	295,474	300,006	13,831	23,680	337,517	145,739	32.9	14.2
Northeast Asia	309,490	307,270	25,458	4,245	336,973	141,102	32.9	8.9
Southern Asia	8,880	9,393	910	981	11,284	3,094	1.1	27.1
Oceania	32,225	30,452	978	2,685	34,115	15,282	3.3	5.9
Northern Europe	51,023	54,285	1,485	590	56,360	23,634	5.5	10.5
Western Europe	86,038	90,694	2,138	3,084	95,916	42,520	9.4	11.5
Central/Eastern Europe	28,731	32,835	465	305	33,605	16,589	3.3	17.0
Southern Europe	17,183	18,910	414	473	19,797	8,655	1.9	15.2
North America	67,005	66,552	1,734	5,803	74,089	34,413	7.2	10.6
Central America	272	204	9	19	232	67	0.0	−14.7
South America	9,988	13,864	53	150	14,067	6,734	1.4	40.8
North Africa	186	173	12	17	202	79	0.0	8.6
Sub-Saharan Africa	1,447	1,398	242	161	1,801	760	0.2	24.5
Middle East	4,468	4,610	88	123	4,821	1,931	0.5	7.9

Source: Tourism Statistic Report (2017)

is highly categorised by ethnically segmented customers (Carter et al. 2015). Earlier studies on customer's purchase decisions explored ethnic background as a significant predictor of shopping decisions (Ihtiyar and Shannon 2016). It is, therefore, significant for academicians and practitioners to identify and understand cultural individuality to deliver value to a specific segment (Ihtiyar and Ahmad 2015, 2014). Besides other contexts in marketing, multiculturalism has great significance and consistent application in the society. However, there is limited consideration on the subject while empirical studies are marginal, particularly with regard to intercultural communication competence's efficiency in general, especially for ethnically different customers.

This study is primarily of an exploratory nature, as it purposes to establish causal relationships between variables (Saunders et al. 2009). The aim of this exploratory research is to examine the structural relationships between multicultural personality, intercultural communication, customer satisfaction and post-purchase stage dimensions which are, namely, word of mouth, intention to pay more and intention to revisit among the Cambodian Generation Y. This study is structured in several sections. Initially, introduction as a comprehensive overview of the research, research objectives, foundation and importance of the study were discussed. Then, the theoretical background and an inclusive literature review are stipulated to investigate the influences of antecedents of customer satisfaction and its impact on customer satisfaction and total effect on post-purchase intentions. Thirdly, a validation of the indicated research methodology such as selected statistical tests, measurement of constructs, research design, sampling plan and data collection methods, common method bias and variance are delivered. It is then continued by the research results and discussions. Finally, conclusion, recommendations and research limitations are deliberated.

8.2 Literature Review

8.2.1 An Integrated Framework

The projected conceptual framework for analysing customer satisfaction with the antecedent of intercultural communication is presented in Fig. 8.1. The framework incorporates and synthesises research within an overall model based on role theory, interdependence theory and cognitive consistency theory. The model concentrates on the customer perceptions and whether the proposed dimensions affect their perception positively or negatively. Furthermore, personality (MP) and perceived cultural distance (PCD) are assumed constructs of intercultural communication, and customer satisfaction (CS) is contemplated as a consequence of customers' shopping experience as highlighted in the study of Oliver (1981).

Moreover, the study was conducted in the context of hospitality industry of Cambodia, whereby limited studies related to specifically trendy coffee shops have examined various topics in the hospitality context in Cambodia. In recent years, hospitality industry in Cambodia has experienced rapid growth. Focusing on travel-

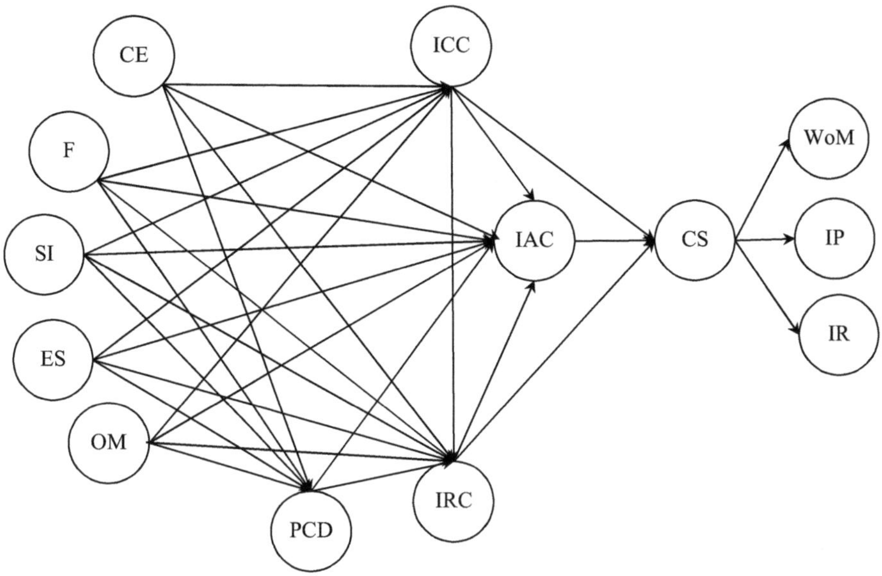

Fig. 8.1 Research framework

ler (as a customer) satisfaction has become a major goal in the industry, particularly in trendy coffee shops, and, thus, a complete understanding of its antecedents is essential for service providers and practitioners (Ihtiyar and Ahmad 2015). Although there are various studies on interrelationships of customer satisfaction with product categories, industry players, pricing, store atmosphere and so on, however, limited studies have investigated interrelationship of intercultural communication (with antecedents and consequences) and customer satisfaction within the trendy coffee shops in Cambodian tourism market.

Furthermore, recent studies have called for attachment of intercultural communication in different service settings and markets (Tam et al. 2014; Sharma et al. 2012). The majority of the studies that investigate the Cambodia context are related to employee (domestic perspective) (Morrow 2015), education (Masnan and Nagjib 2016; Miller et al. 2016), technology usage (Murphy et al. 2016), public relation (Hashim and Mahpuz 2011) and heath communication (Satawedin 2017). According to their suggestion (included earlier literature review), empirical studies examining emotional, cognitive, and behavioural factors of intercultural communication simultaneously are still limited in the context of Cambodia. It is, therefore, significant for academicians and practitioners to identify and understand the impact of intercultural communication on customer satisfaction to deliver value to a specific segment (Sharma et al. 2012).

Another potential strength of the study lies in its capability to recommend approaches to reduce communication problems and to increase awareness among service encounters who possess dissimilar languages, ethnic beliefs, religious, values, norms and cultural backgrounds. The competencies here could include knowing the values of the different societies, their religion, ethnic beliefs, culture, language and practices in culture, as well as their verbal and nonverbal communication. This study will make recommendations for managerial, policymaker and practitioner level to enhance the intercultural service encounter's communication skills effectively. This is particularly significant for retailers, which consider the service encounters have different cultural backgrounds.

Considering these gaps, this study contemplates multicultural personality (MP) as an antecedent of intercultural communication and interaction comfort (IAC), interrole congruence (IRC), and customer satisfaction (CS), as the consequence to examine integrated view of intercultural communication. Accordingly, a comprehensive framework is established, and it is contemplated as a momentum point to better understand the intercultural communication of service encounters in Cambodia tourism industry.

8.3 Exploring Theoretical Background

This review of literature will focus on the theoretical background of the research, which focuses on role theory, interdependence theory and cognitive consistency theory. It will also explain the research determinants of the proposed research framework by understanding its embedded theory and past literature. Then, it will suggest the research propositions that explain the interrelationships between the determinants.

8.3.1 Role Theory

One of the significant topics of the current concentration in marketing is the importance of the person-to-person interaction in terms of intercultural contexts and its overall influence on basic marketing outcomes such as the level of customer satisfaction and loyalty (Ihtiyar et al. 2013; Sharma et al. 2012; 2009). Contrary to *mixed product/service and pure service,* situations are much more tangible than *mixed product/service* situations, and customer satisfaction and positive word of mouth are determined by the communication competence of employees (Sharma et al. 2012; Solomon et al. 1985). Successful person-to-person interaction is becoming a crucial indicator to decrease dissatisfaction levels and create powerful strategies for many *pure service*-oriented or *mixed product/service* providers (Sharma et al. 2009; Paswan and Ganesh 2005). To an extent, all employees who are involved in customer contact are a service provider and each employee represents the company, explains the product, directly promotes it to customers and is first to receive the feedback (Solomon et al. 1985). Therefore, the employee becomes a central point of the interaction for customers and service providers. However, even if the role of employee in interaction is significant, the interaction is not defined as linear. It is a reciprocal and interactive experience among the encounters. In other words, the person-to-person interaction between customer and employee is a dyadic interaction (Negi 2009; Evans 1963), and as mentioned in interdependence theory (Surprenant et al. 1983), each part of the interaction has an impact on another part because the behaviour of the second part is influenced by the outcomes of the first part. With respect to this definition, explaining and clarifying the role of encounters (employee and customer) in this interaction play a significant role in the buying or selling process (Nicholls 2011; Sharma et al. 2009; Solomon et al. 1985).

After the brief introduction on nature of interaction, this study introduces to role theory for constructing theoretical background. The role theory is one of the fundamental concepts for different disciplines that included studies in psychology and sociology (Opler 1957), dyadic social exchanges (Emerson 1976) in anthropology, socialisation of individuals (Moreno 1940), multi-person interactions (Bales 1955) in social psychology (Solomon et al. 1985), motivation (Oliver 1981), dyadic quality of personal selling (Weitz 1981; Webster 1968) and management of service encounters in marketing literature (Broderick 1999).

The theory developed in the 1970s leading Biddle (2013) and Solomon et al. (1985) to conclude that it was a bridge between the processes of social participation and intrapsychic life (Ackerman 1958). According to this concept, the theory is formulised by various interdisciplinary studies to better understand the behaviour of the encounters (Sizoo et al. 2004; Broderick 1999). The well-known definition of the theory proposed by Biddle (2013) is "….A science concerned with the study of behaviours that are characteristic of persons within contexts and with processes that produce, explain or are affected by these behaviours". In accordance with this definition, social cohesions and environments affect individuals' personality and individuality as an integrative model of behaviour (Solomon et al. 1985). Therefore, the concentration of the theory among marketers is essentially on the recognition of interaction of service encounters on anticipated customer behaviours.

In the context of this research, personality is defined as critical indicator that intentionally or unintentionally/concisely or unconsciously influences dyadic interaction among service encounters. Therefore, this influence may affect positive or negative communication competence, perceived cultural distance, interaction comfort, interrole congruence and finally customer satisfaction (Sharma et al. 2012; Paswan and Ganesh 2005).

8.3.2 Interdependence Theory

The perspective of interdependence theory is established on the logical analysis of the structure of interpersonal interactions (Sharma et al. 2012; Paswan and Ganesh 2005). The theory implies a logical and explicable process to better understand specific situations, problems, motivations and opportunities via the similarity-attraction paradigm and concept of transformation. The theory hypothesises that the ability of a person to interact in a social environment is partly due to the anticipation of the social reward or social comforts (Rusbult and Van Lange 2003).

The interdependence theory also identifies that the most important characteristics of interpersonal situations are not just from intrapersonal perspectives but also through the interpersonal process or a comprehensive analysis of situation structure (Rusbult and Lange 2008). A few studies on the theory resulted in interesting findings. For example, an individual in a multicultural team who has a foreign accent is frequently more culturally distant than another who can speak the official language fluently (Rao Hill and Tombs 2011). Another finding from the study that indicates that the linguistic ability of the employee in a service setting (i.e. American, British and Indian accent) influences the customers' perceptions and interpretations during the service experience in the call centre setting (Wang et al. 2009). Additionally, the theory can also illuminate widely known assumptions on core interpersonal and intrapersonal processes in the concept of personality and social psychology. Herewith, the processes of interaction in terms of internal processes and external processes, background of interaction and its possible impact on other dimensions are indicated.

Thus, in this study, clarifying, describing and decoding the interaction among the encounters is a crucial process to understand the consequences and possible outcomes of encounters better.

8.3.3 Cognitive Consistency Theory

An additional theory for explaining perspectives and to predict differences of service encounters is cognitive consistency theory. The theory assumes that behaviours or attitudes of individuals change when there are differences between expected and perceived situations, particularly when outcomes are significant to them (Pekerti and Thomas 2015). In accordance with the theory, individuals who are comfortable with differences between realised situations do not attempt to change their opinions or behaviours in others. These behaviours are entitled inconsistency-support behaviours. Contrary to this definition, individuals who are not comfortable attempt to change opinions or behaviours. These behaviours are entitled inconsistency-reduction behaviours (Iwao 1997; Kelman and Baron 1968).

In short, intentionally or unintentionally, cognitive differences among service encounters and its reflection on purchasing and post-purchasing process are significant for service providers. Therefore, particularly, customer satisfaction and post-purchasing constructs are linked to cognitive consistency orientation and communication behaviour (Vater and Schröder-Abé 2015).

8.4 Personality

Theories of personality concentrate on the dimensions of human characteristics that can be categorised under cognitive and affective patterns such as thoughts and emotions for explaining the behavioural aspects of human beings (Ihtiyar and Ahmad

2015; Boag and Tiliopoulos2011). According to traditional definitions of personality, it is a self-motivated psychophysical system that creates an individual's characteristic patterns of behaviour, thoughts and feelings, and these patterns give direction to the individual's life (Boag and Tiliopoulos 2011). The popular trait psychology approach materialised a theory called the five-factor model with determinants of neuroticism, extraversion, openness, agreeableness and conscientiousness (Hofstede and McCrae 2004). Although the model has been researched from numerous scholars in various fields, there are debates concerning its limitations in explaining "how culture shapes personality", "how personality traits and culture interact to shape the behaviour of individuals and social groups" (Hofstede and McCrae 2004) and "what extent do the culture and sub-cultures in which people are immersed shape their personality" (Dumont et al. 2010). Furthermore, the Big Five personality assessment does not explain the personality issues related to intercultural adaptation or competency (Rossier et al. 2007; Leone et al. 2005).

Another criticism on the Big Five personality assessment is that it has broad dimensions. In accordance with Paunonen et al. (1999), to achieve an effective accuracy level of prediction, multiple narrow dimensions are required to represent the broad constructs. In parallel with this explanation, the Big Five personality assessment model may fail to provide the situations that require specificity and to be narrowed to emphasise the influence of personality on specific behaviours, manners, attitudes or circumstances such as intercultural interaction and perceived differences between the encounters, role conflict and so on (Judge et al. 2013; Van der Zee et al. 2013; Leone et al. 2005; Van der Zee and Van Oudenhoven 2000; Paunonen et al. 1999; Spreitzer et al. 1997). Therefore, evaluations or measurements for personality in this field have been seriously disadvantaged due to the limited consideration on specifically tailored measurements to predict individual differences of multicultural behaviours, and, although the majority of evaluations were not specifically established for this research, measures like the Big Five have been used (Leone et al. 2005). Contrary to the conventional perspective on personality and the subsequent criticism, instead of using the broader model for understanding the impact of personality on perceived intrapersonal or interpersonal differences and its cumulative impact on interaction between (intercultural) encounters, the study suggests understanding personality through cultural indicators via multicultural personality model.

In addition to understand personality through cultural indicators, personality has a significant role to explain past and previous experiences and communication process of individuals (Lee and Ciftci2014; Harrison-Walker 2012), and strengthened (intercultural) communication process reduces perceived risk, increases satisfaction and improves relational exchange (Ihtiyar and Shannon 2016; Vater and Schröder-Abé 2015; Lee and Ciftci 2014; Harrison 2012; Sharma et al. 2012). Therefore, considering to measure the interrelationship of customers' personality traits (with cultural indicators) and (intercultural) communication process contributes to better understand and predict the customers' behaviours in a multicultural service environment (Castillo 2017). In other words, in order to increase positive experiences of customers in the service environment, service providers may consider customers' personalities and communication process (Castillo 2017; Vater and Schröder-Abé 2015; Lee and Ciftci 2014; Harrison 2012; Sharma et al. 2012).

According to the multicultural personality model, there are five different indicators for the model, namely, *cultural empathy*, the capability to emphasise the behaviours, feelings and thoughts of individuals who are culturally different from the host country; *flexibility*, the inclination to adapt himself/herself to new circumstances in the multicultural social environment efficiently; *social initiative*, the capability to take an active role in social situations; *emotional stability*, the capability to remain calm in stressful and complicated situations; and *open-mindedness*, the capability to present objective and approachable behaviour towards culturally different groups and social systems (Van der Zee et al. 2013).

According to earlier explanation on personality, in the study, trait theory of personality within the multicultural personality perspective will assist in explaining how individuals' personality affects intercultural communication competence, perceived cultural distance and interaction comfort.

8.4.1 Hypothesis for Multicultural Personality

H1a,b,c,d: Cultural empathy influences perceived cultural distance, interaction comfort, interrole congruence and intercultural communication competence.

H2a,b,c,d: Flexibility influences perceived cultural distance, interaction comfort, interrole congruence and intercultural communication competence.

H3a,b,c,d: Social initiative influences perceived cultural distance, interaction comfort, interrole congruence and intercultural communication competence.

H4a,b,c,d: Emotional stability influences perceived cultural distance, interaction comfort, interrole congruence and intercultural communication competence.

H5a,b,c,d: Open-mindedness influences perceived cultural distance, interaction comfort, interrole congruence and intercultural communication competence.

8.5 Intercultural Communication

8.5.1 Interaction Comfort

Earlier studies in social psychology, sociology of culture and customer behaviour literature have provided several validated reasons of increasing customer comfort during service encounters. Strengthened interaction comfort reduces the perceived risk, increases confidence and satisfaction and improves relational exchange (Sharma et al. 2012; Lloyd and Luk 2011; Paswan and Ganesh 2005).

As stated in the interdependence theory (Surprenant and Solomon 1987), each part of the interaction has an impact on another part because the behaviour of the second part is influenced by the outcomes of the first part. Interaction comfort is defined as the likeliness of individuals sharing common norms, values, languages and other factors related to culture, and this will improve predictability of individual's expectations and behaviour, decrease uncertainty and create effective communication. On the other spectrum, perceived dissimilarities in behavioural values or norms such as in language, religion and so on will lead to a sense of discomfort (Ihtiyar and Ahmad 2015). In this study, interaction comfort is a critical dimension for mirroring the impact of ICC and IRC in the overall research framework.

8.5.2 Interrole Congruence

A favourable interaction outcome depends on role clarity, and each other's roles and perceptions during communication should be understood (Solomon et al. 1985). Unfortunately, even when service encounters transpire among between people of similar cultures, they may have different perceptions about each other, or they may not always be able to act within their expected roles in interaction (Huang et al. 2016). The interrole congruence would be more complicated in intercultural interactions where participants must consider the different roles among retailers and consumers.

According to role theory (Solomon et al. 1985), the level of conflict and misunderstanding that are caused by the role perception (the degree of understanding and agreement between both sides on each other's role in an interaction) may involve breaking the communication between both sides, and it may lead to a dissatisfied shopping or service experience for customers.

8.5.3 Intercultural Communication Competence

Intercultural communication competence (ICC) is the aptitude to communicate effectively and correctly with people from culturally diverse social environments (Messner and Schäfer 2012). It deals with the capability to think positively, discriminate the differences, internalise various cultural settings, manage the experiences properly and build effective communication with individuals from different cultures (Ihtiyar 2017 and Samovar et al. 2014). ICC influences cross-cultural interactions (Sharma et al. 2012) where people with stronger ICC have greater propensity to learn foreign languages and norms or values of other culture (Thomas and Peterson 2017). Furthermore, revenue contribution, service concentration, interpersonal skills and social and job satisfaction are influenced by intercultural sensitivity (Sizoo et al. 2005). However, most of the encounter's expectations and reactions have been examined without ICC (Ihtiyar and Ahmad 2015).

Individuals with lower ICC are likely to feel less comfortable and have weak interrole congruence in intercultural encounters compared to those with higher levels of ICC (Friedman and Antal 2005). Those with higher ICC exhibit more empathy and respect for individuals from various cultures, respond to curious circumstances and behaviours in a non-judgemental way without showing visible or perceivable discomfort and enthusiastically use their knowledge and experience to predict various expectations in numerous situations (Samovar et al. 2014). Individuals with higher ICC may not only be more aware of cross-cultural differences in service roles and perceptions, but they are also more likely to agree with these differences. Furthermore, they have greater experience and knowledge about other cultures and can use this experience and knowledge efficiently with people from other cultures, compared to those with lower ICC (Thomas and Peterson 2017). Hence, higher ICC may also contribute to reducing discomfort and uncertainty associated with service encounters. The study thus proposes that ICC predicts IAC, IRC and CS.

8.5.4 Hypotheses for Intercultural Communication

H6 a,b,c: Intercultural communication competence influences customer satisfaction, interaction comfort and interrole congruence.
H7: Interaction comfort influences customer satisfaction.
H8a,b: Interrole congruence influences customer satisfaction and interaction comfort.

8.6 Perceived Cultural Distance

Dissimilarities or familiarities about the cultural background of culturally diverse customers indicate significant changes in their expectations and/or perceptions of service or goods and shopping experiences (Sharma et al. 2012). In the literature of sociology, social psychology, positive psychology and associated fields, dissimilarities or familiarities of individuals have been investigated under cultural distance. Most studies utilise cultural distance using various indexes. For instance, Hofstede's cultural dimensions are uncertainty, avoidance, individualism-collectivism, masculinity-femininity and power distance (Sharma et al. 2012). Instead of this general approach, the present study will consider applying Sharma et al.'s (2012) method entitled "perceived cultural distance" to measure overall differentiation among individuals from different cultures.

Perceived cultural distance (PCD) measures "an individual difference of the perceived discrepancies between social and physical aspects of home and host culture environments" (Sharma et al. 2012). According to this definition, individuals' personality and their social environment may influence perceived cultural distance. Although Sharma et al.'s (2012) model, which is assumed as a core model for the present study, did not reflect how personality influenced perceived cultural distance, the present study seeks to fill this gap by integrating personality into the current model.

In addition to Sharma et al. (2012) and Suanet and Van de Vijver (2009), the present study assumes that personality has a significant impact on perceived cultural distance.

8.6.1 *Hypotheses for Perceived Cultural Distance*

H9a,b: Perceived cultural distance influences interaction comfort and interrole congruence.

8.7 Customer Satisfaction and Post-purchase Intention

Related to customer satisfaction, customers perceive that they made a good choice and that the product satisfies their needs and wants, with the services that they expected; this will have a positive impression on their forthcoming purchasing intention. In other words, any unhappiness and/or feeling on not affective commitment, including communication issues, in any causes of customer satisfaction may reduce forthcoming purchasing intention of customers.

Instead of several indicators for defining dissatisfaction of customers, the study is concentrating on communication and individual's personality, particularly importance of personal interaction and their personalities as suggested by Sharma et al. (2012) and Wu and Liang (2009). They have addressed the significance of personal interaction between service encounters. According to their findings, customers are more satisfied when they receive particular attention, and they are more likely to recommend the stores they shop to friends and/or families. Therefore, personal attention has a positive impact on both customer satisfaction and loyalty. Moreover, Nadiri and Gunay (2013) pointed out the loyalty turns for profitability. It is well established that cumulative customer loyalty is positively associated with cumulative profitability.

The consequences of customer satisfaction in service industries are in the form of intention to recommend (word of mouth), intention to pay more and intention to repurchase/revisit. Hence, the present study contemplates the consequences of customer satisfaction as post-purchase intention to be word of mouth, intention to pay more and intention to revisit.

8.7.1 *Customer Satisfaction and Word of Mouth*

Word of mouth (WOM) has interested an excessive deal of consideration among marketers and scholars. WOM is one of the distinguished characteristics of post-purchase theme (Kim et al. 2009). For instance, although several coffee brands indicated WOM as a feasible alternative to old-fashioned marketing communication tools, however, some practitioners are specifically interested with WOM as traditional forms of communication appear to be losing effectiveness (Forrester Research 2005). This study efforts to predict *how customers' service experience in coffee shops influences their satisfaction and WOM.* Customers will be enthused to encourage their social environment to involve in the behaviour, when their service experience is distinctively beneficial and/or enjoyable. In other words, WOM relates both to constructive and contrary evaluations of service encounters, and it affects other individual's (re)purchase and/or behaviour. Hence the study proposes another hypothesis to investigate the relationship between WOM and customer satisfaction.

8.7.2 Customer Satisfaction and Intention to Pay More

Alternative way to examine customers' behavioural intentions is to concentrate on their intention to pay more as in post-purchase path. Customers' intention to pay more is defined as *the maximum price for a customer is intending to pay* (Byrd et al. 2016; Barber et al. 2012). Furthermore, it is also related to the organisations' profitability. For instance, according the study of Anderson et al. (1994, p. 63), which was based on Swedish Customer Satisfaction Index, "firms that actually achieve high customer satisfaction also enjoy superior economic returns". In other words, intention to pay more relates both to constructive and contrary evaluations of service encounters, and it affects organisations' profitability.

Although recent research supports the notion that there is a positive relationship between customer satisfaction and willingness to pay more (Byrd et al. 2016; Hultman et al. 2015; Cheng and Lu 2013; Barber et al. 2012; Kim and Han 2010;), however, there are limited studies in specific markets and industries, such as Cambodian hospitality industry. Therefore, the study efforts to predict *how customers' service experience in coffee shops influences their satisfaction and intention to pay more as a consequence of purchasing path.* Hence the study proposes another hypothesis to investigate the relationship between intention to pay more and customer satisfaction.

8.7.3 Customer Satisfaction and Intention to Revisit

Purchase intentions indicate to the degree of perceptual persuasion of a customer to repurchase a particular goods and/or services and/or to repurchase from particular service provider. The embedded statement is that such behaviour will reflect forthcoming transaction behaviours. Research in social psychology recommends that intentions are the greatest interpreters of an individual behaviour because they allow each individual to individualistically incorporate all applicable factors that may impact the recent behaviour. Armstrong et al. (2000) associated buying intentions and sales forecasting tools. According to their findings, they found that purchase intentions signify a precise measure of forthcoming sales and that it provides enhanced predictions than previous sales trends.

The major contribution of retaining a base of long-term customers is generally recognised by service providers, and the cost of holding a current customer is less expensive than mining for a new customer (Filieri and McLeay 2014). Therefore, highlighted in the several studies in food and beverage sector indicates that customer satisfaction is a significant predictor of customer intent to return, repurchase and revisit intention (Antón et al. 2017; Kim et al. 2009; Hui et al. 2007). Thus, the study establishes a hypothesis to test the relationship between customer satisfaction and intention to revisit.

8.7.4 Hypothesis for Customer Satisfaction and Post-purchase Intention

H10a,b,c: Customer satisfaction influences word of mouth, intention to pay more and intention to revisit.

Based on the concepts explained above, the dimensions stimulate intercultural communication and customer satisfaction. Concepts of personality and intercultural communication are further divided into several areas to be effectively measured. Personality is measured by *cultural empathy, flexibility, social initiative, emotional stability and open-mindedness*; intercultural communication is measured by *intercultural communication competence, interaction comfort, interrole congruence and perceived cultural distance.*

8.8 Coffee Shop Hospitality in Cambodia

Over the past few years, coffee culture has been changing in Southeast Asia. Coffee drinkers have moved away from traditional coffee shops to international chains and the trendy cafés. And in Cambodia, coffee shops which are home-grown brand and/or international coffee shop chains are becoming the prominent recreation centres for Cambodian hospitality industry. International chains such as Starbucks, Gloria Jean's, Costa Coffee and Coffee Bean & Tea Leaf, beside with small independent and niche cafés, alleviated to grow the Cambodian coffee industry by serving to local and international travellers, expats and the upper and/or mid-upper social classes.

Coffee shops in Cambodia construct with philosophy of Western or American cafés interior design, serving type, store atmosphere, etc. while amalgamating and reflecting Cambodian cultural codes. Cambodia has started developing and opening up to worldwide since 1990s; there was a cumulative and steady increasing need for spaces where young generations could hang out in a cool however socially and culturally proper environment and experience the global lifestyle in Western and American style design and with technological infrastructure.

With the arrival of other brands such as Starbucks in 2015, Amazon (from Thailand) and other global F&B brands, there is an increase in competition in the market; hence, coffee shops are required to provide and serve their customers with distinctive and unique shopping experiences for survival in the market. Therefore, the managerial aim of the study is contemplated to provide practical and productive outcomes for the industry players, customers and also scholars.

8.9 Research Method

The present study focuses on travellers, foreign students and expatriates as customers, and research was carried out in Phnom Penh during the spring of 2017. The economy of the country is heavily dependent on the services, and Cambodia in particular has a multicultural consumer market as a result of the large number of travellers, foreign students and expatriates with respect to population of Cambodia (around 15,600,000 for Cambodia and over 2, 000,000 for Phnom Penh populations). Therefore, the sample of the study consists of travellers and expatriates in Phnom Penh. According to Tourism Statistics Report (2017), over 220,000 tourists arrived in Phnom Penh in February 2017, and the residential expatriates' population in Cambodia is over 100,000. Other than national coffee shops, Gloria Jean's, Starbucks, Coffee Bean and Amazon are the well-known international coffee shop chain that operates in Phnom Penh.

8.9.1 Research Framework

The study proposes a research framework and research design to empirically examine the interactions among the determinants discussed above based on the theories proposed in Fig. 8.1.

In designing this research, the nature of the research problems and the objectives of the study, which are translated from the conceptual framework, served as a basis to indicate the type of design to be applied. This research tests whether the proposed model can conceptualise and explain the endogenous and exogenous variables related to ICC in a multicultural service environment. To that end, a survey was designed based on the intercultural sensitivity scale for measuring intercultural communication competence (ICC) by Chen and Starosta (2000), the multicultural personality scale by Van Oudenhoven and Van der Zee (2002) and items derived from Sharma et al. (2012), Zeithaml et al. (1996), Phau and Ferguson (2013) and Huddleston et al. (2009).

8.9.2 Research Methodology

To provide an empirical assessment of the proposed research model shown in Fig. 8.1, a quantitative technique was performed using the cross-sectional data collection approach. For the purpose of this study, customers who were from Phnom Penh were targeted to test and examine the model statistically. Due to the time limitations, reduction of cost, increased ease of distribution, improved data accuracy and difficulties to collect the data, paper-pencil questionnaires were used to collect the primary data from the target population (Kays et al. 2012).

The questionnaires were categorised into two sections. The first part of the questionnaire intends to obtain the information relating to the demographic profile of the respondents, such as age, gender, education, income and purchasing frequency (Table 8.2). However, due to the sensitivity on "ethnicity" question as a demographic indicator, the question was not added in the questionnaire. The second sec-

Table 8.2 Frequency distribution of demographic indicators

Demographic indicators	Status	Percentages
Gender	Female	58.37%
	Male	41.63%
Age	18–25	14.64%
	26–32	25.36%
	33–39	37.19%
	40+	22.81%
Education	Diploma	24.57%
	Degree	75.43%
Purchasing frequency (weekly)	1–4	19.42%
	5–9	52.84%
	10–14	16.89%
	15≤	10.85%
Amount of purchasing in one time	≤ USD 19	14.94%
	USD 20–USD 39	39.30%
	USD 40–USD 59	21.05%
	USD 60 ≤	24.71%
Income	≤ USD 1000	9.34%
	USD 1001–USD 3999	19.65%
	USD 4000–USD 6999	47.44%
	USD 10000≤	23.57%
Stores	Brown	21.15%
	Starbucks	18.42%
	Gloria jeans	15.76%
	Tea Coffee Bean	14.33%
	Costa	12.17%
	Amazon	8.39%
	Others	9.78%

tion aims to collect the information regarding the four main research concepts: personality, communication, customer satisfaction and post-purchase behaviour. The questionnaire was developed in only English. Respondents were asked to use a 5-point Likert-type scale (where 5-point scales anchored 1, "strongly disagree", and 5, "strongly agree") to record their perceptions. Samples were selected on the basis of a non-probability convenience sampling technique (Bryman and Bell 2015). Additionally, to increase the readability of the questionnaire and reduce the wording errors, a pretest (N 37) and pilot test (N 143) were undertaken.

Following the successful results of the pretest and pilot test, the study proceeded with the actual data collection. In March 2017, questionnaires were distributed to potential respondents in Phnom Penh. By the end of May 2017, 900 questionnaires were distributed, and 742 responded questionnaires were received. Of these, 26 were not completed and thus rejected for this study. The study gathered 679 completed questionnaires, which are appropriate for further assessments.

8.9.3 Missing Value Treatment

Missing values are a unanimous problem in surveys, which leads to difficulties in the multivariate data analyses in behavioural and social sciences (Rezaei 2015). To manage the missing values, Gold and Bentler (2000) suggested using the expectation maximisation method (EMM). In accordance with the Hair et al. (2013), missing data of up to 5% was not large and does not cause problematic results. In terms of the study, the number of collected responses was 679 with 89 missing data points from 44, 814 points, and the percentage was 0.00198, which was not a significant value.

The second phase of the analysis was to adjust the missing values. EMM was utilised to input the values for further assessment. Furthermore, EMM generated Little's MCAR test statistics via SPSS 21.0, and according to the assumption of the test, the test result should generate an insignificant chi-square result for randomisation of missing values. According to the chi-square result, the data represented insignificant values, and the values for the model were chi-square, 28039.631; DF, 1953; and Sig., 0.424.

8.9.4 Common Method Bias

Common method bias (CMB) is a critical challenge in quantitative studies, which influences the validity of the findings on the results of constructs, item reliabilities, structural relationships and the covariation between latent constructs (MacKenzie and Podsakoff 2012). To decrease the probability of CMB, Hair et al. (2013) suggested conducting Harman's one-factor test. Another step for the assessment model is calculating the potential common method bias. The study applied Harman's one-factor test to determine whether the data included any potential common method bias (Hashim 2012). The criterion for common method bias is that the accounted covariance for a single factor should be lower than 40%. For this study, the statistical results indicate that common method biases are not a concern in the study.

8.9.5 *Non-response Bias*

The last assumption in the study was to ensure that non-response bias is not a concern in the study and there was no difference between early and late distributions among respondents in terms of key constructs in the model, demographic indicators and responses collected individually (Hair et al. 2013). Non-response bias is a *critical issue* during the data collection via survey methods. The main concern of researchers is that non-response bias affects the generalizability of the research findings (Hair et al. 2013). Therefore, researchers should seek to reduce non-response bias. The criterion is that the p-value for the t-statistics should be greater than 0.05 (Hair et al. 2013). In the study, the study was done in one period, and there was not any early and late response conflict.

8.9.6 *Structural Model Analysis*

Following the data collection period, selecting an appropriate statistical analysis remains a challenge for studies in business, management and social sciences (Sarstedt et al. 2014). The final decision of which statistical technique or method is applicable to operate for the study is a critical phase for researchers. In regard to these matters, the study is preferred to apply structural equation model (SEM).

Structural equation model assesses all constructs of the model and the hypothesised structural relations among variables simultaneously via measurement model and structural model analysis (Hair et al. 2013; Urbach and Ahlemann 2010). SEM is a flexible modelling tool for administrating many multivariate statistical analyses that included factor, regression, path and correlation analysis and is employed to test indirect and direct relationships between independent latent variables and dependent latent variables (Urbach and Ahlemann 2010). In terms of principal statistical assumptions and nature of fit statistics, the two different approaches are the component-based approach such as partial least square (PLS-SEM) and a covariance-based approach (CB-SEM) (Soh et al. 2017; Hair et al. 2013; Marcoulides et al. 2009; Wetzels et al. 2009).

In regard to this selection, the study is contemplated to follow the suggestions of Hair et al. (2013) and Henseler et al. (2009). In accordance with these authors, a few factors such as objectives of research, structural modelling, measurement modelling, characteristics of data, type of constructs and evaluation of model may help to make decision on selection of PLS-SEM and CB-SEM for further analysis. For instance, covariance-based approach (CB-SEM) reduces the differences that are between the sample covariance and predicted model by maximum likelihood function; however, in order to use the maximum likelihood, the observed variables have to be independent and follow a normal distribution (Hair et al. 2013; Urbach and Ahlemann 2010). On the other hand, the main objective of PLS-SEM is to maximise the covariance score that is between dependent and predictor latent variable and PLS-SEM to provide an estimation on single- and multicomponent models and canonical correlation (Hair et al. 2013). Furthermore, instead of using maximum likelihood (ML) estimation procedure, PLS-SEM is based on ordinary least squares (OLS) regression-based technique and the goal of it maximises the R^2 values of the endogenous constructs (Hair et al. 2013). Therefore, these features provided particular characteristics for using theory development and prediction of the constructs (Hair et al. 2013; Barroso et al. 2010).

Although there are many objections on examining the relationships of latent variables with PLS-SEM such as data examining procedure (Rouse and Corbitt 2008), the positive suggestions and recommendations to use in various literature of PLS-SEM have been increased dramatically (Henseler et al. 2009). Additionally, many scholars have been accepting to use PLS-SEM when distributional assumptions of CB-SEM cannot be met; unrealistic inquiries for informational or distributional are demanded (Hair et al. 2011; Ringle et al. 2010).

In addition to these advantages, PLS-SEM are distribution of data and sample size (Wong 2013; Hair et al. 2013). Contrary to CB-SEM method, PLS-SEM does not demand a large sample size and due to the calibration mechanism of PLS-SEM, it does not require normality distribution (Wong 2013; Hair et al. 2013). However, the CB-SEM method is extremely sensitive to normality of data, interdependence of observation, large sample size and uniformity of variable metric (Sosik et al. 2009). The last phase for the model is determining the construct type as reflective or as formative. Even though identification problem is a major concern for the researchers, CB-SEM is not an appropriate technique for formative constructs due to the explanation problem of covariance of all indicators (Hair et al. 2013; Henseler et al. 2009). On the other hand, PLS-SEM can be operated for the model that consists of a reflective or formative model and the model that includes both reflective and formative constructs simultaneously (Wong 2013; Hair et al. 2013; Urbach and Ahlemann 2010; Henseler et al. 2009). In regard to these matters, the study is assumed to apply PLS-SEM for further statistical analysis.

In accordance with Wong (2013) and Hair et al. (2013), evaluating the model consists of a two-stage approach, namely, a measurement model and structural model. According to Wong (2013) and Hair et al. (2013), the measurement model must meet the minimum requirements in terms of construct reliability, outer loadings, indicator reliability and average variance extracted. The structural model assesses the size and significance of path coefficients, coefficients of determination, predictive relevance, model fit (GoF and SRMR) and effect size (f^2 and q^2) (Wong 2013; Hair et al. 2013) by examining the bootstrapping procedure of 5000 resamples. Hence, bootstrapping and blindfolding were examined to assess the measurement and structural model via SmartPLS software 3.0.

8.10 Results

8.10.1 Assessment of Measurement Model

To examine the measurement model, the study applies the criteria proposed by Wong (2013) and Hair et al. (2013). According to them, composite reliability, outer loadings, Cronbach's alpha, average variance extracted (AVE for convergent validity) and discriminant validity, which is determined by heterotrait-monotrait data ratio, VIF values and Fornell-Larcker criteria, were assessed to examine the measurement models. As shown in Table 8.3, the majority of the outer loadings of the constructs are well above the minimum threshold value of 0.70. However, according to Neupane et al. (2014), Wong (2013) and Lew and Sinkovics (2013), if the research is exploratory, the loading scores should be at least 0.40 or greater. According to these authors, the loading scores are well above the minimum value. Composite reliability and Cronbach's α-values for internal consistency reliability are also above the minimum threshold level of 0.70. Furthermore, the AVE values for convergent validity are well above the minimum threshold level of 0.50, thereby demonstrating convergent validity for all constructs.

To assess discriminant validity tests as to whether the items do unintentionally measure or not (Urbach and Ahlemann 2010), VIF values and Fornell-Larcker's (Table 8.4), HTMT criterion (Table 8.5) and VIF values (Table 8.6) were used. The study examined the collinearity problem using SmartPLS 3.0, and the results indicate that the values of the predictor constructs were lower than the tolerance level, which is VIF value of 5.00. The summary of validity results to evaluate a measurement model is presented in Tables 8.4, 8.5 and 8.6. Furthermore, we refer to Henseler et al. (2014) and Sutton (2014) to check the heterotrait-monotrait data ratio (HTMT) criteria for comparing the PLS-SEM discriminant validity scores, as indicated in Table 8.5. Comparing the loadings across the columns, Tables 8.4, 8.5 and 8.6 illustrate discriminant validity between all the constructs.

8.10.2 Assessment of Structural Model

The following phase of the measurement model is the confirmation step for the model, which examines the collinearity, capabilities of the model's predictive ability, predictive relevance and interrelationships of the constructs (Wong 2013 and Hair et al. 2013). Following this step, the structural model was assessed, and Table 8.7 shows the results of the hypothesis testing and structural relationships (Fig. 8.2).

Table 8.3 Construct validity

Items	Loading scores	CA	rho_A	CR	AVE
CS1	0.738	0.902	0.916	0.921	0.593
CS2	0.798				
CS3	0.803				
CS4	0.697				
CS5	0.789				
CS6	0.669				
CS7	0.864				
CS8	0.785				
IAC1	0.888	0.929	0.931	0.947	0.781
IAC2	0.887				
IAC3	0.895				
IAC4	0.922				
IAC5	0.822				
ICCIC5	0.820	0.899	0.905	0.918	0.745
ICCIC6	0.817				
ICCIC7	0.641				
ICCIC8	0.692				
ICCIE1	0.799				
ICCIE4	0.740				
ICCRDC10	0.711				
ICCRDC9	0.711				
ICCIE3	0.751				
IRC1	0.880	0.939	0.941	0.954	0.806
IRC2	0.837				
IRC3	0.939				
IRC4	0.936				
IRC5	0.894				
WoM1	0.761	0.819	0.857	0.892	0.734
WoM2	0.912				
WoM3	0.889				
IR1	0.910	0.817	0.824	0.916	0.845
IR2	0.929				
IPM1	0.916	0.825	0.829	0.919	0.851
IPM2	0.929				
CE1	0.611	0.725	0.738	0.830	0.551
CE2	0.800				
CE3	0.773				
CE4	0.771				
ES1	0.898	0.849	0.866	0.908	0.767
ES2	0.900				
ES3	0.827				
F1	0.924	0.835	0.835	0.924	0.858
F2	0.929				

(continued)

Table 8.3 (continued)

Items	Loading scores	CA	rho_A	CR	AVE
OM1	0.833	0.801	0.811	0.882	0.713
OM2	0.868				
OM3	0.832				
SI2	1.000	1.000	1.000	1.000	1.000
PCD1	0.851	0.902	0.905	0.928	0.721
PCD2	0.841				
PCD3	0.867				
PCD4	0.928				
PCD5	0.747				

The values for R^2 of the endogenous latent constructs were obtained using the PLS algorithm procedure. According to the results, customer satisfaction (CS), intention to pay more (IP) and perceived cultural distance (PCD) indicate small effect sizes; interaction comfort (IAC) and intercultural communication competence (MP) present large effect sizes, and finally, interrole congruence (IRC) and intention to revisit (IR) and word of mouth (WoM) show medium effect sizes (Table 8.8).

8.11 Conclusion

The most significant contribution of the study is that it is a theoretically and empirically integrated model developed from social psychology, sociology of culture and marketing communication and consumer behaviour in marketing. The boundaries of these disciplines provide a comprehensive framework for understanding antecedents and the impact of customer behaviour in hospitality marketing in Asia, particularly Cambodia. The study has extended the understanding of major and confirmed the relationships among these constructs and the proposed hypotheses.

The relationships of perceived cultural distance, interrole congruence (IRC) and interaction comfort (IAC) were illustrated in this study. In accordance with Sharma et al. (2012, 2009), perceived cultural distance (PCD) determines IAC and IRC; according to outcomes of this study, hypotheses for the model were admitted as *supported* for interrole congruence, however *not supported* for interaction comfort. In other words, perceived cultural distance has a statistically significant impact on only IRC, and findings of the study were matched with findings of Sharma et al. (2012, 2009).

Concerning interrole congruence, interaction comfort and customer satisfaction (CS), the constructs have not been tested in Cambodian hospitality literature. The aim of establishing a direct interrelationship instead of an indirect perspective, as mentioned in Sharma et al. (2012) and Paswan and Ganesh (2005), is that customers and employees identified a service encounter by role theory and interdependence

Table 8.4 Discriminant validity – Fornell-Larcker criterion

Items	CE	CS	ES	F	IAC	ICC	IP	IR	IRC	OM	PCD	SI	WoM
CE	0.551												
CS	0.110	0.593											
ES	−0.201	−0.060	0.767										
F	0.417	0.107	−0.040	0.858									
IAC	0.484	0.165	−0.340	0.214	0.781								
ICC	0.601	0.201	−0.356	0.318	0.714	0.745							
IPM	−0.035	0.346	0.103	0.055	−0.004	−0.005	0.851						
IR	0.107	0.513	−0.058	0.105	0.122	0.171	0.449	0.845					
IRC	0.373	0.065	−0.211	0.134	0.616	0.563	−0.043	0.107	0.806				
OM	0.449	0.080	0.054	0.429	0.302	0.368	0.060	0.097	0.234	0.713			
PCD	0.073	−0.038	0.069	0.072	0.054	0.091	0.117	0.067	0.170	0.086	0.721		
SI	0.380	0.111	−0.126	0.210	0.240	0.337	−0.013	0.042	0.246	0.208	0.070	1.000	
WoM	0.090	0.554	0.008	0.083	0.052	0.104	0.450	0.662	0.076	0.068	0.074	0.081	0.734

Table 8.5 Discriminant validity – HTMT values

Items	CE	CS	ES	F	IAC	ICC	IP	IR	IRC	OM	PCD	SI
CE												
CS	0.135											
ES	0.248	0.080										
F	0.529	0.126	0.062									
IAC	0.586	0.178	0.379	0.245								
ICC	0.741	0.224	0.403	0.373	0.776							
IPM	0.087	0.394	0.120	0.067	0.038	0.066						
IR	0.133	0.578	0.069	0.126	0.138	0.195	0.549					
IRC	0.450	0.080	0.235	0.151	0.658	0.609	0.056	0.120				
OM	0.574	0.103	0.105	0.526	0.344	0.428	0.097	0.117	0.264			
PCD	0.093	0.066	0.082	0.099	0.059	0.102	0.136	0.076	0.181	0.107		
SI	0.453	0.116	0.132	0.231	0.248	0.354	0.026	0.045	0.254	0.233	0.072	
WoM	0.118	0.616	0.046	0.100	0.064	0.120	0.558	0.803	0.088	0.084	0.087	0.083

Table 8.6 Discriminant validity – VIF values

Items	CE	CS	ES	F	IAC	ICC	IP	IR	IRC	OM	PCD	SI	WoM
CE					1.899	1.569			1.895		1.569		
CS							1.000	1.000					1.000
ES					1.217	1.075			1.216		1.075		
F					1.347	1.331			1.336		1.331		
IAC		2.362											
ICC		2.146			2.225				1.838				
IPM													
IR													
IRC		1.697			1.519								
OM					1.457	1.411			1.454		1.411		
PCD					1.046				1.024				
SI					1.204	1.179			1.199		1.179		

theory. Both theories explain that service encounters are dependent on each other because the behaviour of customer or employees influences the perceptions of others. In other words, the interaction between service encounters is clearly interrelated with the definition of scripts and roles during the interaction of encounters. As highlighted in the studies of Ali et al. (2016) and Armenski et al. (2011), interaction comfort among encounters will lead to satisfaction. Therefore, any fluctuations, misunderstanding and confusion from this relationship lead to a dissatisfactory shopping experience, and hence, role clarification for enhanced interaction between customers and employees is key component for many service settings (Ranjan et al. 2015; Zhang et al. 2016; Sharma et al. 2012). Although, in contrast to the findings of

Table 8.7 Result of hypothesis testing and structural relationships

Constructs	Original sample	Sample mean	Standard deviation	T statistics	P values	Decisions
H1a CE - > IAC	0.071	0.070	0.038	1.853	0.064	Supported***
H1b CE - > ICC	0.422	0.423	0.037	11.311	0.000	Supported
H1c CE - > IRC	0.052	0.051	0.044	1.182	0.237	Not supported
H1d CE - > PCD	0.039	0.042	0.049	0.797	0.425	Not supported
H2a F - > IAC	−0.021	−0.020	0.032	0.651	0.515	Not supported
H2b F - > ICC	0.045	0.044	0.033	1.360	0.174	Not supported
H2c F - > IRC	−0.087	−0.088	0.042	2.057	0.040	Supported
H2d F - > PCD	0.031	0.032	0.053	0.585	0.558	Not supported
H3a SI - > IAC	−0.037	−0.037	0.028	1.294	0.196	Not supported
H3b SI - > ICC	0.102	0.102	0.034	3.027	0.002	Supported
H3c SI - > IRC	0.054	0.053	0.036	1.495	0.135	Not supported
H3d SI - > PCD	0.051	0.051	0.044	1.154	0.249	Not supported
H4a ES - > IAC	−0.101	−0.101	0.028	3.639	0.000	Supported
H4b ES - > ICC	−0.265	−0.265	0.029	9.045	0.000	Supported
H4c ES - > IRC	−0.028	−0.029	0.032	0.878	0.380	Not supported
H4d ES - > PCD	0.082	0.084	0.041	1.988	0.047	Supported
H5a OM - > IAC	0.052	0.052	0.032	1.604	0.109	Not supported
H5b OM - > ICC	0.152	0.153	0.034	4.517	0.000	Supported
H5c OM - > IRC	0.042	0.043	0.040	1.055	0.291	Not supported
H5d OM - > PCD	0.040	0.041	0.048	0.839	0.402	Not supported
H6a ICC - > CS	0.194	0.195	0.057	3.371	0.001	Supported
H6b ICC - > IAC	0.461	0.462	0.045	10.298	0.000	Supported
H6c ICC - > IRC	0.505	0.506	0.044	11.482	0.000	Supported
H7 IAC - > CS	0.087	0.085	0.055	1.585	0.113	Not supported
H8a IRC - > CS	−0.098	−0.097	0.051	1.905	0.057	Supported***
H8b IRC - > IAC	0.315	0.315	0.040	7.827	0.000	Supported
H9a PCD - > IAC	−0.040	−0.040	0.027	1.469	0.142	Not supported
H9b PCD - > IRC	0.121	0.121	0.037	3.239	0.001	Supported
H10a CS - > WoM	0.554	0.556	0.027	20.224	0.000	Supported
H10b CS - > IPM	0.346	0.348	0.038	9.203	0.000	Supported
H10c CS - > IR	0.513	0.515	0.031	16.684	0.000	Supported

*p**: 0.01; 2.58, *p***: 0.05; 1.96, *p****:0.10; 1.65

earlier literature, in this study, interaction comfort does not lead to customer satisfaction, however, interrole congruence and intercultural communication have positive impact on interaction comfort. Furthermore, interrole congruence and intercultural communication have positive impact on customer satisfaction, and both hypotheses are supported. In other words, interaction quality on satisfaction might not be ignored. Therefore, results on interaction comfort in this study cannot be generalised

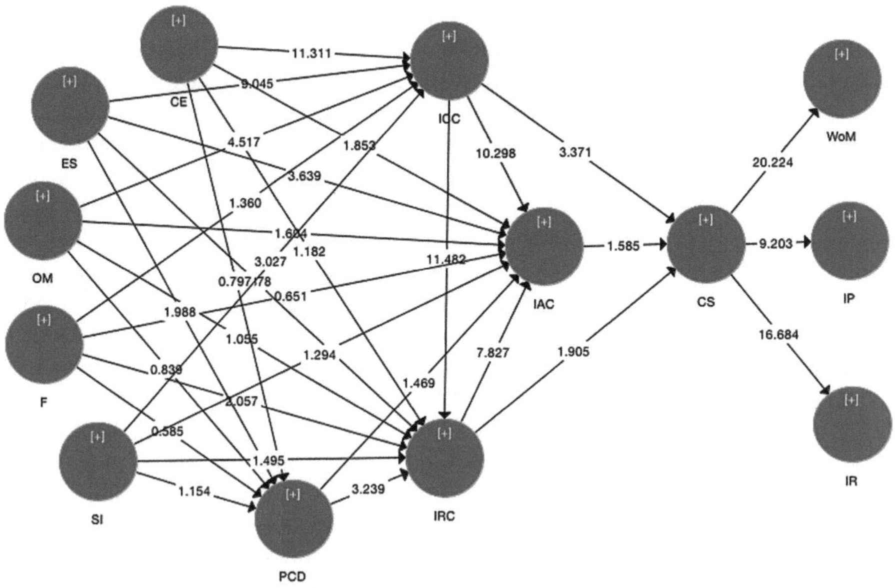

Fig. 8.2 Structural model

in other countries or industries. According to outcomes, interrole congruence has negative impact on customer satisfaction (level 0.10/1.65), while the hypothesis of interaction comfort was not supported statistically. As highlighted in the study of Sharma and Wu (2015), Sharma et al. (2012:2009) Barker and Hartel (2004), these studies mention that there is positive interrelation with interaction comfort and customer satisfaction; however, findings in this study are not matched with these studies in terms of relation of interaction comfort and customer satisfaction. Therefore, these findings might not generalise and it is assumed as market limitation.

The model also examined and partly justified the indicators of personality (MP) for the model. In explaining related hypotheses, the relationships of between constructs of multicultural personality and IAC, ICC, IRC and PCD, respectively, and associations with assumed theories were examined. As highlighted in the studies of William (2005), Van Der Zee and Brinkmann (2004), Chen and Starosta (1996), Chen (1989) and Brein and David (1971), an individuals' personality, which individual is aware of him/her multiple identities and is able to maintain a multicultural coexistence in order to transforms himself or herself from a mono cultural into a multicultural person. Therefore, personality is a crucial indicator to enlighten intercultural communication. In other words, personality is a crucial dimension for understanding the complicated process of communication and its impact on other constructs. Individual differences and its background in terms of behavioural aspects such as sociocultural experiences influence interaction involvements and perceptions on service encounters (Harrison

2012). Although conventional theories of personality have concentrated on the dimensions of human characteristics that can be categorised under cognitive and affective patterns such as thoughts and emotions for explaining the behavioural aspects of human beings (Ihtiyar and Ahmad 2015; Boag and Tiliopoulos 2011). In contrast, the study suggests understanding personality through cultural indicators. In other words, a personality is involved in various patterns of individual differences and affects how people interact with one another (Lee and Ciftci 2014). Therefore, personality has a significant role on the perceptions and communication process of individuals (Lee and Ciftci 2014; Harrison-Walker 2012), and strengthened (intercultural) communication that is based on mentioned variables process reduces perceived risk, positively contributes to shopping experience, increases satisfaction and improves relational exchange (Ihtiyar and Shannon 2016; Vater and Schröder-Abé 2015; Lee and Ciftci 2014; Harrison 2012; Sharma et al. 2012). In regard to examination, although there were several "supported" hypotheses, however, the majority of the relationships were not supported. However, except emotional stability, the other dimensions of personality are not related with perceived cultural distance. In contrast to this finding, except flexibility the rest of the dimensions are related with intercultural communication competence. According to empirical results for personality, cultural empathy only has a statistical relationship with intercultural communication competence, while the rest of the hypotheses, which were related with cultural empathy, were not supported. In contrast to cultural empathy, emotional stability has negative impact on intercultural communication competence and interaction comfort while positive impact on perceived cultural distance. However, there is no statistically proven relationship between interrole congruence and emotional stability. Furthermore, among the personality indicators, only, flexibility has statistically relationship with interrole congruence and according to the outcomes; flexibility has negative impact on interrole congruence. Furthermore, open-mindedness and social initiative have similarly positive impact on intercultural communication competence and have no statistical relationship with interaction comfort, interrole congruence and perceived cultural distance. Additionally, as highlighted in Table 8.7, cultural empathy, emotional stability, open-mindedness and social initiative have empirically proven relationship with intercultural communi-

Table 8.8 Result of R^2

Constructs	R^2	Effect Size
CS	0.043	Small
IAC	0.587	Large
ICC	0.449	Large
IPM	0.118	Small
IR	0.263	Medium
IRC	0.335	Medium
PCD	0.011	Small
WoM	0.306	Medium

*Value effect sizes are, namely, level of 0.02 is small, 0.15 is medium, and 0.35 is large

cation competence. Therefore, even though overall the majority of dimension are related with concept of intercultural communication, 9 of 20 hypothesises were rejected.

In analysing the model, the model employed the PLS-SEM model with new implementations. Sharma et al. (2012) employed the model with a covariance-based approach (CB-SEM), whereas the present study applied the PLS-SEM perspective.

Intercultural communication poses one of the considerable challenges for service providers in the multicultural service environment, and Cambodia is included. Particularly, when the communication strategy of the service providers is not correctly executed, intercultural communication conflicts would heat up and generate incurable problems among service encounters such as difficulties in relational exchange, misunderstanding, dissatisfaction, etc. Hence, the findings recommend that managers and entrepreneurs in service settings should consider these suggestions to resolve intercultural communication conflicts and increase intercultural communication awareness and training in communication skills, encourage to improve general knowledge of other cultures, as well as set up a unique organisational culture that integrates both cultures.

8.12 Limitations and Future Research Directions.

The study was controlled by several limitations and generalisability to larger populations, or other diverse markets may be influenced by these limitations.

The first limitation for the study is related with the definition of construct types for the model. In line with the preceding discussions such as in the studies of Ringle et al. (2012), Becker et al. (2012) and Wetzels et al. (2009), there is limited information on the reflective hierarchical latent variable models. Therefore, regarding these complications on the selection of second-order or first-order model, the study examined the constructs as first-order reflective model.

This study only focuses on service encounters' personality, intercultural communication, satisfaction and post-purchase stage towards luxury trendy coffee shops; it would be interesting to extend the theoretical model in other industries such as the luxury fashion, automobile, electronics industry or healthcare industry. Furthermore, the study is limited with Cambodian perspective, and it might be extended to other ASEAN countries for obtaining more comprehensive results. Finally, future studies may concentrate on additional antecedents or consequences such as perceived quality or physical environment of the stores. These suggestions will promote to generalisability and applicability of the model to different service settings and culturally diverse markets.

Acknowledgement The research was supported by Zaman University, Research Office via Exploratory Research Grant Scheme of Zaman University, grant no. ERGS/2/2017/ZURO/09/17.

Research Constructs and Measurement Items

Research constructs	Acronym	Measurement items	Item source
Word of mouth	WoM	WOM1. It would be difficult to change my preferences towards the store	Zeithaml et al. (1996)
		WOM2. I will recommend the store to someone who seeks my advice	
		WOM3. I will encourage friends and relatives to buy more from the store	
Intention revisit	IR	IR1. I consider the store my first choice to buy groceries	
		IR2. I intend to keep purchasing from the store in the near future	
Intention pay more	IPM	IPM1. I will continue to buy products from the store, even if its prices increase slightly	
		IPM2. I do not mind paying more for the benefits I currently receive from the store	
Customer satisfaction	CS	CS1. The stores' employees are treating me as a valued customer	Phau and Ferguson (2013)
		CS2. The stores' employees are polite to me	Huddleston et al. (2009)
		CS3. The stores' employees are helpful and informative	
		CS4. The stores' employees look presentable	Sharma et al. (2012)
		CS5. The stores' employees are competent	Phau and Ferguson (2013)
		CS6. The store has enough number of employees available to assist me	Huddleston et al. (2009)
		CS7. In general, based on my experience with the store, I am satisfied	
		CS8. Compared to other stores, I am still satisfied with this store	

Intercultural communication competence	ICC	ICC1. I enjoy interacting with people from different cultures	Chen and Starosta (2000)
		ICC2. I avoid those situations where I will have to deal with people from different cultures	
		ICC3. I often give positive responses when interacting with people from different cultures	
		ICC4. I often show my understanding through verbal or nonverbal cues to people from different cultures	
		ICC5. I am sure of myself when interacting with people from different cultures	
		ICC6. I feel confident when interacting with people from different cultures	
		ICC7. I always know what to say when interacting with people from different cultures	
		ICC8. I can be as sociable as I want to be with people from different cultures	
		ICC9. I respect the values of people from different cultures	
		ICC10. I respect the life styles of people from different cultures	
		ICC11. I think people from different cultures and me have different opinions	
		ICC12. I would prefer to interact with people from same culture of mine	

(continued)

Research constructs	Acronym	Measurement items	Item source
Interaction comfort	IAC	IAC1. I feel comfortable dealing with people from different race or ethnicity	Sharma et al. (2012)
		IAC2. I feel comfortable dealing with people from different nationality	
		IAC3. I feel comfortable dealing with people from different language	
		IAC4. I feel comfortable dealing with people from different customs and culture	
		IAC5. I feel comfortable dealing with people from different religious beliefs	
Interrole congruence	IRC	IRC1. I have many friends with different race or ethnicity than mine	
		IRC2. I have many friends with different nationality	
		IRC3. I have many friends with different language than mine	
		IRC4. I have many friends with different customs and culture than mine	
		IRC5. I have many friends with different religious beliefs than mine	
Perceived cultural distance	PCD	PCD1. Most of employees servicing in the store are different from me in terms of race or ethnicity	
		PCD2. Most of employees servicing in the store are different from me in terms of nationality	
		PCD3. Most of employees servicing in the store are different from me in terms of language	
		PCD4. Most of employees servicing in the store are different from me in terms of customs and cultures	
		PCD5. Most of employees servicing in the store are different from me in terms of religious belief	

Cultural empathy	CE	CE1. I am considering the emotions when interacting with people from different cultures	Van der Zee et al. (2013)
		CE2. I am a good listener during the interaction with people from different cultures	
		CE3. I can feel when others get nervous	
		CE4. I can get to share profoundly with people from different cultures	
Emotional stability	ES	ES1. I have worries when interacting with people from different cultures	
		ES2. I am feeling distressed when interacting with people from different cultures	
		ES3. I am feeling lonely when interacting with people from different cultures	
		ES4. I keep calm when things do not go well during the interaction with people from different cultures	
Flexibility	F	F1. I am behaving according to strict rules or regulations in the multicultural society	
		F2. I am behaving according to expectations in the multicultural society.	
Open-mindedness	OM	OM1. I am trying to find various solutions for problems of people from different cultures	
		OM2. I am looking for new ways to achieve her or his goal	
		OM.3 I can feel what is appropriate in different cultures	
Social initiative	SI	SI.1. Sometimes, I find it difficult to interact with people from different cultures	
		SI2. I would like to speak obviously with people from different cultures	

*Notes: Three questions were removed from the construct. SI (Sometimes, I find it difficult to interact with people from different cultures), ICC 2 (I avoid those situations where I will have to deal with people from different cultures), ICC11 (I think people from different cultures and me have different opinions) and ICC12 (I would prefer to interact with people from same culture of mine) due to low loading (below 0.5)

References

Ackerman, N.W. (1958). The psychodynamics of family life. (Retrieved by Broderick, A. 1999).

Ali, F., Amin, M., & Cobanoglu, C. (2016, May 18). An integrated model of service experience, emotions, satisfaction, and price acceptance: An empirical analysis in the Chinese hospitality industry. *Journal of Hospitality Marketing & Management, 25*(4), 449–475.

Anderson, E.W., Fornell, C., & Lehmann, D.R. (1994). Customer satisfaction, market share, and profitability: Findings from Sweden. The Journal of marketing (pp. 53–66).

Antón, C., Camarero, C., & Laguna-García, M. (2017). Towards a new approach of destination loyalty drivers: Satisfaction, visit intensity and tourist motivations. *Current Issues in Tourism, 20*(3), 238–260.

Armenski, T., Dragičević, V., Pejović, L., Lukić, T., & Djurdjev, B. (2011, January). Interaction between tourists and residents: Influence on tourism development. *Polish Sociological Review., 1*, 107–118.

Armstrong, J. S., Morwitz, V. G., & Kumar, V. (2000, September 30). Sales forecasts for existing consumer products and services: Do purchase intentions contribute to accuracy? *International Journal of Forecasting., 16*(3), 383–397.

Asiah Omar, N., & Musa, R. (2011, September 6). Measuring service quality in retail loyalty programmes (LPSQual) Implications for retailers' retention strategies. *International Journal of Retail & Distribution Management., 39*(10), 759–784.

Awang-Rozaimie, A.S., Affidah, M., Aiza, J., & Musdiana, M.S. (2011) *"Multicultural awareness among Malaysian undergraduates': A scale validation" eProceedings for 2011 Int. Res. Con. and Coll.* 119–130.

Bales, R. F. (1955). Task roles and social roles in problem-solving groups. In E. M. T. Newcomb & E. Hartley (Eds.), *Readings in social psychology* (pp. 437–447). New York: Holt.

Barber, N., Kuo, P. J., Bishop, M., & Goodman, R., Jr. (2012, June 22). Measuring psychographics to assess purchase intention and willingness to pay. *Journal of Consumer Marketing., 29*(4), 280–292.

Barker, S., & Härtel, C. E. (2004, March 1). Intercultural service encounters: An exploratory study of customer experiences. *Cross Cultural Management: An International Journal., 11*(1), 3–14.

Barroso, C., Carrión, G. C., & Roldán, J. L. (2010). Applying maximum likelihood and PLS on different sample sizes: Studies on SERVQUAL model and employee behavior model. In *Handbook of partial least squares* (pp. 427–447). Berlin: Springer.

Becker, J. M., Klein, K., & Wetzels, M. (2012). Hierarchical latent variable models in PLS-SEM: guidelines for using reflective-formative type models. *Long Range Planning, 45*(5), 359–394.

Biddle, B. J. (2013, October 22). *Role theory: Expectations, identities, and behaviors*. New York: Academic Press.

Boag, S., & Tiliopoulos, N. (2011). *Personality and individual differences: Theory, assessment, and application*. Hauppauge: Nova Science Publisher.

Bryman, A., & Bell, E. (2015). *Business research methods*. Oxford: Oxford University Press.

Brein, M., & David, K. H. (1971, September). Intercultural communication and the adjustment of the Sojourner. *Psychological Bulletin, 76*(3), 215.

Broderick, A. (1999). Role theory and the management of service encounters. *The Service Industries Journal, 19*(2), 117–131.

Byrd, E. T., Canziani, B., Hsieh, Y. C., Debbage, K., & Sonmez, S. (2016, February 29). Wine tourism: Motivating visitors through core and supplementary services. *Tourism Management., 52*, 19–29.

Carter, R. W., Thok, S., O'Rourke, V., & Pearce, T. (2015, May 28). Sustainable tourism and its use as a development strategy in Cambodia: A systematic literature review. *Journal of Sustainable Tourism., 23*(5), 797–818.

Castillo, J. (2017). The relationship between big five personality traits, customer empowerment and customer satisfaction in the retail industry. *Journal of Business and Retail Management Research., 11*(2), 11–29.

Chang, S. H. (2015, December 11). *Raising mixed race: Multiracial Asian children in a post-racial world*. New York: Routledge.
Chen, G. M. (1989, March 1). Relationships of the dimensions of intercultural communication competence. *Communication Quarterly., 37*(2), 118–133.
Chen, G. M., & Starosta, W. J. (1996, January 1). Intercultural communication competence: A synthesis. *Annals of the International Communication Association, 19*(1), 353–383.
Chen, G. M., & Starosta, W. J. (2000). The development and validation of the intercultural sensitivity scale. *Human Communication, 3*, 1–15.
Cheng, T. M., & Lu, C. C. (2013, October 1). Destination image, novelty, hedonics, perceived value, and revisiting behavioral intention for island tourism. *Asia Pacific Journal of Tourism Research., 18*(7), 766–783.
Dumont, F., Michaels, H. G., Katu, G., & McKeown, M. (2010). *A history of personality psychology. Theory, science and research from Hellenism to the twenty-first century*. Cambridge: Cambridge University Press.
Emerson, R. M. (1976). Social exchange theory. *Annual Review of Sociology, 2*, 335–362.
Evans, F. (1963). Selling as a dyadic relationship- a new approach. *The American Behavioral Scientist, 6*(9), 76–79.
Filieri, R., & McLeay, F. (2014, January). E-WOM and accommodation: An analysis of the factors that influence travelers' adoption of information from online reviews. *Journal of Travel Research., 53*(1), 44–57.
Forrester Research. (2005). What's the buzz on word-of-mouth marketing? Social computing and consumer control put momentum into viral marketing. Accessed Aug 2017. May. http://www.forrester.com
Friedman, V. J., & Antal, A. B. (2005, March). Negotiating reality: A theory of action approach to intercultural competence. *Management Learning., 36*(1), 69–86.
Gold, M. S., & Bentler, P. M. (2000, July 1). Treatments of missing data: A Monte Carlo comparison of RBHDI, iterative stochastic regression imputation, and expectation-maximization. *Structural Equation Modeling, 7*(3), 319–355.
Hair, J. F., Jr., Hult, G. T., Ringle, C., & Sarstedt, M. (2013). *A primer on partial least squares structural equation modeling (PLS-SEM)*. Thousand Oaks: Sage Publications.
Hair, J. F., Ringle, C. M., & Sarstedt, M. (2011, April 1). PLS-SEM: Indeed a silver bullet. *Journal of Marketing theory and Practice., 19*(2), 139–152.
Harrison, N. (2012, March 31). Investigating the impact of personality and early life experiences on intercultural interaction in internationalised universities. *International Journal of Intercultural Relations., 36*(2), 224–237.
Hashim, K. F. (2012). Understanding the determinants of continuous knowledge sharing intention within business online communities. Doctoral dissertation, Auckland University of Technology.
Hashim, M. A., & Mahpuz, M. (2011, March 1). Tackling multiculturalism via human communication: A public relations campaign of Malaysia. *International Journal of Business and Social Science., 2*(4), 70–81.
Haug, W. F., & Bock, R. (1986). *Critique of commodity aesthetics: Appearance, sexuality and advertising in capitalist society*. Cambridge: Polity Press.
Henseler, J., Ringle, C. M., & Sinkovics, R. R. (2009, March 6). The use of partial least squares path modeling in international marketing. In *New challenges to international marketing* (pp. 277–319). Bingley: Emerald Group Publishing Limited.
Henseler, J., Dijkstra, T. K., Sarstedt, M., Ringle, C. M., Diamantopoulos, A., Straub, D. W., Ketchen, D. J., Jr., Hair, J. F., Hult, G. T., & Calantone, R. J. (2014, April). Common beliefs and reality about PLS: Comments on Rönkkö and Evermann (2013). *Organizational Research Methods., 17*(2), 182–209.
Hofstede, G., & McCrae, R. R. (2004, February). Personality and culture revisited: Linking traits and dimensions of culture. *Cross-Cultural Research., 38*(1), 52–88.
Huang, M. H., Huang, M. H., Cheng, Z. H., & Cheng, Z. H. (2016, July 11). Strategies to enhance consumers' identification with a service firm. *Journal of Services Marketing., 30*(4), 449–461.

Huddleston, P., Whipple, J., Nye Mattick, R., & Jung, L. S. (2009, January 30). Customer satisfaction in food retailing: Comparing specialty and conventional grocery stores. *International Journal of Retail & Distribution Management., 37*(1), 63–80.

Hui, T. K., Wan, D., & Ho, A. (2007, August 31). Tourists' satisfaction, recommendation and revisiting Singapore. *Tourism management., 28*(4), 965–975.

Hultman, M., Kazeminia, A., & Ghasemi, V. (2015, September 30). Intention to visit and willingness to pay premium for ecotourism: The impact of attitude, materialism, and motivation. *Journal of Business Research., 68*(9), 1854–1861.

Ihtiyar, A. (2017). A conceptual framework to explain transition of intercultural communication conflict, social environment and personality. *Journal of Management, Economics and Industrial Organization., 1*, 37–50.

Ihtiyar, A., & Ahmad, F. S. (2014 Sep 15). Intercultural communication competence as a key activator of purchase intention. *Procedia-Social and Behavioral Sciences., 150*, 590–599.

Ihtiyar, A., & Ahmad, F. S. (2015, April 1). The role of intercultural communication competence on service reliability and customer satisfaction. *Journal of Economic and Social Studies., 5*(1), 145.

Ihtiyar, A., & Shannon, R. (2016). The role of social environment on interaction of intercultural service encounters and its effects on customer satisfaction. Retailing in emerging markets: Sustainability and regulatory challenges the eighth conference on retailing in the Asia.

Ihtiyar, A., Ahmad, F. S., & Baroto, M. B. (2013, November 6). Impact of intercultural competence on service reliability and customer satisfaction in the grocery retailing. *Procedia-Social and Behavioral Sciences., 99*, 373–381.

Iwao, S. (1997, November 1). Consistency orientation and models of social behavior: Is it not time for west to meet east. *Japanese Psychological Research., 39*(4), 323–332.

Jameson F. (1998). *The cultural turn: Selected writings on the postmodern, 1983–1998*. Verso.

Jean Harrison-Walker, L. (2012, April 20). The role of cause and effect in service failure. *Journal of Services Marketing., 26*(2), 115–123.

Judge, T. A., Rodell, J. B., Klinger, R. L., Simon, L. S., & Crawford, E. R. (2013). Hierarchical representations of the five-factor model of personality in predicting job performance: Integrating three organizing frameworks with two theoretical perspectives. *Journal of Applied Psychology, 98*(6), 875–925.

Kays, K., Gathercoal, K., & Buhrow, W. (2012, January 31). Does survey format influence self-disclosure on sensitive question items? *Computers in Human Behavior., 28*(1), 251–256.

Kelman, H. C., & Baron, R. M. (1968). Inconsistency as a psychological signal. In *Theories of cognitive consistency. A source book*. Chicago: Rand McNally.

Kim, Y., & Han, H. (2010, October 7). Intention to pay conventional-hotel prices at a green hotel–a modification of the theory of planned behavior. *Journal of Sustainable Tourism., 18*(8), 997–1014.

Kim, W. G., Ng, C. Y., & Kim, Y. S. (2009, March 31). Influence of institutional DINESERV on customer satisfaction, return intention, and word-of-mouth. *International Journal of Hospitality Management., 28*(1), 10–17.

Kuo, Y. F., Wu, C. M., & Deng, W. J. (2009, July 31). The relationships among service quality, perceived value, customer satisfaction, and post-purchase intention in mobile value-added services. *Computers in human behavior., 25*(4), 887–896.

Lee, J. Y., & Ciftci, A. (2014, January 31). Asian international students' socio-cultural adaptation: Influence of multicultural personality, assertiveness, academic self-efficacy, and social support. *International Journal of Intercultural Relations., 38*, 97–105.

Leone, L., Van der Zee, K. I., van Oudenhoven, J. P., Perugini, M., & Ercolani, A. P. (2005, April 30). The cross-cultural generalizability and validity of the multicultural personality questionnaire. *Personality and Individual Differences., 38*(6), 1449–1462.

Lew, Y. K., & Sinkovics, R. R. (2013, April 30). Crossing borders and industry sectors: Behavioral governance in strategic alliances and product innovation for competitive advantage. *Long Range Planning., 46*(1), 13–38.

Lloyd, A. E., & Luk, S. T. (2011, May 24). Interaction behaviors leading to comfort in the service encounter. *Journal of Services Marketing., 25*(3), 176–189.

MacKenzie, S. B., & Podsakoff, P. M. (2012, December 31). Common method bias in marketing: Causes, mechanisms, and procedural remedies. *Journal of Retailing., 88*(4), 542–555.

Marcoulides, G. A., Chin, W. W., & Saunders, C. (2009, March 1). A critical look at partial least squares modeling. *Mis Quarterly., 33*(1), 171–175.

Masnan, A. H., & Ngajib, S. H. (2016). The dilemmas of teaching English in Cambodia kindergarten. *International Journal of Academic Research in Business and Social Sciences., 6*(12), 190–196.

Matos, C. A., & Leis, R. P. (2013, May 1). The antecedents of complaint behaviour for Brazilian and French consumers of services. *International Journal of Consumer Studies., 37*(3), 327–336.

Messner, W., & Schäfer, N. (2012). *The ICCA Facilitator's manual: Intercultural communication and collaboration appraisal.* London: GloBus Research Limited.

Miller, C., Brisbane, B., & McAloon, S. (2016, September). Is your school looking for a meaningful global partnership? *Ethos, 24*(3), 16.

Moreno, J. L. (1940). Psychodramatic treatment of psychoses. *Sociometry, 3*(2), 115–132.

Morrow, J. S. (2015). An effective ESP program design for enhanced employment and income in tourism: The case of Siem Reap, Cambodia. *Course and Material Design for Active ESP Journalism English Teaching, 12*

Murphy, A., Jones, H., & Farley, H. (2016). Mobile learning in the Asia-Pacific region: Exploring challenges hindering the sustainable design of mobile learning initiatives. In33rd international conference of innovation, practice and research in the use of educational technologies in tertiary education (p. 461).

Nadiri, H., & Gunay, G. N. (2013, February 1). An empirical study to diagnose the outcomes of customers' experiences in trendy coffee shops. *Journal of Business Economics and Management, 14*(1), 22–53.

Negi, R. (2009). User's perceived service quality of mobile communications: Experience from Ethiopia. *International Journal of Quality & Reliability Management, 26*(7), 699–711.

Neupane, A., Soar, J., Vaidya, K., & Yong, J. (2014, October 14). Willingness to adopt e-procurement to reduce corruption: Results of the PLS Path modeling. *Transforming Government: People, Process and Policy., 8*(4), 500–520.

Nicholls, R. (2011). Customer-to-Customer Interaction (CCI): A cross-cultural perspective. *International Journal of Contemporary Hospitality Management, 23*(2), 209–223.

Oliver, R. L. (1981). Measurement and evaluation of satisfaction processes in retail settings. *Journal of retailing., 57*, 25–48.

Opler, M. K. (1957). Culture and personality: Culture, psychiatry and human values: The methods and values of a social psychiatry. *American Anthropologist, 59*(1), 192–194.

Paswan, A. K., & Ganesh, G. (2005, December 19). Cross-cultural interaction comfort and service evaluation. *Journal of International Consumer Marketing, 18*(1–2), 93–115.

Paunonen, S. V., Rothstein, M. G., & Jackson, D. N. (1999, May). Narrow reasoning about the use of broad personality measures for personnel selection. *Journal of Organizational Behavior., 1*, 389–405.

Pekerti, A. A., & Thomas, D. C. (2015, August). The role of self-concept in cross-cultural communication. *International Journal of Cross Cultural Management., 15*(2), 167–193.

Phau, I., & Ferguson, G. (2013, August 31). Validating the Customer Satisfaction Survey (CSS) Scale in the Australian fast food industry. *Australasian Marketing Journal (AMJ)., 21*(3), 147–154.

Polsa, P., Fuxiang, W., Sääksjärvi, M., & Shuyuan, P. (2013, January 4). Cultural values and health service quality in China. *International journal of health care quality assurance., 26*(1), 55–73.

Ranjan, K. R., Sugathan, P., & Rossmann, A. (2015, February 9). A narrative review and meta-analysis of service interaction quality: New research directions and implications. *Journal of Services Marketing, 29*(1), 3–14.

Rao Hill, S., & Tombs, A. (2011, November 15). The effect of accent of service employee on customer service evaluation. *Managing Service Quality: An International Journal., 21*(6), 649–666.

Rezaei, S. (2015, January 31). Segmenting consumer decision-making styles (CDMS) toward marketing practice: A partial least squares (PLS) path modeling approach. *Journal of Retailing and Consumer Services., 22*, 1–5.

Ringle, C. M., Sarstedt, M., & Straub, D. W. (2012). Editor's comments: A critical look at the use of PLS-SEM in MIS quarterly. *MIS Quarterly, 36*(1), iii–xiv (Retrieved by Becker et al. 2012).

Ringle, C. M., Wende, S., & Will, A. (2010). Finite mixture partial least squares analysis: Methodology and numerical examples. In *Handbook of partial least squares* (pp. 195–218). Heidelberg: Springer.

Rossier, J., Aluja, A., García, L. F., Angleitner, A., De Pascalis, V., Wang, W., Kuhlman, M., & Zuckerman, M. (2007, November 1). The cross-cultural generalizability of Zuckerman's alternative five-factor model of personality. *Journal of personality assessment., 89*(2), 188–196.

Rouse, A., & Corbitt, B. (2008, January 1). *There's SEM and" SEM": A critique of the use of PLS regression in information systems research.* ACIS 2008 Proceedings, vol. 81.

Rusbult, C. E., & Van Lange, P. A. (2003, Feb). Interdependence, interaction, and relationships. *Annual review of psychology., 54*(1), 351–375.

Rusbult, C. E., & Van Lange, P. A. (2008, September 1). Why we need interdependence theory. *Social and Personality Psychology Compass., 2*(5), 2049–2070.

Samovar, L. A., Porter, R. E., McDaniel, E. R., & Roy, C. S. (2014). *Intercultural communication: A reader*. Boston: Cengage Learning.

Sarstedt, M., Ringle, C. M., Henseler, J., & Hair, J. F. (2014, June 30). On the emancipation of PLS-SEM: A commentary on Rigdon (2012). *Long range planning., 47*(3), 154–160.

Satawedin, P. (2017). Health Communication Issues among Migrant Workers in Thailand: A Systematic Review for Health Communication Practices. *BU Academic Review., 16*(1), 87–100.

Saunders, M., Lewis, P., & Thornhill, A. (2009). *Research methods for business students*. New York: Pearson Education.

Sharma, P., & Wu, Z. (2015, April 13). Consumer ethnocentrism vs. intercultural competence as moderators in intercultural service encounters. *Journal of Services Marketing., 29*(2), 93–102.

Sharma, P., Tam, J. L., & Kim, N. (2009, November). Demystifying intercultural service encounters: Toward a comprehensive conceptual framework. *Journal of Service Research., 12*(2), 227–242.

Sharma, P., Tam, J. L., & Kim, N. (2012, October 5). Intercultural service encounters (ICSE): an extended framework and empirical validation. *Journal of Services Marketing., 26*(7), 521–534.

Sizoo, S., Iskat, W., Plank, R., & Serrie, H. (2004, January 21). Cross-cultural service encounters in the hospitality industry and the effect of intercultural sensitivity on employee performance. *International journal of hospitality & tourism administration., 4*(2), 61–77.

Sizoo, S., Plank, R., Iskat, W., & Serrie, H. (2005). The effect of intercultural sensitivity on employee performance in cross-cultural service encounters. *Journal of Services Marketing, 19*(4), 245–255.

Soh, C. Q., Soh, C. Q., Rezaei, S., Rezaei, S., Gu, M. L., & Gu, M. L. (2017, June 19). A structural model of the antecedents and consequences of Generation Y luxury fashion goods purchase decisions. *Young Consumers., 18*(2), 180–204.

Solomon, M. R., Surprenant, C., Czepiel, J. A., & Gutman, E. G. (1985, January). A role theory perspective on dyadic interactions: the service encounter. *The Journal of Marketing., 1*, 99–111.

Sosik, J. J., Kahai, S. S., & Piovoso, M. J. (2009, Feb). Silver bullet or voodoo statistics? A primer for using the partial least squares data analytic technique in group and organization research. *Group & Organization Management., 34*(1), 5–36.

Souiden, N., & Pons, F. (2009, April 17). Product recall crisis management: the impact on manufacturer's image, consumer loyalty and purchase intention. *Journal of Product & Brand Management., 18*(2), 106–114.

Spreitzer, G. M., McCall, M. W., & Mahoney, J. D. (1997, Feb). Early identification of international executive potential. *Journal of applied Psychology., 82*(1), 6.
Suanet, I., & Van de Vijver, F. J. (2009, May 1). Perceived cultural distance and acculturation among exchange students in Russia. *Journal of Community & Applied Social Psychology., 19*(3), 182–197.
Surprenant, C. F., Michael, R. S., & Gutman, E. G. (1983). Service encounters are human interactions. *Journal of Business Research, 20*, 13–21.
Surprenant, C. F., & Solomon, M. R. (1987, April). Predictability and personalization in the service encounter. *The Journal of Marketing., 51*, 86–96.
Sutton, C. C. (2014). Are all shares equal? A measure of secondary agency conflict and an examination of its influence on investor reactions to acquisition announcements.
Tam, J., Sharma, P., & Kim, N. (2014, May 6). Examining the role of attribution and intercultural competence in intercultural service encounters. *Journal of services marketing., 28*(2), 159–170.
Tham, D. S., Bremner, J. G., & Hay, D. (2017, March 31). The other-race effect in children from a multiracial population: A cross-cultural comparison. *Journal of experimental child psychology., 155*, 128–137.
Thomas, D. C., & Peterson, M. F. (2017, January 17). *Cross-cultural management: Essential concepts*. Thousand Oaks: Sage Publications.
Tourism Statistics Report. (2017). Available at: http://www.catacambodia.com/Cambodia%20Tourism%20Statistic-2017-February.pdf
Urbach, N., & Ahlemann, F. (2010, June 1). Structural equation modeling in information systems research using partial least squares. *JITTA: Journal of Information Technology Theory and Application., 11*(2), 5.
Van Der Zee, K. I., & Brinkmann, U. (2004, September 1). Construct validity evidence for the intercultural readiness check against the multicultural personality questionnaire. *International Journal of Selection and Assessment., 12*(3), 285–290.
Van Der Zee, K. I., & Van Oudenhoven, J. P. (2000, July 1). The Multicultural Personality Questionnaire: A multidimensional instrument of multicultural effectiveness. *European journal of personality., 14*(4), 291–309.
Van der Zee, K., Van Oudenhoven, J. P., Ponterotto, J. G., & Fietzer, A. W. (2013, January 1). Multicultural Personality Questionnaire: Development of a short form. *Journal of personality assessment., 95*(1), 118–124.
Van Oudenhoven, J. P., & Van der Zee, K. I. (2002, November 30). Predicting multicultural effectiveness of international students: The Multicultural Personality Questionnaire. *International Journal of Intercultural Relations., 26*(6), 679–694.
Vater, A., & Schröder-Abé, M. (2015, March 1). Explaining the link between personality and relationship satisfaction: Emotion regulation and interpersonal behaviour in conflict discussions. *European Journal of Personality., 29*(2), 201–215.
Wang, C. Y., & Mattila, A. S. (2010, July 13). A grounded theory model of service providers' stress, emotion, and coping during intercultural service encounters. *Managing Service Quality: An International Journal., 20*(4), 328–342.
Wang, Z., Arndt, A., Singh, S., & Biernat, M. (2009). The impact of accent stereotypes on service outcomes and its boundary conditions. *ACR North American Advances, 36*, 940–941.
Webster, F. E., Jr. (1968). Interpersonal communication and salesman effectiveness. *Journal of Marketing, 32*(3), 7–13.
Weitz, B. A. (1981). Effectiveness in sales interactions: A contingency framework. *Journal of Marketing, 45*(1), 85–103.
Wetzels, M., Odekerken-Schröder, G., & Van Oppen, C. (2009, March). Using PLS path modeling for assessing hierarchical construct models: Guidelines and empirical illustration. *MIS Quarterly., 1*, 177–195.
Williams, T. R. (2005, December). Exploring the impact of study abroad on students' intercultural communication skills: Adaptability and sensitivity. *Journal of studies in international education., 9*(4), 356–371.

Wong, K. K. (2013, January 1). Partial least squares structural equation modeling (PLS-SEM) techniques using SmartPLS. *Marketing Bulletin., 24*(1), 1–32.

Wu, C. H., & Liang, R. D. (2009, December 31). Effect of experiential value on customer satisfaction with service encounters in luxury-hotel restaurants. *International Journal of Hospitality Management., 28*(4), 586–593.

Zeithaml, V. A., Berry, L. L., & Parasuraman, A. (1996, April). The behavioral consequences of service quality. *The Journal of Marketing, 1*, 31–46.

Zhang, M., Jin, B., Wang, G. A., Goh, T. N., & He, Z. (2016, March 1). A study of key success factors of service enterprises in China. *Journal of Business Ethics., 134*(1), 1–4.

Chapter 9
Structural Equation Modeling with Path Analysis: Antecedents of Corporate Commitment to Sustainable Tourism

Jung Wan Lee

Abstract The chapter demonstrates an application of structural equation modeling (SEM) and path analysis (PA) in tourism research, and the associated statistics are described. First, we discuss the basic concepts of SEM, followed by an explanation of the key statistics and terms associated with this procedure. Then we describe the procedure for conducting SEM, including second-order confirmatory factor analysis (CFA). Finally, we describe the related technique of path analysis. In doing so, this chapter provides an example of structural equation modeling with a path model, of which path analysis assumes that all variables are measured without error so that it has a more restrictive set of assumptions than general structural equation models. The study examines antecedents of corporate commitment to sustainable tourism and corporate environmental responsibility in a country in Asia. The study employs factor analysis (i.e., exploratory factor analysis, confirmatory factor analysis, and internal consistency reliability tests) and structural equation modeling analysis and path analysis (i.e., the analysis of moment structures and regression analysis) using 386 samples collected from tourism employees in South Korea.

Keywords Structural equation model · Path analysis · Factor analysis · Internal consistency reliability tests · Corporate environmental responsibility · Organizational behavior · Stakeholder theory · Tourism research · Korea

9.1 Introduction of Methodology

Chapter Objectives

- To explain the basic concepts of structural equation modeling (SEM) such as theory, model, path diagram, exogenous versus endogenous constructs, dependence and correlational relationships, model fit, and model identification

J. W. Lee (✉)
Administrative Sciences Department, Metropolitan College, Boston University, Boston, MA, USA
e-mail: jwlee119@bu.edu

S. Rezaei (ed.), *Quantitative Tourism Research in Asia*, Perspectives on Asian Tourism, https://doi.org/10.1007/978-981-13-2463-5_9

- To introduce the basic statistics associated with SEM
- To describe the process of conducting SEM and explain the various steps involved
- To explain the concept of model fit and the differences between absolute, incremental, and parsimony fit indices
- To describe the procedure for conducting SEM with path analysis (PA)
- To explain path analysis and its relationship to SEM

9.1.1 Theory, Model, and Path Diagram

The structural equation modeling (SEM) is based on a theory that all relationships must be specified before the SEM model can be estimated. In SEM, models are often constructed to test certain hypotheses derived from the theory. A SEM model consists of two models: the measurement model and the structural model.

The researcher uses knowledge of the theory, empirical research, or both to postulate the relationship pattern a priori and then tests the hypothesis statistically. Indicator variables are selected on the basis of theory, and confirmatory factor analysis (CFA) is used to see if they load as predicted on the expected number of factors. The terms construct and factor are used interchangeably. In other words, in testing for measurement model, the researcher has complete control over which indicators describe each construct. On the other hand, a structural model shows how the constructs are interrelated to each other, often with multiple dependence relationships. It specifies whether a relationship exists or does not exist. If a relationship is hypothesized by the theory, then an arrow is drawn. If a relationship is not hypothesized, then no arrow is drawn.

A measurement model is designed and constructed by following norms. Constructs are represented by ovals or circles, while measured variables are represented by squares. Straight arrows are drawn from the constructs to the measured variables. Dependence relationships are portrayed by straight arrows and correlational relationships by curved arrows. A dependence relationship is shown by straight arrows. The arrows flow from the antecedent (independent) to the subsequent effect (dependent) measured variable or latent construct. In a measurement model, the straight arrows are drawn from the construct to its measured variables. In a structural model, the dependence occurs between constructs, and so straight arrows are drawn between constructs.

A path diagram typically involves a combination of dependence and correlational relationships among endogenous and exogenous constructs, as stipulated by theory. An exogenous construct is the latent, multi-item equivalent of an independent variable in traditional multivariate analysis. Multiple observed variables or items are used to represent an exogenous construct that acts as an independent variable in the model. An exogenous construct is determined by factors outside of the model, and it cannot be explained by any other construct or variable in the model. Graphically, an exogenous construct does not have any paths (single-headed arrows) coming into it from any other construct or variable in the model; it will only have paths (single-headed arrows) going out of it. In contrast, an endogenous construct is the latent, multi-item equivalent of a dependent variable. It is determined by constructs or variables within the model, and thus it is dependent on other constructs. Graphically, an endogenous construct has one or more paths (single-headed arrows) coming into it from one or more exogenous constructs or from other endogenous constructs.

The process of conducting SEM is as follows: The steps involved in conducting SEM are (1) define the individual constructs, (2) specify the measurement model, (3) assess measurement model reliability and validity, (4) specify the structural model if the measurement model is valid, (5) assess structural model validity, and (6) draw conclusions and make recommendations if the structural model is valid. For further readings, the principles of SEM such as theory, model, and path diagram are well documented in the book (Malhotra 2010).

9.1.2 Model Fit and Model Identification

SEM analyzes correlation or covariance. SEM determines how well the proposed theory explains the observed correlation or covariance matrix among measured variables. Based on the proposed measurement and structural models, it is possible to estimate the covariance matrix between the observed variables, Σ_k. Model fit is then determined by comparing how closely the estimated covariance matrix, $\Sigma_{k,}$ matches the observed covariance matrix S, i.e., the fit statistics are based on $|S- \Sigma_k|$. A residual in SEM is the difference between the observed value and the estimated value of a covariance.

Model identification concerns whether there is enough information in the covariance matrix to enable us to estimate a set of structural equations. The researcher can estimate one model parameter for each unique variance or covariance among the observed variables. If there are p observed variables, then up to a maximum of $(p(p + 1))/2$ parameters can be estimated. Note that this number is the sum of all the unique covariance $(p(p - 1)/2)$ and all the variances, p. Thus,

$$\left(p\left(p+1\right)\right)/2 = p\left(p-1\right)/2 + p.$$

If the actual number of estimated parameters, k, is less than $(p(p + 1))/2$, the model is overidentified. In that case, we have positive degrees of freedom. Conversely, if k is greater than $(p(p + 1))/2$, the model is underidentified, and a unique solution cannot be found. As a general guideline, having at least three observed variables for each latent construct helps in model identification, i.e., results in an overidentified model. This practice is, therefore, recommended.

9.1.3 Sample Size and Scale Design

The sample size required for SEM depends upon several considerations, including the complexity of the model, estimation technique, amount of missing data, amount of average error variance among the indicators or measured variables, and multivariate distribution of the data. For further readings, see the discussion in the book (Malhotra 2010).

The principles of scale design and development are well documented in the book (Nunnally and Bernstein 1994), and they describe methods of item selection, content and construct validations, reliability assessment, scaling, and analysis. The sensitivity of data in measuring individuals' perceptions and behavioral intentions in different cultural contexts poses a problem for the adoption of a single superior scale due to limited data comparability (Bartoshuk et al. 2005; Dawes 2008). For this reason, different researchers have employed different scales in their measurement of individuals' perceptions and behavioral intentions as one size does not fit all.

9.1.4 Factor Analysis and Internal Consistency Reliability

Evidence of the effectiveness of a scale for its purpose must be examined. Many methods of validation rely heavily on the analysis of inter-item or inter-scale correlations. Construct validity embraces a variety of techniques for assessing the degree to which an instrument measures the concept that it is designed to measure. This may include testing dimensionality and homogeneity.

Factor analysis is an often-used key technique in this process. In order to ensure the construct validity of the measurement instrument, factor analysis is often employed in a two-stage process. First, exploratory factor analysis with a varimax rotation procedure is employed to identify underlying predictors based on an eigenvalue cutoff of 1. Secondly, confirmatory factor analysis using structural equation modeling techniques is employed to confirm that the identified predictors are fitted to the items correctly and reliably.

Internal consistency reliability is a measure of how well a test addresses different constructs and delivers reliable scores. Among others, the most common method for assessing internal consistency is Cronbach's alpha. This form of intra-class correlation is closely related to convergent validity, i.e., the extent to which the items in a scale are all highly intercorrelated. For example, in a series of questions that ask the subjects to rate their response between one and seven, Cronbach's alpha gives a score between zero and one, with 0.7 and above being reliable. The test also takes into account both the size of the sample and the number of potential responses. The Cronbach's alpha test is preferred in tourism research due to the benefit of averaging the correlation between every possible combination of split halves and allowing multilevel responses. For example, the survey items are divided into four constructs.

The internal consistency reliability test provides a measure so that each of these particular constructs is measured correctly and reliably.

For further readings, see the more comprehensive description of scale development and reliability (Dunn 1989) and scale development and validation (Bartholomew 1996; Basilevsky 1994).

9.1.5 Path Analysis and Its Relationship to SEM

Path analysis (PA) can be viewed as a special case of SEM. The researcher could think of PA as SEM with only single indicators for each of the variables in the causal model. In other words, path analysis is SEM with a structural model, but no measurement model. Path analysis may also be viewed as an extension of the regression model. The PA model is depicted in a rectangle-and-arrow figure in which single-headed arrows indicate causation. A regression is done for each variable in the model as a dependent on others, which the model indicates are causes. The regression weights estimated by the model are compared with the observed correlation matrix for the variables, and a goodness-of-fit statistic is calculated. Path analysis calculates the strength of each relationship using only a correlation or covariance matrix as input.

9.2 Literature Review and Hypotheses

9.2.1 Research Background

Corporate environmental responsibility views the corporation as members of society. As a part of society, corporations more or less would be affected by environmental expectation and responsibility. Traditionally, environmental protection has been considered to be internal in the public interest and external to private life. Public entities, viewed as a main source of environmental expectation, have assumed principal responsibility for assuring environmental protection and have focused on creating and preserving a safe environment. In this approach, some have contended that unrestricted private sector behavior has been considered as presenting environmental problems. Corporate environmental sustainability has grown to be a dominant theme today in different private industries. Leaders of major corporations are increasingly facing the challenge of managing organizations that meet the expectations of a wide range of stakeholders in addition to delivering a financial return to shareholders. As a result, the practice of corporate environmental responsibility is now an essential ingredient for a company's long-term success.

However, according to the report of the Science and Technology Policy Institute of Korea (2011), only 53 hotels out of 184 samples, about 29.1%, expressed to accept corporate environmental responsibility in guidance of governmental regulations and as an environmental obligation. Accordingly, there was a discussion that whether or not the tourism industry should be awarded for accepting corporate environmental responsibility by providing economic incentives and benefit. Under such situation, it is essential to examine various factors that are driving the tourism industry to accept corporate environmental responsibility. As applying new environmental criteria to tourism operations requires new resource combinations, green practice adoption includes implementing new or modified processes, techniques, and systems to reduce environmental harms. Much literature has proposed various explanations as to what factors influence tourism firms' adoption of green practices. Stakeholder interests, environmental regulation, organizational support, managers' characteristics, and human resources are relevant environmental and organizational variables frequently appearing in literature. Although organizational characteristics have been taken into account in previous literature, much remains to be learned empirically about how organizational characteristics influence green practice adoption in the tourism industry.

To fill the research gap, the current paper aims to identify antecedents influencing organizational acceptance of corporate environmental responsibility in the tourism industry of Korea. This paper also makes an effort to integrate components of corporate environmental responsibility, stakeholder theories, and organizational behavior theories in accepting corporate environmental responsibility.

9.2.2 Sustainable Tourism and Corporate Environmental Responsibility

Examples of corporate environmental responsibility in the tourism industry include carbon emission reduction policies, green supply chain policies, and energy and water efficiency strategies. In this pathway, as local governments lead the way toward tourism leadership in energy environmental design certifications (Heisterkamp 2009), green hotels have been concentrating on most common practices like saving water and energy and reducing solid waste to shield the environment. In addition, hoteliers deliberately intensify their efforts on issues like energy-efficient lighting, installing hinge-activated lighting, and using electronic methods such as email, electronic check-in, and other more. Benavides-Velasco et al. (2014) report that the adoption of corporate environmental responsibility improves the capacity of hotels to create benefits for their stakeholders, and so these results have a positive effect on hotel performance in the Andalusian region (Spain). It is said that employees' perceptions of corporate environmental responsibility influence organizational citizenship through organizational commitment in the Chinese tourism industry (Fu et al. 2014), in Hong Kong (Kucukusta et al. 2013;

Tsai et al. 2012), and in the United States (Rahman et al. 2012). However, Grosbois (2012) reports while a large number of hotels in the world tourism industry report commitment to corporate environmental responsibility goals, a smaller number of them provide details of specific initiatives undertaken to contribute to these goals and even less of them report actual performance achieved.

Green practices for the tourism industry include commitments to environmental practices, recycling and reuse, energy efficiency and conservation, water efficiency and conservation, landscape, transportation, and purchasing. It was reported that hotel operators generally recognize the need for environmental protection and are involved in a number of activities in the tourism industry in Europe (Bohdanowicz 2006) and in the United States (Singh et al. 2014). In accordance to green practices, reuse of room amenities like linens and towels has become a common policy, while many hotels have installed low-flow showers and toilets and occupancy sensor controls for guestroom thermostats. In addition, many hotels have created green mission statements. Accordingly, the benefits that green practice offers will serve as motivation for tourism companies to accept corporate environmental responsibility.

Chan et al. (2014) report that three green triggers – environmental knowledge, environmental awareness, and environmental concern – are positively associated with environmental behavior, and this behavior is positively associated with intention to implement green practices in tourism. Chan and Hawkins (2012) find that a top-down approach to implementing environmental management systems was more suitable for the tourism industry in China. An assortment of organizational characteristics such as quality of human resources, management leadership, organizational support, and organizational size has been discussed on their influences on the environment management strategy. In general, sufficient organizational resources are important organizational characteristics toward adopting the green practice (Lin and Ho 2011). For accepting corporate environmental responsibility, organizational support is essential because the resources required for adopting green practices will be more easily available and the employees will be motivated to implement green practices. In this regard, top management plays an important role in organizational support and commitment to environmental responsibility. The central task of top management is to acquire resources and allocate them efficiently so that the company is able to adopt green practices to achieve environmental competitive advantage (González-Benito and González-Benito 2006). Based on the evidence from previous literature, we propose the following hypotheses:

Hypothesis 1: The perceived corporate environmental responsibility has a positive effect on acceptance of corporate environmental responsibility.

Hypothesis 2: The perceived corporate environmental responsibility has a positive effect on organizational resources to corporate environmental responsibility.

Hypothesis 3: Organizational resources to the environment responsibility have a positive effect on acceptance of corporate environmental responsibility.

9.2.3 Stakeholder Approaches Toward Corporate Environmental Responsibility

Kuvan and Akan (2012) suggest that the two biggest stakeholders in tourism are local residents and the managers of tourism facilities. Their research findings indicate that while both local residents and managers of tourism facilities believe tourism benefits in providing more jobs and enhancing local economics, different opinions toward environmental issues have been observed. Managers of tourism facilities believe they have only little negative environmental impact, but local residents do not believe so. Vasi and King (2012) categorize stakeholders in tourism into two categories: primary stakeholders who tend to directly relate to corporates and affect corporates' decision-making such as shareholder and management board and secondary stakeholders who tend to spur public emotion and involve individuals and groups into protest or other public activities. But their result does not indicate that secondary stakeholders do not contribute to the environmental progress within corporates. In contrast, secondary stakeholders might increase corporates' level of awareness of environmental responsibility. Uecker-Mercado and Walker (2012) categorize stakeholders into other two groups: internal and external. They suggest that external stakeholders are considered relatively bigger source of pressure on the organization to conduct positive environmental change than internal stakeholders. This result is completely different from Vasi and King's (2012) finding. Park and Kim (2014) report that stakeholder pressure is the most dominant predictor of management attitudes toward adoption of green practices, followed by economic benefits and top managers' personal environmental concern. Several stakeholders such as environmental activist groups, governments, employees, and consumers have been discussed in literature of environmental management (Etzion 2007; Kim and Lee 2011; Kim and Yoo 2012). Among various groups of stakeholders, customers and regulators are arguably viewed as a company's most important stakeholders. A series of research revealed the positive relationships between firms' environmental practices and customer and regulatory pressure (e.g., Christmann 2004; Lee 2008).

Kang et al. (2012) report that hotel guests with higher degrees of environmental concerns show a higher willingness to pay a premium for environmentally friendly and sustainable practices of the US tourism industry. On the other hand, a study collects data from 239 hotels in Spain and finds that if a manager is acting more like a proactive environmental manager than passive, he or she will negatively perceive stakeholder pressure of the competitive opportunity inherent in sustainability (López-Gamero et al. 2011). The study also finds that proactive managers prefer voluntary norm to environmental legislation. In this case, any extreme expression of environmental attitude will not help a manager implement green practices. However, Thijssens et al. (2015) indicated that when the level of perceived legitimacy increased in management, stakeholders who promote environmental reform will benefit most. Delgado-Ceballos et al. (2012) examine data from a sample of 73 managers and find that stakeholder integration has positive impact toward the development of proactive environmental strategies. Huang and Kung (2010) introduce an intermediary stakeholder: accounting firms. They suggest that since accounting firms are responsible for auditing and disclosing information, they can become a source of stakeholder pressure, which will increase the level of perceived environmental risks.

When a company has corporate environmental responsibility initiatives, employees take more pride in their organizations (Brammer et al. 2007). A study also finds that employees' perception of their organizations' concern for the environment was linked to their level of organizational commitment and it eventually became an organizational culture (Stawiski et al. 2010). The perceived corporate environmental responsibility will also drive stakeholder's interest. Arikan et al. (2015) observe that firms engaging in environmental responsibility-related activities will trigger positive signals to both internal and external stakeholders and shape cognitive and behavioral stakeholder outcomes. This result is related to the findings of Sarkis et al. (2010), which means that when corporates perceive much environmental responsibility and are willing to provide employee green practices, stakeholders will pay more attention to environmental issues. Guay et al. (2004) find that nongovernmental organizations have opportunities to influence corporate conduct via direct, indirect, and interactive influences on the environmentally responsible investment, with attendant consequences for corporate governance and environmental performance. Based on the evidence from previous literature, we propose the following hypotheses:

Hypothesis 4: The perceived corporate environmental responsibility has a positive effect on stakeholder interests of environmental responsibility.

Hypothesis 5: Stakeholder interests of the environmental responsibility have a positive effect on acceptance of corporate environmental responsibility.

9.3 Research Methods

9.3.1 Survey and Sample Characteristics

A mail survey was conducted in 2012 with the Korea Hotel Association. 1000 sets of questionnaires were actually mailed to 200 hotels from the member directory of the Korea Hotel Association or 5 questionnaires per hotel (1253 member hotels were registered to the Association as of December 31, 2011). Of 424 filled-in and returned questionnaires, 38 cases were removed from the dataset because they have missing data or outliers. The final sample size was of 386 cases that have no missing data and were used for the following analyses. The concern that the mail survey might induce a self-selection bias would be less serious on this sample because mail survey participants in this kind of institutional mail surveys would be more likely to respond. Table 9.1 provides the sample statistics.

Table 9.1 Survey and sample characteristics

Characteristics	Classifications	Frequency[a]	Percent
Gender	Male	275	71.2
	Female	111	28.8
Hotel location	Metropolitan city (Seoul)	314	81.3
	Metropolitan city (Busan)	30	7.8
	Large city (Kwangju)	20	5.2
	Large city (Daejeon)	2	0.5
	Others	20	5.2
Job title	Low level	131	33.9
	Middle level	116	30.1
	Manager	96	24.9
	Top management	43	11.1
Age group	Under 30 years old	123	31.9
	31–40 years old	195	50.5
	41–50 years old	66	17.1
	Over 51 years old	2	0.5
Work experience in the tourism industry	Under 5 years	190	49.2
	6–10 years	97	25.1
	11–15 years	71	18.4
	More than 16 years	28	7.3

[a]Sample size = 386

Given that the model embeds complex relationships in the path of acceptance of corporate environmental responsibility, this study collected self-reported hoteliers' perceptions using a questionnaire. An initial structured questionnaire was developed based on a study of existing literature (e.g., Bastič and Gojčič 2012; Chan et al. 2014), and the model's hypotheses were initially examined with 15 participants of focus group interviews. The initial questionnaire included 23 items related to various constructs discussed in this study and 5 items that capture information pertaining to respondent demographic characteristics such as gender, age, hotel location, job title, and work experience in the tourism industry. The questionnaire was refined based on the feedback and the initial analysis. A final questionnaire retained 18 items related to the various constructs and 5 items for demographic information. A 5-point scale is used in this study to be better consistent with research in a different cultural context at its best. The variation of responses in this study is ranging from (1) strongly disagree to (3) neutral and to (5) strongly agree.

9.3.2 *Factor Analysis and Internal Consistency Reliability*

To identify underlying antecedents of acceptance of corporate environmental responsibility, factor analysis with a varimax rotation procedure was employed. The component factor analysis was used to uncover the underlying structure of a large set of items and identified four components: component one with four items (eigenvalue = 2.743), component two with four items (eigenvalue = 2.550), component three with four items (eigenvalue = 2.334), and component four with three items (eigenvalue = 2.125). This resulted in the retention of 15 items out of 18, which represented the four components. Afterward, the four components were used for the following analysis.

To test the appropriateness of factor analysis, two measures – the Kaiser-Meyer-Olkin and Bartlett's test – were used. The Kaiser-Meyer-Olkin overall measure of sampling adequacy is 0.802, which falls within the acceptable significant level ($p < 0.01$). Bartlett's test of sphericity is 2292.269 at 105 degree of freedom, which shows a highly significant correlation among the survey items ($p < 0.01$). The sum of squared loadings is the cumulative value of 64.289% by explaining the total variance of the data. The results of exploratory factor analysis using principal component analysis extraction method are reported in Table 9.2.

The results of internal consistency reliability tests for the four constructs are reported in Table 9.2: perceived corporate environmental responsibility (4 items, $\alpha = 0.793$), organizational resources (3 items, $\alpha = 0.716$), stakeholder interests (4 items, $\alpha = 0.812$), and acceptance of corporate environmental responsibility (4 items, $\alpha = 0.765$). The results of internal consistency reliability tests, including item-total correlation coefficient values, are also reported in Table 9.2. The confirmatory factor analysis using structural equation modeling techniques was employed to confirm that the identified predictors fitted the items correctly and reliably. The results of confirmatory factor analysis indicated that a single-factor solution fitted the items acceptably. The corrected item-total correlation value of each item to the construct is presented in Table 9.2.

Table 9.2 Results of factor analysis and reliability statistics

Item code	Factor loadings	Eigenvalue	Extracted variance	Construct name	Corrected item-total correlation	Cronbach α
X201	0.756	2.733	18.223%	Perceived corporate environmental responsibility	0.604	0.793
X202	0.803				0.638	
X203	0.670				0.579	
X204	0.740				0.605	
X207	0.580	2.125	13.051%	Organizational resources	0.509	0.716
X208	0.747				0.564	
X209	0.780				0.558	
X210	0.515	2.550	17.003%	Stakeholder interests	0.512	0.812
X211	0.828				0.638	
X212	0.885				0.752	
X213	0.754				0.719	
Y201	0.734	2.334	15.562%	Acceptance of corporate environmental responsibility	0.588	0.765
Y202	0.724				0.609	
Y203	0.730				0.558	
Y204	0.743				0.565	

9.4 Results

9.4.1 *Structural Equation Modeling Estimates and Path Analysis*

The results of the analysis of moment structures are reported in Table 9.3, which the indexes generally achieve acceptable goodness-of-fit measures. For example, the index of the goodness-of-fit index (= 0.947) indicates that the fit of the proposed model is about 95% of the saturated model (the perfectly fitting model). The normed fit index (= 0.933) indicates that the fit of the proposed model is about 93%. The other goodness-of-fit measures are as follows:

Model fit summary: The minimum value of the sample discrepancy chi-square (CMIN) = 440.781, degree of freedom (DF) = 84, and CMIN/DF = 5.247.
Model fit measures: The goodness-of-fit index (GFI) = 0.947, the adjusted goodness-of-fit index (AGIF) = 0.927, the parsimony goodness-of-fit index (PGFI) =0.918, the root mean square residual (RMR) = 0.043, and the root mean square of approximation (RMSEA) = 0.034.
Baseline comparisons measures: The Bentler-Bonett normed fit index (NFI) = 0.933, Bollen's relative fit index (RFI) = 0.924, the Tucker-Lewis coefficient index (TLI) = 0.947, and the comparative fit index (CFI) = 0.958.
Parsimony-adjusted measures: The parsimony ratio (PRATIO) = 0.927, the parsimony normed fit index (PNFI) = 0.903, and the parsimony comparative fit index (PCFI) = 0.909.
Other measures: The estimate of the non-centrality parameter (NCP) = 356.781, the Akaike information criterion (AIC) = 512.781, the Browne-Cudeck criterion (BCC) = 515.903, and the Bayes information criterion (BIC) = 655.191.

Table 9.3 displays the estimates of the structural equation model. In testing hypothesis 1, where there is relationship between perceived corporate environmental responsibility and acceptance of corporate environmental responsibility, the test

Table 9.3 Results of structural equation modeling estimates and path analysis

Independent variable	Path	Dependent variable	Coefficients
H1: Perceived corporate environmental responsibility	→	Acceptance of corporate environmental responsibility	0.383***
H2: Perceived corporate environmental responsibility	→	Organizational resources	0.692***
H3: Organizational resources	→	Acceptance of corporate environmental responsibility	0.178**
H4: Perceived corporate environmental responsibility	→	Stakeholder interests	0.499***
H5: Stakeholder interests	→	Acceptance of corporate environmental responsibility	0.128**

Numbers in the cells are standardized coefficient values, which are regression estimates
Probability values for rejection of the null hypothesis of zero coefficient are employed at 0.05 level (** $p < 0.05$ and *** $p < 0.001$)

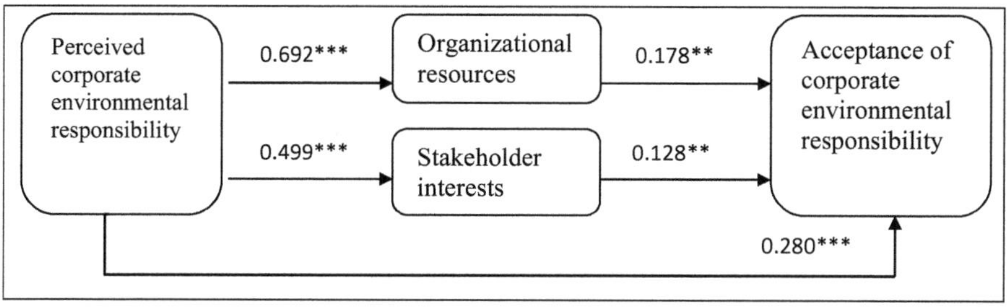

Fig. 9.1 Path diagram of acceptance of corporate environmental responsibility
Coefficient is statistically significant at 95% confidence level. (** $p < 0.05$ and *** $p < 0.001$)

result shows a significant positive relationship at the 95% confidence level (H1, $p < 0.001$). This meant that perceived corporate environmental responsibility has a positive propensity toward acceptance of corporate environmental responsibility.

In testing hypotheses 2 and 3, where there is relationship between perceived corporate environmental responsibility and organizational resources and between organizational resources and acceptance of corporate environmental responsibility, the test result shows that there are positive relationships between constructs. Each pair shows to be statistically significant at a 95% confidence level (H2, $p < 0.001$, and H3, $p < 0.05$). This meant that perceived corporate environmental responsibility has a positive and direct effect on organizational resources, and so organizational resources have a positive propensity toward acceptance of corporate environmental responsibility.

The hypotheses 4 and 5 test a relationship between perceived corporate environmental responsibility and stakeholder interests and between stakeholder interests and acceptance of corporate environmental responsibility. The results of testing hypotheses 4 and 5 show that they are statistically significant at a 95% confidence level (H4, $p < 0.001$, and H5, $p < 0.05$). This meant that perceived corporate environmental responsibility has a positive and direct effect on stakeholder interests, and so stakeholder interests directly and positively influence acceptance of corporate environmental responsibility.

Overall, perceived corporate environmental responsibility, organizational resources, and stakeholder interests serve as important antecedents of acceptance of corporate environmental responsibility. In the path, perceived corporate environmental responsibility has the highest impact on both organizational resources and acceptance of corporate environmental responsibility. Figure 9.1 displays a path diagram of acceptance of corporate environmental responsibility.

9.5 Discussion and Policy Implications

The results from the structural equation modeling and path analysis support all of the hypotheses. The current study shows that perceived corporate environmental responsibility as the most important antecedent influences on organizational resources to acceptance of corporate environmental responsibility and green practices. The positive impact of organizational resources and stakeholder interests on acceptance of corporate environmental responsibility is of special interest. The results indicate that perceived corporate environmental responsibility, organizational resources, and stakeholder interests should be viewed as important antecedents explaining the acceptance of corporate environmental responsibility and green practices.

Regarding the influence of perceived corporate environmental responsibility, hypotheses 1, 2, and 4 are supported. The hypotheses supported are highlighting the role of perceived corporate environmental responsibility in driving the acceptance of corporate environmental responsibility and green practices in the tourism industry of Korea. It meant that the perceived corporate environmental responsibility of employees of the tourism industry is a critical factor in influencing organizational resources, stakeholder interests, and the acceptance of corporate environmental responsibility. It gives the impression that their organizational resources and stakeholder interests drive more acceptance of corporate environmental responsibility in the tourism industry of Korea.

Based on these findings, tourism business stakeholders and regulatory bodies should put more efforts to encourage tourism employees in accepting environmental responsibility. Energy and water consumptions are probably the most critical environmental issues of the tourism industry in Korea. There is, therefore, a significant opportunity for tourism employees to make substantial contributions in reducing their energy and water consumptions and other environmental impacts and then help in improving the operational efficiency and environmental sustainability. In doing so, they should be aware of the reality that most tourism businesses in Korea are yet small and medium sized; thus they may suffer from the lack of financial, technical, and qualified human resources. Hence, policy makers may offer economic incentives and provide required resources for the tourism industry. Although most of tourism businesses in Korea, regardless of their size and business magnitude, understand the importance of their making environmental resources, they are less likely to put an amount of resources on adopting new technologies and green practices since they have a proclivity to focus on a short-term interest about their return on investments rather than a long-term environmental sustainability goal. In addition, small tourism businesses in highly uncertain business environments would put more resources on improving their primary business activities rather than allocating resources to environmental sustainability initiatives.

Concerning the organizational resources, hypothesis 3 is also supported. The hypothesis supported is highlighting the role of organizational resources in driving the acceptance of corporate environmental responsibility. It meant that the organizational resources are a critical factor in influencing the acceptance of corporate environmental responsibility. It suggests that their organizational support, the quality of human resources, and the understanding of environmental initiatives drive the acceptance of environmental responsibility. The awareness of corporate environmental responsibility and organizational resources gives tourism employees motivation and required resources to exert environmental initiatives and green practices. In doing so, organizational commitment to environmental responsibility and green practices may include:

- Tourism business from an environmental committee that is responsible for developing an environmental green plan for energy, water, and solid waste use
- Tourism businesses that manage their environmental performance by monitoring the use of energy, water, and waste management information on a monthly and annual basis
- Tourism businesses that actively communicate their environmental efforts with all stakeholders, shareholders, and the public

Pertaining to the stakeholder interests, hypothesis 5 is also supported. The hypothesis highlights the role of stakeholder interests in driving the acceptance of corporate environmental responsibility. It meant that the stakeholder interest of tourism business is a critical factor in influencing the acceptance of corporate environmental responsibility. It indicates that consumers and governmental regulatory interests drive the acceptance of corporate environmental responsibility. The stakeholder interests give tourism business leadership motivation to exert environmental initiatives and green practices. In this regard, government actions will force a green agenda on the tourism industry in Korea first, in a top-down approach. Although this may be the least desirable outcome for tourism businesses, it is already evident that government intervention and legislation are reaching ever more directly for environmental issues. For example, pricing is one aspect of government intervention. Although there is a clear trend in policy guidelines to make the users pay the full costs of using the tourism infrastructure, many tourism businesses have largely escaped these initiatives. As an alternative, requiring certification is seen as evidence on tourism businesses' commitment to the environment and is frequently used as a public relation and marketing promotion. That having said, corporate environmental responsibility of tourism businesses is more likely to be accepted if tourism businesses are provided with required resources for green practices, subject to stakeholders and regulatory interests.

9.6 Conclusions

The paper examined the antecedents of accepting corporate environmental responsibility in the tourism industry in Korea. Unlike research done so far about this topic, this paper integrated the components of corporate environmental responsibility in conjunction with organizational behavior and stakeholder theories in accepting environmental responsibility in their tourism business. The results show that perceived corporate environmental responsibility, organizational resources, and stakeholder interests are important antecedents of accepting corporate environmental responsibility. Moreover, the higher job titles are, the more positively significant they have on the propensity in acceptance of environmental responsibility.

The tourism industry of Korea is consequently under interest from both outside and inside parties to be more proactive and accountable for a wide range of corporate environmental responsibility and to report publicly on their environmental performance. Such anxiety, whether rational or not, should be seen in the context of the tourism business' power. However, the role of tourism business has been changing radically in accepting corporate environmental responsibility around the world in general and particularly in Korea. Today, protecting the environment with diverse tourism operations is a core responsibility of tourism business operators.

Future research to generalize these findings may examine the same business from different countries as potential sources of variation in the antecedents of acceptance of corporate environmental responsibility. Additional research is required to determine whether there is a consistent pattern or not. In addition, further studies should validate the selected survey items, which are used to elucidate corporate environmental responsibility, organizational resources, and stakeholder interests in tourism.

References

Arikan, E., Kantur, D., & Telci, E. E. (2015). Investigating the mediating role of corporate reputation on the relationship between corporate environmental responsibility and multiple stakeholder outcomes. *Quality and Quantity, 50*(1), 129–149.

Bartholomew, D. J. (1996). *The statistical approach to environmental measurement*. San Diego: Academic.

Bartoshuk, L. M., Fast, K., & Snyder, D. J. (2005). Differences in our sensory worlds: Invalid comparisons with labeled scales. *Current Directions in Psychological Science, 14*(3), 122–125.

Basilevsky, A. (1994). *Statistical factor analysis and related methods*. New York: Wiley.

Bastič, M., & Gojčič, S. (2012). Measurement scale for eco-component of hotel service quality. *International Journal of Hospitality Management, 31*(3), 1012–1020.

Benavides-Velasco, C. A., Quintana-García, C., & Marchante-Lara, M. (2014). Total quality management, corporate environmental responsibility and performance in the hotel industry. *International Journal of Hospitality Management, 41*(August), 77–87.

Bohdanowicz, P. (2006). Environmental awareness and initiatives in the Swedish and Polish hotel industries – Survey results. *International Journal of Hospitality Management, 25*(4), 662–682.

Brammer, S., Millington, A., & Rayton, B. (2007). The contribution of corporate environmental responsibility to organizational commitment. *International Journal of Human Resource Management, 18*(10), 1701–1719.

Chan, E. S. W., & Hawkins, R. (2012). Application of EMSs in a hotel context: A case study. *International Journal of Hospitality Management, 31*(2), 405–418.

Chan, E. S. W., Hon, A. H. Y., Chan, W., & Okumus, F. (2014). What drives employees' intentions to implement green practices in hotels? The role of knowledge, awareness, concern and ecological behavior. *International Journal of Hospitality Management, 40*(July), 20–28.

Christmann, P. (2004). Multinational companies and the natural environment: Determinants of global environmental policy standardization. *Academy of Management Journal, 47*(5), 747–760.

Dawes, J. G. (2008). Do data characteristics change according to the number of scale points used? An experiment using 5-point, 7-point and 10-point scales. *International Journal of Market Research, 50*(1), 61–77.

Delgado-Ceballos, J., Aragón-Correa, J. A., Ortiz-de-Mandojana, N., & Rueda-Manzanares, A. (2012). The effect of internal barriers on the connection between stakeholder integration and proactive environmental strategies. *Journal of Business Ethics, 107*(3), 281–293.
Dunn, G. (1989). *Design and analysis of reliability studies: The statistical evaluation of measurement errors*. London: Edward Arnold Publishers.
Etzion, D. (2007). Research on organizations and the natural environment, 1992 – Present: A review. *Journal of Management, 33*(4), 637–664.
Fu, H., Ye, B. H., & Law, R. (2014). You do well and I do well? The behavioral consequences of corporate environmental responsibility. *International Journal of Hospitality Management, 40*(July), 62–70.
González-Benito, J., & González-Benito, O. (2006). A review of determinant factors of environmental proactivity. *Business Strategy and the Environment, 15*(2), 87–102.
Grosbois, D. (2012). Corporate environmental responsibility reporting by the global hotel industry: Commitment, initiatives and performance. *International Journal of Hospitality Management, 31*(3), 896–905.
Guay, T., Doh, J. P., & Sinclair, G. (2004). Non-governmental organizations, shareholder activism, and environmentally responsible investments: Ethical, strategic, and governance implications. *Journal of Business Ethics, 52*(1), 125–139.
Heisterkamp, M. (2009). Guests expect more from green programs. *Hotel Management*. Available from: http://www.hotelmanagement.net/heisterkamp. Accessed 18 Jan 2015.
Huang, C. L., & Kung, F. H. (2010). Drivers of environmental disclosure and stakeholder expectation: Evidence from Taiwan. *Journal of Business Ethics, 96*(3), 435–451.
Kang, K. H., Stein, L., Heo, C. Y., & Lee, S. (2012). Consumers' willingness to pay for green initiatives of the hotel industry. *International Journal of Hospitality Management, 31*(2), 564–572.
Kim, Y. M., & Lee, K. N. (2011). A study on the driving factors of green physical distribution and its subsequent impact on environmental and economic performance. *Journal of the Korean Academy of International Commerce, 26*(3), 91–109.
Kim, T. H., & Yoo, S. G. (2012). A study on the green growth and government support on green physical distribution. *Journal of the Korea Association for International Commerce and Information, 14*(1), 315–344.
Kucukusta, D., Mak, A., & Chan, X. (2013). Corporate environmental responsibility practices in four and five-star hotels: Perspectives from Hong Kong visitors. *International Journal of Hospitality Management, 34*(September), 19–30.
Kuvan, Y., & Akan, P. (2012). Conflict and agreement in stakeholder attitudes: Residents' and hotel managers' views of tourism impacts and forest-related tourism development. *Journal of Sustainable Tourism, 20*(4), 571–584.
Lee, S. (2008). Drivers for the participation of small and medium sized suppliers in green supply chain initiatives. *Supply Chain Management: An International Journal, 13*(3), 185–198.
Lin, C. Y., & Ho, Y. H. (2011). Determinants of green practice adoption for logistics companies in China. *Journal of Business Ethics, 98*, 67–83.
López-Gamero, M. D., Claver-Cortés, E., & Molina-Azorín, J. F. (2011). Environmental perception, management, and competitive opportunity in Spanish Hotels. *Cornell Hospitality Quarterly, 52*(4), 480–500.
Malhotra, N. K. (2010). *Marketing research: An applied orientation* (6th ed.). Upper Saddle River: Pearson.
Nunnally, J. C., & Bernstein, I. H. (1994). *Psychometric theory* (3rd ed.). New York: McGraw-Hill.
Park, J., & Kim, H. J. (2014). Environmental proactivity of hotel operations: Antecedents and the moderating effect of ownership type. *International Journal of Hospitality Management, 37*(February), 1–10.
Rahman, I., Reynolds, D., & Svaren, S. (2012). How "green" are North American hotels? An exploration of low-cost adoption practices. *International Journal of Hospitality Management, 31*(3), 720–727.

Sarkis, J., Gonzalez-Torre, P., & Adenso-Diaz, B. (2010). Stakeholder pressure and the adoption of environmental practices: The mediating effect of training. *Journal of Operation Management, 28*(2), 163–176.

Singh, N., Cranage, D., & Lee, S. (2014). Green strategies for hotels: Estimation of recycling benefits. *International Journal of Hospitality Management, 43*(October), 13–22.

Stawiski, S., Deal, J. J., & Gentry, W. (2010). *Employee perceptions of corporate environmental responsibility: The implications for your organization*. Greensboro: The Center for Creative Leadership.

The Science and Technology Policy Institute of Korea. (2011). *A study on the practice and adoption of green innovation in business sectors* (Science, Technology & Innovation Policy Report No. DT-395N-2010090). Available from: http://www.stepi.re.kr/app/report/list.jsp?cmsCd=CM0011 Accessed 18 Jan 2014.

Thijssens, T., Bollen, L., & Hassink, H. (2015). Secondary stakeholder influence on CSR disclosure: An application of stakeholder salience theory. *Journal of Business Ethics, 132*(4), 873–891.

Tsai, H., Tsang, N. K. F., & Cheng, S. K. Y. (2012). Hotel employees' perceptions on corporate environmental responsibility: The case of Hong Kong. *International Journal of Hospitality Management, 31*(4), 1143–1154.

Uecker-Mercado, H., & Walker, M. (2012). The value of environmental responsibility to facility managers: Revealing the perceptions and motives for adopting ESR. *Journal of Business Ethics, 110*(3), 260–284.

Vasi, I. B., & King, B. G. (2012). Environmental movements, risk perceptions, and economic outcomes: The effect of primary and secondary stakeholder activism of firms perceived environmental risk and financial performance. *American Sociological Review, 77*(4), 573–596.

Chapter 10
Partial Least Squares Structural Equation Modeling in Asian Tourism and Hospitality Research: A Systematic Review

Imtiaz Mostafiz, Yeasmin Islam, and Saeed Pahlevan Sharif

Abstract During the recent years, a growing number of tourism and hospitality researchers have used partial least squares structural equation modeling in their studies. However, little is known about the application of PLS-SEM in Asian tourism and hospitality research. This chapter was conducted to fill this research gap by assessing to what extent Asian research in tourism and hospitality has followed the guidelines recommended by the most prominent literature on PLS-SEM. Sixty-four partial least squares structural equation modeling studies conducted by Asian researchers and/or on Asian contexts that were published in 66 tourism and hospitality journals from 2000 to the end of February 2017 were systematically reviewed. The results identified some weaknesses in conducting the analysis which should be addressed in future empirical studies. The results of this systematic review provide tourism and hospitality researchers with best practice guidelines for conducting partial least squares structural equation modeling in their studies.

Keywords PLS-SEM · Partial least squares · Tourism · Asia · Review

10.1 Introduction

Tourism and hospitality research has been growing constantly in the last three decades. Similar to other fields, a variety of data analysis methods such as multiple regression analysis, ordinary least squares, principle component analysis, factor analysis, etc. were used to answer research questions in this field. As the field has grown, research approaches and methods have become more diverse (Ballantyne et al. 2009; Mura et al. 2017). One of the methods that have attracted tourism and hospitality researchers' interest recently is structural equation modeling (SEM).

I. Mostafiz · S. Pahlevan Sharif (✉)
Faculty of Business & Law, Taylor's University, Subang Jaya, Malaysia
e-mail: saeed.sharif@taylors.edu.my

Y. Islam
Independent University, Dhaka, Bangladesh

S. Rezaei (ed.), *Quantitative Tourism Research in Asia*, Perspectives on Asian Tourism, https://doi.org/10.1007/978-981-13-2463-5_10

SEM is an advanced second-generation statistical analysis technique that can be used to test complex and sophisticated research models. There are two approaches to conduct SEM, including covariance-based SEM and variance-based SEM. Several methods of variance-based SEM have been proposed by researchers; however, the most common method is partial least squares-SEM known as PLS-SEM. Each of these methods has its own advantages and drawbacks.

PLS-SEM is recommended for studies that assess new measurement models and can work well with both reflective and formative measurements (e.g., see Yang et al. 2015) and higher-order constructs (e.g., see Huang et al. 2015). PLS-SEM does not assess the theoretical model as a whole (Pahlevan Sharif and Mahdavian 2015). This makes it appropriate for exploratory studies without strong theoretical support. While covariance-based methods are typically used for theory testing, PLS-SEM is referred to as a prediction-oriented approach that increases the predictive power of a model (Sosik et al. 2009). Moreover, PLS-SEM maximizes the variance of the endogenous variables explained by the model (Hair et al. 2016). The advocates claimed that the differences between CB-SEM and PLS-SEM estimates are very small and suggested using PLS-SEM as a viable complementary modeling approach for CB-SEM when the assumptions of covariance-based methods are violated (Hair et al. 2011). For example, PLS-SEM is known as an appropriate SEM method when data deviates from normal distribution and/or sample size is quite small. In this regard, Hair et al. (2011, p. 111) noted that "if correctly applied, PLS-SEM indeed can be a silver bullet in many research situations."

However, the superiority of PLS-SEM relative to CB-SEM methods has been criticized recently (Goodhue et al. 2012; McIntosh et al. 2014; Rönkkö and Evermann 2013; Rönkkö et al. 2016). Rönkkö et al. (p. 1) argued that "although the PLS technique is promoted as a structural equation modeling (SEM) technique, it is simply regression with scale scores and thus has very limited capabilities to handle the wide array of problems for which applied researchers use SEM." In an editorial note, Guide and Ketokivi (2015, p. 7) stated that "claiming that PLS fixes problems or overcomes shortcomings associated with other estimators is an indirect admission that one does not understand PLS." Nevertheless, during the recent years a growing number of tourism and hospitality researchers have used PLS-SEM in their studies (do Valle and Assaker 2016). However, little is known about the application of PLS-SEM in Asian tourism and hospitality research. This paper was conducted to fill this research gap by assessing to what extent Asian research in tourism and hospitality has followed the guidelines recommended by the most prominent literature on PLS-SEM.

10.2 Methods

By following Pahlevan Sharif et al. (Forthcoming) approach, this systematic review addressed how PLS-SEM analysis has been conducted and reported in Asian tourism and hospitality research. This systematic review was conducted in accordance with the Preferred Reporting Items for Systematic Reviews and Meta-Analyses (PRISMA; Liberati et al. 2009). Following Mura and Pahlevan Sharif (2017), the present review systematically searched for empirical research articles published in 66 tourism and hospitality journals, included in the Australian Business Deans Council (ABDC) quality list. "PLS-SEM," "PLS SEM," and "partial least square*" were searched in the full text of the papers through the search engine provided in the website of the journals and in some cases through the advanced search of Google Scholar. The full text of the identified papers were downloaded and screened by the second author of this review to exclude papers that did not perform PLS-SEM although they mentioned one of the search terms in the text or the reference list. Also, as the scope of this systematic review was limited to Asian tourism and hospitality research, only studies conducted by Asian researchers and/or on Asian contexts were included in the review. The search of journal articles was performed in July 2017. In case of any doubt for eligibility of the studies, the first author of this paper reviewed the search process of journal articles and resolved potential disagreements by consensus.

The title, authors' names and affiliations, journal name, year of publication, and context of the included papers were exported into an Excel spreadsheet. Next, the first author assessed the main parts of conducting and reporting PLS-SEM analysis suggested by the most prominent literature on PLS-SEM (Chin 2010; Fornell and Larcker 1981; Hair et al. 2011, 2012a, b, 2016; Ringle et al. 2012). From each included paper, the first author extracted items related to data preparation, measurement model assessment, and structural model assessment (see Table 10.2). The Excel spreadsheet file was extended to record the findings of the review. The second author confirmed the accuracy of the extraction and coding process.

10.3 Findings

In total, 19,880 articles were published in tourism and hospitality journals from 2000 to the end of February 2017. These articles were not including research notes, commentary papers, and review papers. Conducting the search using the defined search terms, this review identified 2394 articles. The results of screening showed that the majority of the articles did not use PLS-SEM as their data analysis method. For example, while some papers had cited PLS-SEM papers for reference purposes, some studies had suggested using PLS-SEM for future research. Finally, 64 articles from 50 journals met the eligibility criteria of this systematic review. More specifically, 51 papers by Asian researchers on Asian contexts, 4 papers by Asian researchers on non-Asian contexts, and 9 papers by non-Asian researchers on Asian contexts were included in this review. Figure 10.1 shows the flow chart of the article selection process.

The results revealed that the majority of the included papers were published in A* journals (43.75%, 28 papers) followed by A (35.94%, 23 papers), B (17.19%, 11 papers), and C (3.13%, 2 paper) journals based on the ABDC ranking list of 2017.

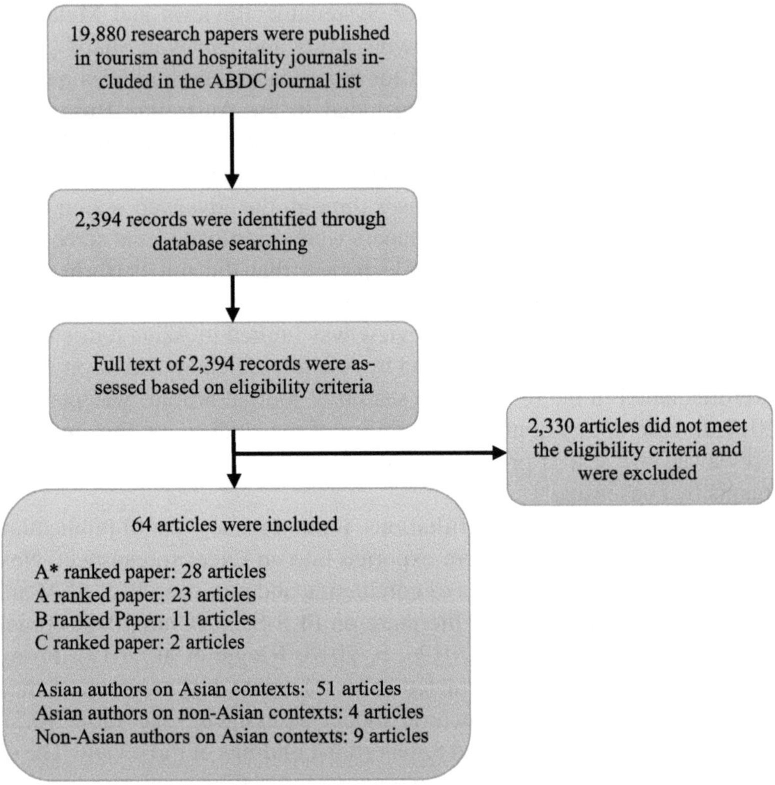

Fig. 10.1 Flow chart of the paper selection process

Even after controlling for the total number of published articles in each of the four ranked groups (total number of articles in A* journals, 5377 articles; A journals, 7282 articles; B journals, 4770 articles; and C journals, 2451 articles), there was a relationship between the quality of journals and using PLS-SEM, so that higher-ranked journals published more PLS-SEM papers. Table 10.1 highlights the overview of the included papers.

Figure 10.2 illustrates the growth of the Asian articles that used PLS-SEM method over time. Findings showed that the application of PLS-SEM attracted researchers' interest in this context after 2011. Approximately 60% of the included papers were published in in the last 3 years. However, the number of Asian PLS-SEM papers dropped in 2016 by around 40%.

Figure 10.3 represents the geographical representation of PLS-SEM research in Asian context. The findings showed that the majority of the included PLS-SEM articles investigated the tourism industry of Malaysia (25%) followed by China (23%) and Taiwan (12%). Countries such as Iran, South Korea, and Hong Kong also received attention by Asian researchers.

Table 10.1 Overview of the included PLS-SEM papers

	Rank	Total number of published papers	Records identified through the search terms	Asian papers by Asian authors	Non-Asian papers by Asian authors	Asian papers by non-Asian authors
Tourism Management	A*	1592	109	8	1	4
Annals of Tourism Research	A*	935	41	1	0	0
Journal of Travel Research	A*	814	337	4	1	2
Journal of Sustainable Tourism	A*	796	185	0	0	0
International Journal of Hospitality Management	A*	1240	32	6	0	1
International Journal of Tourism Research	A	678	41	2	0	0
Current Issues in Tourism	A	544	174	6	0	0
Tourism Recreation Research	A	499	57	0	0	0
Cornell Hospitality Quarterly	A	831	132	1	0	0
Tourism Geographies	A	424	95	0	1	0
International Journal of Contemporary Hospitality Management	A	1154	5	2	0	0
Tourism Analysis	A	612	8	2	1	0
Journal of Hospitality and Tourism Research	A	437	178	3	0	0
Journal of Hospitality Marketing and Management	A	526	130	1	0	0
Journal of Travel and Tourism Marketing	A	832	8	3	0	1
Tourism Economics	A	745	0	0	0	0
Tourism Review	B	293	1	0	0	0
Scandinavian Journal of Hospitality and Tourism	B	274	69	0	0	0
International Journal of Culture, Tourism and Hospitality Research	B	290	1	0	0	0

(continued)

Table 10.1 (continued)

	Rank	Total number of published papers	Records identified through the search terms	Asian papers by Asian authors	Non-Asian papers by Asian authors	Asian papers by non-Asian authors
Journal of Heritage Tourism	B	202	42	1	0	0
Journal of China Tourism Research	B	218	34	0	0	1
Information Technology & Tourism	B	53	3	1	0	0
Tourism and Hospitality Planning and Development	B	295	44	0	0	0
Journal of Ecotourism	B	192	25	0	0	0
Journal of Sport Tourism	B	236	0	0	0	0
Journal of Hospitality and Tourism Education	B	363	55	0	0	0
Asia Pacific Journal of Tourism Research	B	662	4	3	0	0
Journal of Convention Event Tourism	B	246	60	0	0	0
Journal of Hospitality and Tourism Management	B	186	2	1	0	0
Journal of Hospitality and Tourism Technology	B	108	2	1	0	0
Journal of Human Resources in Hospitality and Tourism	B	257	2	0	0	0
Journal of Policy Research in Tourism, Leisure and Events	B	130	21	0	0	0
Journal of Quality Assurance in Hospitality Tourism	B	251	1	0	0	0
Tourist Studies: An International Journal	B	244	35	0	0	0
Tourism and Hospitality Research	B	270	72	3	0	0

(continued)

Table 10.1 (continued)

	Rank	Total number of published papers	Records identified through the search terms	Asian papers by Asian authors	Non-Asian papers by Asian authors	Asian papers by non-Asian authors
Journal of Tourism and Cultural Change	C	209	0	0	0	0
Journal of Tourism History	C	91	0	0	0	0
Advances in Hospitality and Leisure	C	115	1	0	0	0
Asian Journal of Tourism and Hospitality Research	C	24	0	0	0	0
FIU Hospitality Review	C	189	12	0	0	0
Hospitality & Society	C	91	0	0	0	0
International Journal of Tourism Policy	C	110	0	0	0	0
Journal of Hospitality Application and Research	C	113	4	0	0	0
Journal of Hospitality, Leisure, Sports & Tourism Education	C	74	2	1	0	0
Journal of Teaching in Travel and Tourism	C	240	0	0	0	0
Journal of Travel and Tourism Research	C	296	0	0	0	0
The Journal of Hospitality Financial Management	C	106	25	1	0	0
Tourism in Marine Environments	C	167	0	0	0	0
Tourism Review International	C	286	0	0	0	0
Worldwide Hospitality and Tourism Themes	C	340	1	0	0	0
	Total	19,880	2394	51	4	9

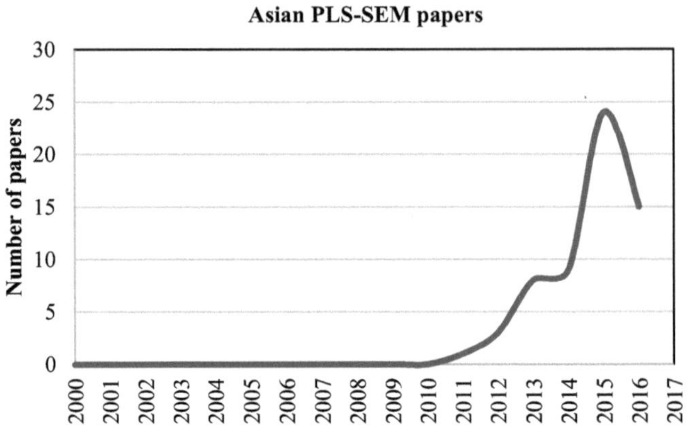

Fig. 10.2 Asian PLS-SEM papers published from 2000 to the end of 2016

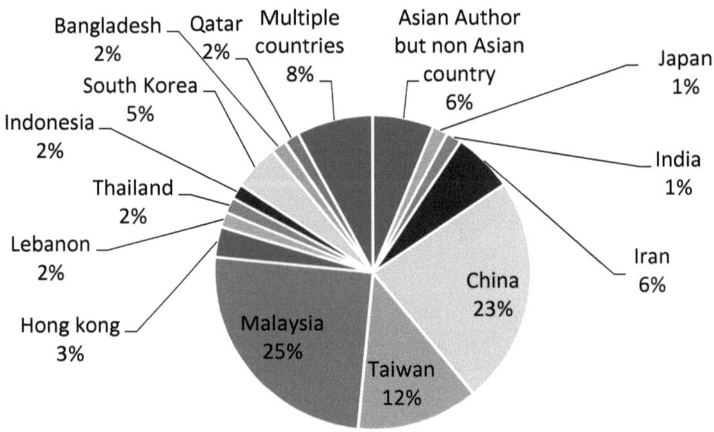

Fig. 10.3 The state of PLS-SEM papers in Asian countries

Table 10.2 presents the summary of the results of assessing the conducting and reporting PLS-SEM analysis in Asian tourism and hospitality research. Results showed that 58 PLS-SEM articles (90.62%) graphically represented the conceptual framework of the investigation. The mean of the sample size used in all PLS-SEM articles was 504.08 (*SD* = 718.24, range = 41 to 5209). The mean of the sample size of the included studies after excluding five studies with the highest sample size and five studies with the lowest sample size was 370.06 (*SD* = 176.07).

Figure 10.4 illustrates the frequency distribution of the sample size used by PLS-SEM articles. More than 42% of the articles (27 papers) used a sample size between 100 and 300 cases. Three cases (4.69%) of the papers used a sample size of less than 100 cases. Among the included studies, 25 papers (39.06%) provided justification to determine the sample size of their study. While 16 papers (25%) reported about missing cases, only 7 papers (10.93%) reported how they handled them. Finding showed

Table 10.2 Summary of systematic analysis

Category and criteria	Time period 2000–2004	2005–2009	2010–2014	2015–2017	2000–2017
Number of publications	0	0	21	43	64
Graphical presentation of conceptual framework	0	0	20	38	58
Data preparation					
Sample size					
Minimum	0	0	154	41	41
Maximum	0	0	2760	5209	5209
Sample size determination	0	0	8	17	25
No. of missing data	0	0	7	9	16
Handle missing data	0	0	1	6	7
Normality	0	0	2	7	9
Outliers	0	0	2	1	3
Justification for using PLS-SEM	0	0	11	29	40
Reflective constructs	0	0	18	43	61
Formative constructs	0	0	9	10	19
Higher-order constructs	0	0	3	7	10
Reflective construct assessment					
Factor loading	0	0	18	41	59
Factor weight			5	5	10
Cronbach's alpha	0	0	15	29	44
Composite reliability	0	0	20	43	63
Other methods to assess reliability	0	0	1	3	4
AVE			21	42	60
Convergent validity	0	0	20	42	62
Discriminant validity	0	0	19	41	60
Other methods to assess discriminant validity	0	0	1	11	12
Common method variance	0	0	2	9	11
Formative construct assessment					
Factor loadings	0	0	6	7	13
Factor weights	0	0	3	2	5
Multicollinearity assessment	0	0	4	11	15
Other validity assessments for formative constructs	0	0	0	2	2
Structural model assessment					
Critical *t*-value	0	0	17	23	40
R^2 value	0	0	21	41	62
Effect size	0	0	2	9	11
Q^2 value	0	0	4	15	19
Goodness-of-fit	0	0	3	10	13

(continued)

Table 10.2 (continued)

Category and criteria	Time period 2000–2004	2005–2009	2010–2014	2015–2017	2000–2017
Bootstrapping					
500 resamples	0	0	4	7	11
1000 resamples	0	0	1	3	4
2000 resamples	0	0	1	2	3
3000 resamples	0	0	0	0	0
5000 resamples	0	0	2	13	15
Addressing limitations of PLS-SEM	0	0	1	7	8
Software package					
Smart PLS	0	0	7	43	50
XLSTAT	0	0	0	4	4
Visual PLS	0	0	1	0	1
PLS-Graph	0	0	5	0	5
Warp PLS	0	0	3	0	3
ADANCO	0	0	0	1	1

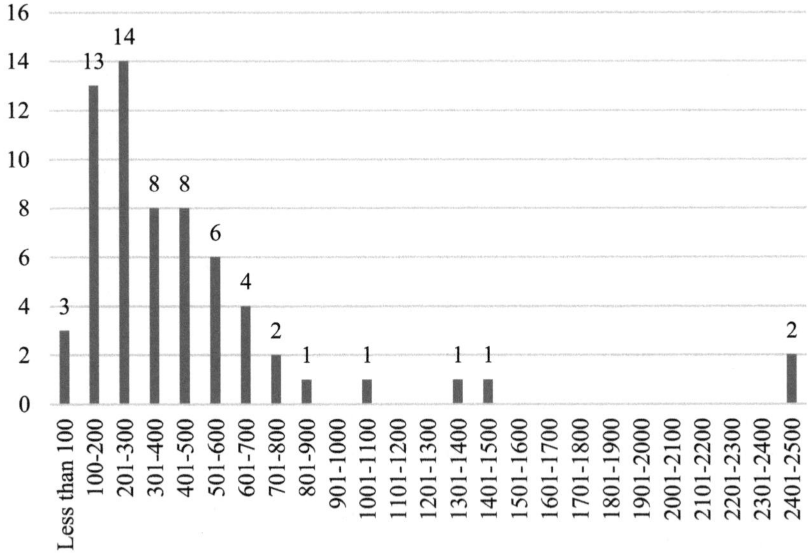

Fig. 10.4 Frequency distribution of sample size used in PLS-SEM papers

that nine articles (14.07%) assessed and reported normality. Moreover, only three papers (4.68%) were transparent about how they detected and handled outliers.

Forty papers (62.5%) provided some explanation to justify using PLS-SEM method although some were not relevant. In terms of measurement, while the majority of the articles used only reflective measurement (44 articles, 68.75%), 17 papers

(26.56%) used both reflective and formative measurements, and 1 paper (1.56%) used only formative measurement. Higher-order constructs were used in 10 papers representing 15.63% of the included studies.

In order to assess the measurement of the reflective constructs, the included papers were mostly consistent. Fifty-nine articles of the 63 papers (93.65%) containing reflective measurements reported items factor loading. In terms of reliability, composite reliability was the most commonly used method reported in all of the 63 articles followed by Cronbach's alpha reported in 44 papers (69.84%). Sixty-two papers (98.41%) assessed convergent validity in which 60 of them (95.23%) computed AVE for this purpose. Sixty articles (95.24%) assessed discriminant validity of reflective constructs, and common method variance was addressed in 12 articles (19.05%). In total, 18 papers used formative constructs. Among them, factor loadings and factor weights were reported in 13 (72.22%) and 5 papers (27.77%), respectively. Fifteen articles (83.33%) assessed multicollinearity issue among the items of formative constructs.

In order to assess structural model, 40 articles (62.5%) reported *t*-value of the statistical tests. Although the absolute majority of the articles (62 papers, 96.88%) computed and reported the coefficient of determination, only 11 studies estimated the effect size (17.19%). The predictive validity of the model was assessed in 19 articles (29.69%), and 13 papers (20.31%) used goodness-of-fit indexes to assess the model fit. The results showed that SmartPLS was the most popular software package used in Asian tourism and hospitality research (50 papers, 78.13%). Eight articles (44.44%) discussed the limitations of PLS-SEM as a statistical analysis method. In conducting bootstrapping, 15 papers (23.44%) used 5000 resamples followed by 11 articles (17.19%) using 500 resamples. However, many papers did not mention about bootstrapping in their report.

10.4 Discussion

The objective of this paper was to provide a methodological review of Asian tourism and hospitality research that used PLS-SEM. This review of 64 papers reveals several weaknesses and areas that researchers should take into consideration while performing PLS-SEM in future research.

Evidence from this review revealed that a very minimum number of researchers reported missing data and how they handled it. Even those who reported about missing data provided very brief information on the extent of missing, how missing cases were handled, and the degree of randomness (Hair et al. 2011; Weston and Gore 2006). Researchers may report whether the missing cases were random, non-random, or completely missing (Hair et al. 2012b). They also may report what method they used to treat missing data. The most common methods of handling missing value include deleting listwise, deleting pairwise, known replacement value, and model-based methods. Moreover, little was reported about normality distribution of data. This may be due to the known distribution-free assumption of

PLS-SEM. However, not all researchers support this view. For example, Rönkkö et al. (2016) stated that "although normality is not required for consistency, unbiasedness, or efficiency of the OLS estimator, normality is assumed when using inferential statistical tests. Furthermore, an estimator cannot at the same time have fewer distributional assumptions and work better with smaller samples, because this notion violates the basics of information theory." A very few studies handled and reported outliers. Identifying and removing both univariate and multivariate outliers are necessary for the validity of the findings. Indeed, being a "silver bullet" of PLS-SEM is subject to meeting the commonly known standards of collecting adequate sets of empirical data such as identifying and handling outliers and influential cases (Ringle et al. 2012).

The majority of the studies did not explain how they determined the sample size. Future studies should justify the sample size by disclosing more information such as determination criteria of sample size and conducting power analysis. PLS-SEM is known as a method that can work well with small sample size. Some researchers have used this to justify using PLS-SEM rather than covariance-based structural equation modeling analysis which is a large sample size technique (Yang et al. 2015). However, the notion of small sample size has been highly criticized recently (Rönkkö et al. 2016). Hair et al. (2013, p. 2) stated that "some researchers abuse this advantage by relying on extremely small samples relative to the underlying population," and "PLS-SEM has an erroneous reputation for offering special sampling capabilities that no other multivariate analysis tool has." The results of this review showed that unlike other disciplines (e.g., see Ringle et al. 2012), only three studies used a sample size of less than 100 observations. Thus, there is no such concern regarding using small sample size in Asian hospitality and tourism research.

In terms of measurement, around a quarter of the included studies specified their models using both reflective and formative measurements. The majority of the papers used only reflective measurements, while using only formative measurements was limited to one paper. Moreover, around one-fifth of the studies used higher-order constructs. One of the advantages of PLS-SEM over CB-SEM method is coping with both reflective and formative constructs and the ease of assessing models with higher-order measurements. The results of this review revealed that handling formative measurements was provided as a justification for utilizing PLS-SEM in around 15% of the papers. However, surprisingly, 30% of these papers did not use any formative construct in their model.

In order to assess the reliability of the reflective constructs, composite reliability was the most commonly used method followed by Cronbach's alpha. As Cronbach's alpha assumes equal factor loadings or error terms among all indicators of a construct, it computes a less accurate measure of construct reliability than other methods such as composite reliability that takes into consideration the factor loadings of the items (Pahlevan Sharif et al., Forthcoming). To assess convergent validity of reflective constructs, authors computed AVE, as a measure of the average amount of variance in items explained by a construct, using the greater than 0.5 criteria. The absolute majority of the papers reported an AVE of greater than 0.5, while there were a few exceptional cases. However, the papers were not transparent about whether they removed any weak item to improve the AVE of their constructs. AVE of all reflective constructs should be computed and reported even if it does not meet the 0.5 criteria. Indeed, assessing convergent validity using AVE may be too conservative (Pahlevan Sharif, Mostafiz, and Guptan, Forthcoming).

For assessing discriminant validity of reflective constructs, the approach suggested by Fornell and Larcker (1981) was the most commonly used technique, while some recently published papers reported the heterotrait-monotrait ratio of correlations (HTMT) (Henseler et al. 2015). Discriminant validity of a construct requires that "a test not correlate too highly with measures from which it is supposed to differ" (Campbell 1960, p. 548). A lack of discriminant validity indicates that "constructs [have] an influence on the variation of more than just the observed variables to which they are theoretically related," and as a result, "researchers cannot be certain results confirming hypothesized structural paths are real or whether they are a result of statistical discrepancies" (Farrell 2010, p. 324). Some recent studies have questioned the effectiveness of the Fornell-Larcker criterion under certain circumstances (Henseler et al. 2015). To overcome the limitations of the Fornell-Larcker criterion, Henseler et al. (2015) suggest reporting HTMT.

Around half of the papers that used formative measurements did not report factor weight of the items although factor loading of the formative items were reported in most of them. Around one-third of the papers with formative constructs did not assess collinearity among the formative items. To evaluate validity of the formative constructs, only one paper performed nomological validity assessment. Surprisingly, none of the included papers used convergent validity suggested by Hair et al. (2016) also known as redundancy analysis (Chin 1998). Overall, the results showed that assessing formative measurements was mostly limited to estimating the factor loadings and evaluating multicollinearity among the items. Most of the studies neglected factor weight assessment and assessing convergent validity of the construct.

Moreover, the results of this systematic review revealed that less than one-fifth of the included papers computed and reported the effect size, and they only rely on statistical significance of the relationships. The statistical significance measures the probability that the relationship between two variables is due to chance. However, using a large sample size would result in p-value less than the accepted level of 0.05, and even very weak and meaningless relationships become significant (Sullivan and Feinn 2012). Therefore, reporting effect size is often suggested or even required when discussing the results of a study (Kelley and Preacher 2012).

The present study is not without limitations. It is possible that some papers that could meet the inclusion criteria have been missed. Also, reporting different methods of measurement and structural model assessment largely depends on the features of software that researchers use. Moreover, journal format may influence reporting the results. Finally, the authors of this paper acknowledge that the issues that should be addressed in performing PLS-SEM are not limited to those that were discussed in this review.

10.5 Conclusion

A growing number of hospitality and tourism studies have used structural equation modeling analysis. However, little is known about the application of PLS-SEM in Asian tourism and hospitality research. All PLS-SEM studies conducted by Asian researchers and/or on Asian contexts that were published in 66 tourism and hospitality journals from 2000 to the end of February 2017 were systematically reviewed. Studies that use PLS-SEM should address several issues to ensure the validity of their findings. Although performing PLS-SEM analysis is largely dependent on the features that software packages provide, the results of this review identified several weaknesses and inconsistencies in conducting and reporting which should be addressed by researchers in future tourism and hospitality empirical studies.

References

Ballantyne, R., Packer, J., & Axelsen, M. (2009). Trends in tourism research. *Annals of Tourism Research, 36*, 149–152.

Campbell, D. T. (1960). Blind variation and selective retentions in creative thought as in other knowledge processes. *Psychological Review, 67*, 380–400.

Chin, W. W. (1998). The partial least squares approach to structural equation modeling. *Modern Methods for Business Research, 295*, 295–336.

Chin, W. W. (2010). How to write up and report PLS analyses. In *Handbook of partial least squares*. Berlin/Heidelberg: Springer.

do Valle, P. O., & Assaker, G. (2016). Using partial least squares structural equation modeling in tourism research: A review of past research and recommendations for future applications. *Journal of Travel Research, 55*, 695–708.

Farrell, A. M. (2010). Insufficient discriminant validity: A comment on Bove, Pervan, Beatty, and Shiu (2009). *Journal of Business Research, 63*, 324–327.

Fornell, C., & Larcker, D. F. (1981). Evaluating structural equation models with unobservable variables and measurement error. *Journal of Marketing Research, 16*, 39–50.

Goodhue, D. L., Lewis, W., & Thompson, R. (2012). Does PLS have advantages for small sample size or non-normal data? *MIS Quarterly, 36*, 981–1001.

Guide, V. D. R., & Ketokivi, M. (2015). Notes from the editors: Redefining some methodological criteria for the journal. *Journal of Operations Management, 37*, 5–8.

Hair, J. F., Ringle, C. M., & Sarstedt, M. (2011). PLS-SEM: Indeed a silver bullet. *Journal of Marketing Theory and Practice, 19*, 139–152.

Hair, J. F., Sarstedt, M., Pieper, T. M., & Ringle, C. M. (2012a). The use of partial least squares structural equation modeling in strategic management research: A review of past practices and recommendations for future applications. *Long Range Planning, 45*, 320–340.

Hair, J. F., Sarstedt, M., Ringle, C. M., & Mena, J. A. (2012b). An assessment of the use of partial least squares structural equation modeling in marketing research. *Journal of the Academy of Marketing Science, 40*, 414–433.

Hair, J. F., Ringle, C. M., & Sarstedt, M. (2013). Partial least squares structural equation modeling: Rigorous applications, better results and higher acceptance. *Long Range Planning, 46*(1–2), 1–12.

Hair, J. F., Hult, G. T. M., Ringle, C., & Sarstedt, M. (2016). *A primer on partial least squares structural equation modeling (PLS-SEM)*. Thousand Oaks: Sage.

Henseler, J., Ringle, C. M., & Sarstedt, M. (2015). A new criterion for assessing discriminant validity in variance-based structural equation modeling. *Journal of the Academy of Marketing Science, 43*, 115–135.

Huang, S., Weiler, B., & Assaker, G. (2015). Effects of interpretive guiding outcomes on tourist satisfaction and behavioral intention. *Journal of Travel Research, 54*, 344–358.

Kelley, K., & Preacher, K. J. (2012). On effect size. *Psychological Methods, 17*, 137.

Liberati, A., Altman, D. G., Tetzlaff, J., Mulrow, C., Gøtzsche, P. C., Ioannidis, J. P., & Moher, D. (2009). The PRISMA statement for reporting systematic reviews and meta-analyses of studies that evaluate health care interventions: Explanation and elaboration. *PLoS Medicine, 6*, e1000100.

McIntosh, C. N., Edwards, J. R., & Antonakis, J. (2014). Reflections on partial least squares path modeling. *Organizational Research Methods, 17*, 210–251.

Mura, P., & Pahlevan Sharif, S. (2017). Narrative analysis in tourism: A critical review. *Scandinavian Journal of Hospitality and Tourism, 17*, 194–207.

Mura, P., Mognard, E., & Pahlevan Sharif, S. (2017). Tourism research in non-English-speaking academic systems. *Tourism Recreation Research, 42*, 436–445.

Pahlevan Sharif, S., & Mahdavian, V. (2015). *Structural equation modeling by the use of AMOS*. Tehran: Bishe Publications.

Pahlevan, Sharif, S., Mostafiz., I., & Guptan, V. (Forthcoming). Structural equation modeling in nursing research: A systematic review. *Nurse Researcher*. https://doi.org/10.7748/nr.2018.e1577.

Ringle, C. M., Sarstedt, M., & Straub, D. W. (2012). Editor's comments: A critical look at the use of PLS-SEM in "MIS Quarterly". *MIS Quarterly, 36*, 3–14.

Rönkkö, M., & Evermann, J. (2013). A critical examination of common beliefs about partial least squares path modeling. *Organizational Research Methods, 16*, 425–448.

Rönkkö, M., McIntosh, C. N., Antonakis, J., & Edwards, J. R. (2016). Partial least squares path modeling: Time for some serious second thoughts. *Journal of Operations Management, 47–48*, 9–27.

Sosik, J. J., Kahai, S. S., & Piovoso, M. J. (2009). Silver bullet or voodoo statistics? A primer for using the partial least squares data analytic technique in group and organization research. *Group & Organization Management, 34*, 5–36.

Sullivan, G. M., & Feinn, R. (2012). Using effect size—Or why the P value is not enough. *Journal of Graduate Medical Education, 4*, 279–282.

Weston, R., & Gore, P. A. (2006). A brief guide to structural equation modeling. *The Counseling Psychologist, 34*, 719–751.

Yang, E. C. L., Pahlevan Sharif, S., & Khoo-Lattimore, C. (2015). Tourists' risk perception of risky destinations: The case of Sabah's eastern coast. *Tourism and Hospitality Research, 15*, 206–221.

Chapter 11
Structural Equation Modelling with Second-Order Confirmatory Factor Analysis: Critical Factors Influencing Consumer Behavior in Medical Tourism

Jung Wan Lee and Michael Kwag

Abstract The chapter demonstrates an application of structural equation modelling (SEM) with latent variables in tourism research, and the associated statistics are described. First, we discuss the basic concepts of SEM, followed by an explanation of the key statistics and terms associated with this procedure. Then we describe the procedure for conducting SEM, including second-order confirmatory factor analysis (CFA). Finally, we describe the related technique of SEM analysis. In doing so, the chapter provides an application of structural equation modelling with latent variables, of which latent variables are hypothetical constructs that are invoked to explain observed covariation in behavior. The study aims to identify critical factors of consumer acceptance of medical tourism in Asia. The study employs factor analysis (i.e., exploratory factor analysis, confirmatory factor analysis, and internal consistency reliability tests) and structural equation modelling analysis and path analysis (i.e., the analysis of moment structures and regression analysis) using the survey data collected from a sample of 486 international tourists who visited South Korea in 2014.

Keywords Structural equation model · Path analysis · Latent variable · Factor analysis · Internal consistency reliability tests · Medical tourism · Tourism destination · Tangible and intangible attributes · Tourism research · Asia

J. W. Lee (✉)
Administrative Sciences Department, Metropolitan College, Boston University, Boston, MA, USA
e-mail: jwlee119@bu.edu

M. Kwag
School of Hospitality Administration, Boston University, Boston, MA, USA
e-mail: mkwag@bu.edu

S. Rezaei (ed.), *Quantitative Tourism Research in Asia*, Perspectives on Asian Tourism, https://doi.org/10.1007/978-981-13-2463-5_11

11.1 Introduction of Methodology

Chapter Objectives

- To explain the basic concepts of structural equation modelling (SEM) such as theory, model, path diagram, exogenous versus endogenous constructs, dependence and correlational relationships, model fit, and model identification
- To describe the process of conducting SEM and explain the various steps involved
- To explain how to specify a measurement model and assess its validity
- To explain the concept of model fit and the differences between absolute, incremental, and parsimony fit indices
- To describe the procedure for conducting SEM with second-order confirmatory factor analysis (CFA)
- To describe how to specify a structural model and assess its validity

11.1.1 Theory and the Structural Equation Model

The structural equation modelling (SEM) is based on a theory in which all relationships must be specified before the SEM model can be estimated. In SEM, models are often constructed to test certain hypotheses derived from the theory. A SEM model consists of two models: the measurement model and the structural model.

The researcher uses knowledge of the theory, empirical research, or both to postulate the relationship pattern a priori and then tests the hypothesis statistically. Indicator variables are selected on the basis of theory, and confirmatory factor analysis (CFA) is used to see if they load as predicted on the expected number of factors. The terms construct and factor are used interchangeably. In other words, in testing for measurement model, the researcher has complete control over which indicators describe each construct. On the other hand, a structural model shows how the constructs are interrelated to each other, often with multiple dependence relationships. It specifies whether a relationship exists or does not exist. If a relationship is hypothesized by the theory, then an arrow is drawn. If a relationship is not hypothesized, then no arrow is drawn.

A measurement model is designed and constructed by following norms. Construct is represented by ovals or circles, while measured variables are represented by squares. Straight arrows are drawn from the constructs to the measured variables. Dependence relationships are portrayed by straight arrows and correlational relationships by curved arrows. A dependence relationship is shown by straight arrows. The arrows flow from the antecedent (independent) to the subsequent effect (dependent) measured variable or latent construct. In a measurement model, the straight arrows are drawn from the construct to its measured variables.

In a structural model, the dependence occurs between constructs, and so straight arrows are drawn between constructs. The specification of dependence relationships is also related to whether a construct is considered exogenous or endogenous. It should be noted that an endogenous construct can be an antecedent of other endogenous constructs. A correlational relationship, also called covariance relationship, specifies a simple correlation between exogenous constructs. The theory posits that these constructs are correlated, but it is not assumed that one construct is dependent upon another. A correlational relationship is depicted by a two-headed arrow.

The process of conducting SEM is as follows. The steps involved in conducting SEM are (1) define the individual constructs, (2) specify the measurement model, (3) assess measurement model reliability and validity, (4) specify the structural model if the measurement model is valid, (5) assess structural model validity, and (6) draw conclusions and make recommendations if the structural model is valid. For further readings, see the book (Malhotra 2010).

11.1.2 Specification of a Measurement Model

The measurement model is usually represented by a diagram. The assignment of measured variables to each latent construct is graphically equivalent to drawing arrows from each construct to the measured variables that represent that construct. The degree to which each measured variable is related to its construct is represented by that variable's loading. Only the loadings linking each measured variable to its latent construct as specified by the arrows are estimated; all other loadings are set to zero. Also, since a latent factor does not explain a measured variable perfectly, an error term is added. In a measurement model, the researcher often does not distinguish between exogenous and endogenous constructs; they are all treated as being of the same type, similar to that in factor analysis.

In a measurement model, it is common to represent constructs by Greek characters and measured variables by alphabets. The common notations used are:

ζ = latent factors
x = measured variables
λ_x = factor loadings
δ = errors
Φ = correlation between constructs

Specifying the observed variables or indicators for each latent construct requires setting the scale of the latent construct. Because a latent construct is not observed, it has no metric scale (i.e., no range of values). Therefore, this must be provided, and either one of the following two options may be used. First, one of the factor loadings can be fixed, generally to a value of one. Second, the construct variance can be fixed, generally to a value of one. In this case, the relationships between constructs are represented by a correlation matrix.

11.1.3 Factor Analysis and Internal Consistency Reliability

Evidence of the effectiveness of a scale for its purpose must be examined. Many methods of validation rely heavily on the analysis of inter-item or inter-scale correlations. Construct validity embraces a variety of techniques for assessing the degree to which an instrument measures the concept that it is designed to measure. This may include testing dimensionality and homogeneity.

Factor analysis is an often used key technique in this process. In order to ensure the construct validity of the measurement instrument, factor analysis is often employed in a two-stage process. First, exploratory factor analysis with a varimax rotation procedure is employed to identify underlying predictors based on an eigenvalue cutoff of one. Secondly, confirmatory factor analysis using structural equation modelling techniques is employed to confirm that the identified predictors are fitted to the items correctly and reliably.

Internal consistency reliability is a measure of how well a test addresses different constructs and delivers reliable scores. Among others the most common method for assessing internal consistency is Cronbach's alpha. This form of intraclass correlation is closely related to convergent validity, i.e., the extent to which the items in a scale are all highly intercorrelated. For example, in a series of questions that ask the subjects to rate their response between one and seven, Cronbach's alpha gives a score between zero and one, with 0.7 and above being reliable. The test also takes into account both the size of the sample and the number of potential responses. The Cronbach's alpha test is preferred in tourism research due to the benefit of averaging the correlation between every possible combination of split halves and allowing multi-level responses. For example, the survey items are divided into four constructs. The internal consistency reliability test provides a measure so that each of these particular constructs is measured correctly and reliably.

For further readings, see the more comprehensive description of scale development and reliability (Dunn 1989) and scale development and validation (Bartholomew 1996; Basilevsky 1994).

11.1.4 Validity of the Measurement Model

The validity of the measurement model depends on the goodness-of-fit results, reliability, and evidence of construct validity, especially convergent and discriminant validity. The convergent validity measures the extent to which the scale correlates positively with other measures of the same construct. Hence, the size of the factor loadings provides evidence of convergent validity. High factor loadings indicate that the observed variables converge on the same construct. At a minimum, all factor loadings should be statistically significant and higher than 0.5, ideally higher than 0.7. A loading of 0.7 or higher indicates that the construct is explaining 50% or more of the variation in the observed variable (since $((0.71)^2 = 0.5)$. Sometimes, a cutoff level of 0.6 is used.

Another measure that is used to assess convergent validity is the average variance extracted (AVE), which is defined as the variance in the indicators or observed variables that is explained by the latent construct. AVE is calculated in terms of the (completely) standardized loadings. AVE varies from 0 to 1, and it represents the ratio of the total variance that is due to the latent variable. An AVE of 0.5 or more indicates satisfactory convergent validity, as it means that the latent construct accounts for 50% or more of the variance in the observed variables, on the average. If AVE is less than 0.5, the variance due to measurement error is larger than the variance captured by the construct, and the validity of the individual indicators, as well as the construct, is questionable. Note that AVE is a more conservative measure than composite reliability (CR), which is defined as the total amount of true score variance in relation to the total score variance. On the basis of CR alone, the researcher may conclude that the convergent validity of the construct is adequate, even though more than 50% of the variance is due to error. One should also interpret the standardized parameter estimates to ensure that they are meaningful and in accordance with theory.

In order to establish discriminant validity, the researcher must show that the construct is distinct from other constructs and thus makes a unique contribution. First, individual observed variables should load on only one latent construct. Cross loadings indicate lack of distinctiveness and present potential problems in establishing discriminant validity. One formal way to show distinctiveness is to set the correlation between any two constructs as equal to one, i.e., we are specifying that observed variables measuring the two constructs might as well be represented by only one construct. Evidence of discriminant validity is obtained if the fit of the two-construct model is significantly better than the fit of the one-construct model. However, this actually turns out to be a weak test as significant fit differences may be obtained even when the correlations between the two constructs are very high.

An alternative test of discriminant validity is based on the logic that a construct should explain its observed variables better than it explains any other construct. This test is conducted by showing that the average variance extracted is greater than the square of the correlations. Equivalently, discriminant validity is achieved if the square root of the average variance extracted is larger than correlation coefficients.

11.1.5 Specification and Assessment of the Structural Model

In moving from the measurement model to the structural model, the emphasis shifts from the relationships between latent constructs and the observed variables to the nature and magnitude of the relationships between constructs. Thus, the measurement model is altered based on the relationships among the latent constructs. Because the measurement model is changed, the estimated covariance matrix based on the set of relationships examined will also change. However, the observed covariance matrix, based on the sample data, does not change as the same data are used to estimate the structural model. Thus, in general, the fits statistics will also change, indicating that the fit of the structural model is different than the fit of the measurement model.

If the measurement model is identified, then the structural model is too, provided that it is recursive in that there are no feedback loops or dual dependencies and, in addition, there are no interaction terms. In such cases, generally the structural model is nested within the measurement model and contains fewer estimated parameters. A model is nested within another model if it has the same number of constructs and variables and can be derived from the other model by altering relationships, as by adding or deleting relationships.

The analysis of moment structures is used for an empirical test of a structural model. The maximum likelihood estimation is applied to estimate numerical values for the components in the model. In the process of identifying the best-fit model, multiple models are analyzed because researchers are testing competing theoretical models. From a predictive perspective, the researcher determines which model fits the data best, but sometimes the differences between the models appear small on the basis of the fit indices. When comparing non-nested models, the Akaike information criterion fit index is used as the first choice because the difference in the chi-square values among the models cannot be interpreted as a test statistic (Kline 2005), the root mean square of approximation fit index as the second choice, and then the goodness of fit index as the third choice. For further readings, the concept of model fit and the differences between absolute, incremental, and parsimony fit indices are well documented in the book (Malhotra 2010).

11.2 Literature Review and Hypotheses

11.2.1 Research Background

Medical tourism, which refers to the business of providing medical treatment, hospitality and health services, and associated tour services, is one of the fastest growing business sectors in tourism. This emerging business provides foreign medical tourists with timely, economical, and quality medical treatment and health services in hosting countries. The global medical tourism industry recently achieved annual revenue about $40 billion with around eight million medical tourists from across the world, according to a report of the Institute of Population Health (Labonté et al. 2013). Nowadays, with the increased access to Internet websites, social networking sites, and blogs, this social media brings more medical tourism information to broader population on a timely basis. Thanks to the benefits of the Internet of Things and social networking sites, the medical tourism industry has been directly affected in the way consumers search medical tourism information. The medical tourism websites and social networking sites refer to a community that provides any combination of consulting, sharing, and providing medical tourism information, including medical tour destinations, medical service providers, medical tour packages, medical tour agencies and facilitators, medical tour visa, medical tour prices, and medical dispute and arbitration information. For example, the Visit Medical Korea (2016) website provides much information about medical tour services in Korea in six different languages: English, Japanese, Chinese, Russian, Arabic, and Korean.

Generally, the international medical tourism industry attracts consumers from developed countries such as the United States, the United Kingdom, Japan, and Australia, to developing countries like India, Thailand, Malaysia, Turkey, Mexico, and Brazil. But sometimes and vice versa, the medical tourism industry attracts tourists from the developing countries to the developed countries like the United States, Singapore, and Korea. In particular, the rapidly growing trend of Asian medical tourism industry has promoted their medical tourism to a worldwide competition. For instance, Singapore reported estimated revenue in the medical tourism industry of US$ 1.8 billion in 2012, creating 13,000 new jobs and attracting more than one million medical tourists from across the world (Connell 2006). Furthermore, government officials and tour agencies are increasingly utilizing Internet websites and social networking sites as a specific marketing tool for promoting their medical tourism industry. As a result, Internet websites and social networking sites have been widely used by medical tourism marketers in both developed and developing countries to promote their offerings to prospective medical tourism consumers. On the other hand, word-of-mouth recommendations to friends and relatives are proved to be more effective in promoting medical tourism than commercialized marketing websites, by which tourists generally use them for booking flights and hotels and estimating costs.

Therefore, a better understanding of consumer behavior in accepting medical tourism is a critical area of research. This paper intends to integrate some components of service quality dimensions and consumer needs for medical tours to identify critical factors affecting consumer behavior in accepting medical tourism. This study aims (1) to propose a new simplified framework for a better understanding of consumer behavior in accepting medical tourism, (2) to provide empirical evidence to support the proposed framework, and (3) to help medical tourism marketers in the way what type of information is required and what type of marketing promotion would work better when using Internet websites and social networking sites for the purpose of attracting medical tourism consumers.

11.2.2 Overview of the Medical Tourism Industry

Medical tourism refers to the business of providing medical treatment, health services, and associated tour services. This business is one of the fastest growing business sectors in tourism, especially for cosmetic surgeries: sometimes being considered as a by-product of this growing business. There is much literature on the positive effects of medical tourism on the international tourism industry (e.g., Li et al. 2011; Zhang et al. 2013). The medical tourism promotion by local governments and medical tour operators, for example, promoting their country as an ideal destination for cosmetic surgeries for foreign visitors, is repaying their effort by attracting more inbound tourists into their country. Some Asian countries such as Singapore (Lee 2010), Korea (Lee et al. 2011; Yu and Ko 2012), Thailand (Wongkit and McKercher 2013), Hong Kong (Ye et al. 2011), Malaysia (Yeoh et al. 2013), and Iran (Moghimehfar and Nasr-Esfahani 2011) enjoy seeing a rapid growth of medical tourism. The evidence on growing medical tourism is quite large. Singapore government is known for its immense support to enhance its medical tourism industry. Lee (2010) found empirical evidence that there was a unidirectional long-run interconnection from medical tourism to international tours. The United States has generated $1 billion trade surplus in the medical tourism segment by creating more demand for inbound tourists (Johnson and Garman 2010). Brazil has become a world center of plastic surgeries thanks to international tourist residential experiences (Edmonds 2011).

In Canada, however, a substantial number of medical tourism operators ceased their business. Some reasons may include dwindling consumer base, limited medical service providers, competitions, and limited medical tourism destinations (Turner 2011). It is reported that emerging markets face challenges in the global medical tourism markets, for instance, they all had difficulties attracting international tourists. In addition, medical tourism providers in developed countries are now reacting to the competition they faced from the growth of medical tourism in those emerging markets. The globalization of medical tourism is actualized since international medical tourists were disappointed about their national health system's weaknesses, such as long waiting lists and higher costs (Enderwick and Nagar 2011). Emerging markets could resolve those problems above by virtue of significant cost reduction and comparable ranges of quality medical tourism. A study reports several barriers that would impact medical tourism, including policy and regulation restrictions of the government, costs, infrastructures, capacity problems, and medical service needs of the local community (Heung et al. 2011).

11.2.3 How Consumers Evaluate Quality of Medical Tourism

By engaging in medical tourism, consumers would consider quality of the required medical treatment, location of medical treatment, quality of health services, medical service providers, medical facilities, and safety issues (Cormany and Baloglu 2011; García-Altés 2005). A study reports that consumers look for medical tourism information for their decision-making in opting for or against medical tourism abroad (Runnels and Carrera 2012). Another study reports that the consumer satisfaction level of medical tourism depends on four key factors: patients' background, surgical procedures, surgeon, and medical facilities (Lunt and Carrera 2010). It is often argued that some of the previous studies only concentrated on tourists' personal purposes, ignoring other formal cross-border institutional transfers such as associated tourism activities in seeking fun and leisure (Connell 2013). Some researchers (e.g., Li et al. 2011; Zhang et al. 2013) also argued that cross-cultural consumers would appear to have different preferences and perceptions in choosing their medical tourism destination. For example, Chinese medical tourists under the Korean wave's influence appeared to be more interested in getting cosmetic and plastic surgery in Korea (Yu and Ko 2012). However, Japanese tourists were more concerned about rehabilitation, safety, hospitality services, and cost-effectiveness. Sobo et al. (2011) illustrated that medical tourism operators usually assumed that medical tourists prefer world-class health services for the least cost.

Several studies report that sharing electronic word of mouth among tourists through social networking sites plays increasingly important roles in medical tourism (e.g., Munar and Jacobsen 2014; Luo and Zhong 2015). Some studies demonstrate that website marketing features would substantially affect the effectiveness of message delivering, quality of medical tourism promotion, and the features of medical tourism (e.g., Chang and Caneday 2011; El-Gohary 2012). It is said that website development is the most important form of marketing promotion to attract more medical tourists. Furthermore, the increasing use of mobile devices with web access would become important to better promote medical tourism (Dudensing et al. 2011). A study reports that Internet medical tourism information has a positive effect on the demand creation for medical tourism (Suziedelyte 2012). The Internet is an essential tool for medical tourism marketing since it has created a new distribution channel and opened a new way for tourism research. It is often argued, however, that Internet marketing platforms to obtain medical tourism information, such as professional diagnosis, self-diagnosis, aftercare and support, sharing medical tourism experience, posting opinions, and online virtual tours, would be the key driver in medical tourism (Hung and Law 2011; Lunt et al. 2010). Some reasons that explain the rapid growth of Internet marketing in medical tourism are that it is economical, instantaneous, convenient, useful, and playful (Chang and Caneday 2011).

There are differences among medical tourism facilitators in creating website contents. For example, North American agencies are more likely to provide services to multiple countries than any other continent, and they would also provide help in the transference of medical records, international mobile phone support, and financial assistance (Cormany and Baloglu 2011). A study reports that medical tourism facilitators emphasize more on a variety of health-related messages and images rather than cost savings to reach out international medical tourists (e.g., Crooks et al. 2011). Penney et al. (2011) argued that the medical tourism information on Internet websites and social networking sites focuses on promotion only rather than assisting consumers to understand full options of their medical tour or indicating potential risks involved. The number of websites that tourists could share their individual experiences and reviews about hospitals and professionals is ever increasing. It is suggested that medical tourism operators and policy-makers must try to understand why these sharing experiences are contributing in the way and how to better respond to these published opinions, rather than merely ignoring them (Adams 2011).

Horng and Tsai (2010) examined the impact of online introduction and marketing techniques of government-sponsored websites, in particular for medical tourism. They report that establishing on government-sponsored medical tourism information websites is the most essential phase in developing effective marketing strategies for promoting medical tourism. It is said that a large number of photos and written descriptions also would enrich the effect of Internet marketing since the information is randomly accessible (Wang et al. 2007). In addition, a three-step guide for future users on how to facilitate their medical tourism was developed: describing the inefficient home country healthcare system, encouraging consumers to become more active about their own health, and finally introducing low-cost medical tour destinations (Ormond and Sothern 2012). In contrast, it is easily found that most of international medical tour guidebooks are concentrating on persuading consumers about the attractiveness of medical tourism destinations.

11.2.4 Factors Affecting Consumer Acceptance of Medical Tourism

It is said that if medical tourists trust medical tourism operators' service quality, they will choose the service provider over the others. In addition, medical tourists perceive the quality of medical tourism through medical facility and tourism infrastructure of the destination as well (Chomvilailuk and Srisomyong 2014). There is another service quality dimension that might not directly involve with the facility and infrastructure of medical tourism. The associated medical tourism services, on the other hand, work out as an intermediary to support medical tourists by providing tourism information, medical arrangements, choices of travel and transportation, etc. These service quality dimensions also affect how medical tourists made their choices to visit where for their medical tourism purpose. Medical tourism facilitators, the people who contact with medical tourists from planning stages until the medical tourists return home, play also a key role in promoting medical tourism. Medical tourism facilitators generally rely on online communications to facilitate medical tourism services. Through the use of Internet websites and social networking sites, medical tourists can find the required information, including hospital equipment, health facility, certificates and standards of hospital accreditation, listing of medical treatments available, estimated medical treatment costs, and past medical tourist testimonials. Additionally, the auxiliary services associated with the medical tour are also provided on the websites, which include air travel, ground transportation, hotel accommodations, translation services, concierge services, sightseeing options, arrangement of medical appointments, transfer of medical records, and provision of aftercare support services.

According to Lertwannawit and Gulid (2011), quality of medical tourism can be measured through several dimensions such as value, satisfaction, behavioral loyalty, brand trust, etc. The relationships between service quality and several dimensions can be seen in different ways. For example, service quality is positively related to the value that medical tourists perceive in their medical treatment. Also, service quality is positively related to the brand trust of medical tourists toward medical treatment. A study claims that the countries that engage in medical tourism should put effort to develop its own signature brand in which this reflects unique quality (Viladrich and Baron-Faust 2014). For instance, Argentina developed its own brand by advertising its unique culture and applied a body framework to promote their cosmetic surgery quality. Through this branding process, Argentina also presented that Argentina provided superior medical tourism quality than other South American nations. As a result, medical tourists perceived that the cosmetic surgery in Argentina is not only low cost but also high quality. Moreover, Argentina highlighted their nation with the concept "feel at home," especially for European; they would feel like they were at home while visiting Argentina (Viladrich and Baron-Faust 2014, p. 122).

Considering there are wide differences in the perceptions of medical tourism quality and the hierarchy of medical tourism needs from different regions, Cormany and Baloglu (2011) claim that aftercare arrangement is the most important service that allows medical tourism facilitators in Central and South Africa to differentiate them from the others. Accordingly in medical tourism facilitators' websites, there are different presentations emphasizing on their unique competitive advantages. For instance, an Asian website emphasizes on lower cost, while a US website emphasizes on hospital accreditation (Cormany and Baloglu 2011). The medical tourism facilitators in Europe likely offer sightseeing available in many European tourism destinations, which allow them to attract many medical tourists. The medical tourism facilitators in Asia likely offer translation services especially in countries where English is used as a secondary language such as Korea, Thailand, and Malaysia.

11.2.5 A New Framework of Medical Tourism Research

In traditional business environments, consumers who observe service providers' facility and equipment, management, and employees typically generate the perception of tangible attributes. The tangible attributes of medical tourism affect its service evaluation by consumers and their behavioral intentions of having medical tourism there. The examples of tangible attributes of medical tourism include the use of modern medical equipment and technology, professional appearance of service personnel (i.e., doctors, nurses, and clinical support workers), hospital, and healthcare facilities. Tangible attributes are perceived as easy to use, attractive to use, and interesting for consumers to perform tasks on it. Thus, the current study proposes that the tangible attributes of medical tourism are likely positively associated with acceptance of medical tourism.

The intangible attributes are defined as not physical in nature, such as hospital reputation, hospital brand awareness, service quality assessment, and word-of-mouth referrals about medical tourism services. The intangible attributes of medical tourism mostly refer to perceived service quality such as prompt service, positive feedback, reputation, word of mouth, brand, trust, security, safety and privacy, etc. In this sense, in order to draw large number of consumer attention, medical tourism facilitators should make efforts on providing up-to-date information about their tangibles and intangibles of medical tourism. Thus, the current study proposes that the intangible attributes of medical tourism are likely positively associated with acceptance of medical tourism.

Medical tourism facilitators, the people who contact with medical tourists from planning stages until the medical tourists return home, play also a key role in promoting medical tourism. In this regard, the role of medical tourism facilitators is considered important in helping consumers make their medical tourism decision. Medical tourism facilitators used to assume that consumers prefer world-class care for the least cost. Because of this assumption, medical tourism facilitators frequently avoid mentioning the recovery phase or the duration of the recovery on their tourism information to reduce possible consumer loss. To make it worse, the medical tourism website texts highlight nothing regarding the possible dangers of medical treatments. Thus, the current study proposes that the role of medical tourism intermediaries is likely positively associated with acceptance of medical tourism.

Lastly, there is not enough evidence about the relationship between consumer needs for medical tours and acceptance of medical tourism. Consumer needs for medical tours refer to consumer desire for specific benefits of medical tourism, which can be emotional and functional that consumers want to be satisfied in the medical tour. Therefore, consumer needs for medical tours are a necessary factor in accepting medical tourism. Thus, the current study assumes that consumer needs for medical tours are likely positively associated with acceptance of medical tourism.

The hypothesized structural model is described graphically in Fig. 11.1. Figure 11.1 shows measurement components in rectangles and structural components (latent variables) in circles by using thin lines. Big circles represent the latent variables that are unobserved endogenous variables. Rectangles represent the measurement variables that are observed exogenous variables.

In summary, this study proposes hypothesis statements as the following:

Hypothesis 1: Consumer needs for medical tourism are likely related to the tangible attributes of medical tourism.

Hypothesis 2: Consumer needs for medical tourism are likely related to the intangible attributes of medical tourism.

Hypothesis 3: The tangible attributes of medical tourism are likely related to consumer acceptance of medical tourism.

Hypothesis 4: The intangible attributes of medical tourism are likely related to consumer acceptance of medical tourism.

Hypothesis 5: Consumer needs for medical tourism are likely related to consumer acceptance of medical tourism.

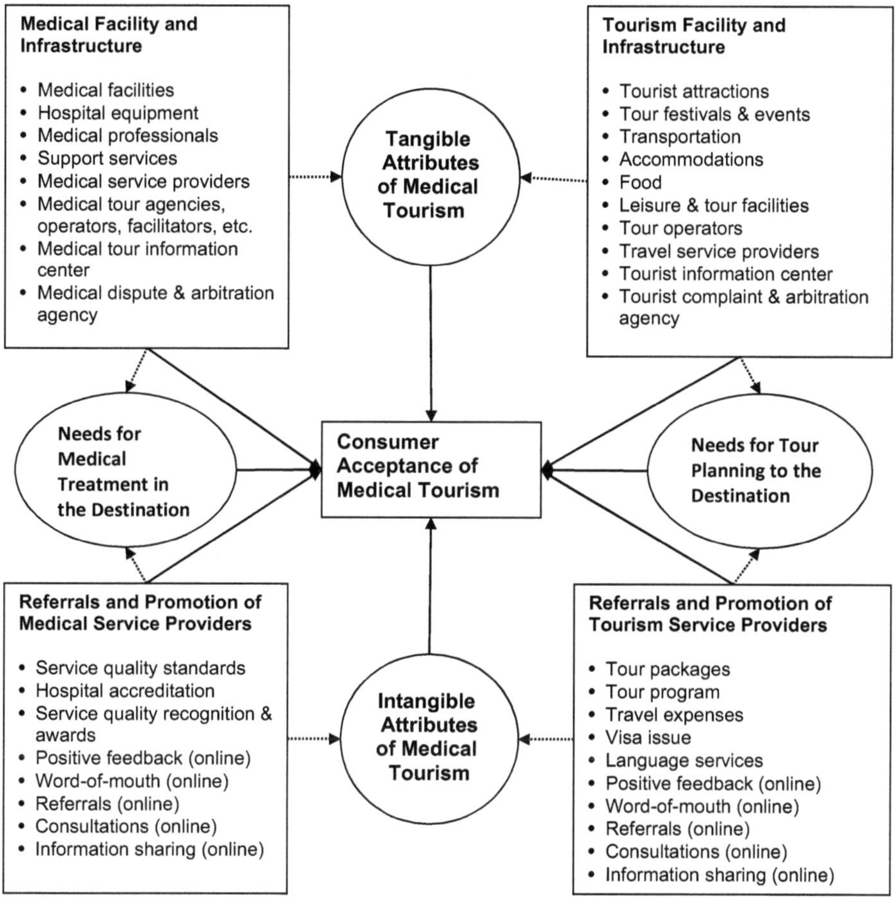

Fig. 11.1 A framework for understanding consumer behavior in medical tourism research

11.3 Research Methods

11.3.1 Survey and Sample Characteristics

Given that the model embeds complex relationships of consumer acceptance of medical tour destinations, this study collected self-reported consumer perceptions using a questionnaire. A structured questionnaire was developed based on existing literature, including 23 questions related to various constructs discussed in this study and 6 items that capture demographic information of the respondents and experience of medical tourism. A seven-point Likert-type scale is used in this study to be consistent with research in a different cultural context, though the sensitivity of data in measuring consumer perceptions and behavioral intentions in many different cultural contexts poses a problem for the adoption of a single superior scale due to limited data comparability.

The questionnaire was translated into Japanese, Chinese, and English by professional tour guides. Three surveys were simultaneously conducted in Seoul, Busan, and Incheon in Korea from 1 April 2014 to 30 September 2014 by research assistants (these assistants are tour guides, part-time workers, and college students who work part time. Per filled-in questionnaire, the assistants were compensated for their effort). A total of 685 tourists participated in the survey. After excluding 199 respondents that have missing data, outliers, and/or incomplete information, a total of 486 respondents were used for analysis in the study. Table 11.1 describes sample characteristics.

Table 11.1 Sample characteristics

Characteristics		N (486)	% (100)
Gender	Male	178	36.6
	Female	308	63.4
Nationality	China	206	42.4
	Japan	79	16.3
	Taiwan or Hong Kong	38	7.8
	Russia	21	4.3
	Overseas Korean	17	3.5
	Southeast Asia (ASEAN countries)	92	18.9
	Central Asia (Kazakhstan, Uzbekistan, Mongolia, etc.	22	4.6
	Others	11	2.2
Age	Under 20	20	4.1
	21–30	129	26.5
	31–40	181	37.2
	41–50	113	23.3
	Above 51	43	8.9
Travel budget and expenses	Less than US$1000	80	16.4
	US$1001–$3000	135	27.8
	US$3001–$5000	132	27.2
	US$5001–$7000	66	13.6
	US$7001–$9000	50	10.3
	More than US$9001	23	4.7
Education level	Secondary (high school)	141	29.0
	Junior college	84	17.3
	College or university	238	49.0
	Graduate	23	4.7
Occupation	Office worker	143	29.4
	Business person or professional	118	24.3
	Homemaker (housewife)	86	17.7
	Self-employed	75	15.4
	Student	54	11.1
	Others	10	2.1
Experience in medical tourism	Yes [male 43/178; female 226/308 (73.4%)]	269	55.3
	No [male 135/178(75.8%); female 82/308]	217	44.7

Table 11.2 Results of exploratory and confirmatory factor analysis and reliability statistics

Factor and construct name	Question item	Factor loading	Eigenvalue	Extracted variance	Item-total correlation	Cronbach alpha
Tangible attributes of medical tourism	QA1	0.597	2.218	12.325	0.583	0.780
	QA2	0.678			0.612	
	QA3	0.703			0.529	
	QA4	0.755			0.616	
Intangible attributes of medical tourism	QB1	0.784	3.796	21.087	0.726	0.901
	QB2	0.761			0.746	
	QB3	0.815			0.760	
	QB4	0.841			0.770	
	QB5	0.832			0.764	
Consumer needs for medical tourism	QC1	0.781	3.564	19.803	0.743	0.881
	QC2	0.774			0.704	
	QC3	0.789			0.742	
	QC4	0.740			0.712	
	QC5	0.700			0.673	
Consumer acceptance of medical tourism	QD1	0.709	3.202	17.791	0.623	0.893
	QD2	0.872			0.844	
	QD3	0.878			0.863	
	QD4	0.767			0.734	
Total				71.005		

Kaiser-Meyer-Olkin sampling adequacy = 0.894, $p < 0.01$
Bartlett's sphericity = 5800.907, degree of freedom = 153, $p < 0.01$

11.3.2 Results of Factor Analysis

In order to ensure the construct validity of measurement items, factor analysis was employed in a two-stage process. At first an exploratory factor analysis with a varimax rotation procedure was employed to identify underlying constructs based on an eigenvalue cutoff of one. Then again a confirmatory factor analysis using structural equation modelling techniques was employed to confirm that the identified components are fitted in the items correctly and reliably. Table 11.2 displays the results of exploratory and confirmatory factor analysis. Table 11.3 provides the descriptive statistics of question items.

11.4 Empirical Results

The analysis of moment structures was used for an empirical test of the structural model. The maximum likelihood estimation was applied to estimated numerical values for the components in the model. When comparing non-nested models, the Akaike information criterion fit index is used as our first choice, the root mean square of approximation fit index as our second choice, and then the goodness of fit

Table 11.3 Descriptive statistics of question items

Question items	Mean [seven-point scale] (standard deviation)
Tangible attributes of medical tourism:	
QA1. The destination/service provider uses modern equipment and technology	5.825 (1.134)
QA2. The destination/service provider has modern and state-of-the-art facilities	5.125 (1.149)
QA3. The destination/service provider has many excellent medical and healthcare professionals	5.471 (1.083)
QA4. The destination/service provider has official medical tour resources and information centers	5.209 (1.087)
Intangible attributes of medical tourism:	
QB1. The destination/service provider has service quality standards	5.598 (1.113)
QB2. The destination/service provider has quality and accreditation in hospital and healthcare services	5.611 (1.125)
QB3. The destination/service provider has positive feedback, word of mouth, and referrals	5.341 (1.108)
QB4. The destination/service provider provides advertising and promotional marketing	4.341 (1.117)
QB5. The destination/service provider provides medical tour consultations and understands specific needs of their customers	5.405 (1.126)
Consumer needs for medical tourism:	
QC1. Ease of ordering entry visa is important for medical tour	5.419 (1.145)
QC2. Excellent customer service is important for medical tour	5.314 (1.153)
QC3. Costs of medical tour packages are important	6.395 (1.185)
QC4. Privacy and safety issues are important for medical tour	5.584 (1.097)
QC5. Excellent medical treatment is important for medical tour	6.565 (1.167)
Consumer acceptance of medical tourism:	
QD1. Overall, I am satisfied with my medical tour experience	6.004 (1.072)
QD2. I am satisfied with the destination/service provider for medical tour	5.358 (0.982)
QD3. I like to revisit the destination/service provider for medical tour	5.323 (1.009)
QD4. If asked, I like to recommend the destination/service provider for medical tour	5.236 (1.064)

index as out third choice. Table 11.4 displays the results of structural equation modelling estimates with latent variables.

In testing hypothesis 1 that consumer needs for medical tourism are likely related to the tangible attributes of medical tourism, the results in Table 11.4 show that consumer needs for medical tourism are positively associated with the tangible attributes of medical tourism. The relationship is significant at the 0.05 level ($p < 0.001$).

In testing hypothesis 2 that consumer needs for medical tourism are likely related to the intangible attributes of medical tourism, Table 11.4 shows that consumer needs for medical tourism are positively associated with the intangible attributes of medical tourism. The relationship is significant at the 0.05 level ($p < 0.001$).

Table 11.4 Results of structural equation model estimates and latent variables

Hypo	Path	Estimates [critical ratio]	Results	Total effect	Direct effect	Indirect effect
H1	Consumer needs ➔ Tangible attributes	0.595 [5.231]***	Supported	0.595	0.595	
H2	Consumer needs ➔ Intangible attributes	0.451 [3.766]***	Supported	0.451	0.451	
H3	Tangible attributes ➔ Acceptance of medical tourism	0.615 [7.383]***	Supported	0.691	0.615	0.076
H4	Intangible attributes ➔ Acceptance of medical tourism	0.433 [3.437]***	Supported	0.579	0.433	0.146
H5	Consumer needs ➔ Acceptance of medical tourism	0.489 [4.745]***	Supported	0.730	0.489	0.241

Numeric values in the cells of regression estimates are standardized coefficient values
When the critical ratio in square brackets [ ] is > 1.96 for a regression weight, that path is significant at the 0.05 level (*** $p < 0.001$)
Numeric values in square brackets [ ] are critical ratio values
Model fit summary: CMIN = 871.868, DF = 131, CMIN/DF = 6.655
Model fit measures: GFI = 0.934, AGFI = 0.909, PGFI = 0.895, RMR = 0.043, RMSEA = 0.036
Baseline comparisons measures: NFI = 0.928, RFI = 0.902, TLI = 0.938, CFI = 0.946
Parsimony-adjusted measures: PRATIO = 0.905, PNFI = 0.872, PCFI = 0.895
The Akaike information criterion = 951.868, NCP = 740.868

In testing hypothesis 3 that the tangible attributes of medical tourism are likely related to consumer acceptance of medical tourism, the results in Table 11.4 indicate that the tangible attributes of medical tourism are positively associated with consumer acceptance of medical tourism. The relationship is significant at the 0.05 level (p < 0.001).

In testing hypothesis 4 that the intangible attributes of medical tourism are likely related to consumer acceptance of medical tourism, Table 11.4 shows that the intangible attributes of medical tourism are positively associated with consumer acceptance of medical tourism. The relationship is significant at the 0.05 level (p < 0.001).

Lastly, in testing hypothesis 5 that consumer needs for medical tourism are likely related to consumer acceptance of medical tourism, the results in Table 11.4 confirm that consumer needs for medical tourism are positively associated with consumer acceptance of medical tourism. The relationship is also significant at the 0.05 level (p < 0.001).

Overall, consumer needs for medical tourism are the most important antecedent in the process of consumer acceptance of medical tourism. The total effect of consumer needs for medical tourism was 0.730, the highest one in which the direct and indirect effect was 0.489 and 0.241, respectively (see Table 11.4).

Figure 11.2 displays a path diagram of consumer acceptance of medical tourism. Overall, consumer needs for medical tourism are the most important factor in the process of accepting medical tourism. Consumer needs for medical tourism also serve as an important antecedent in the process of accepting medical tourism, and both tangible and intangible attributes of medical tourism serve as important intermediaries in the process of consumer decision-making on acceptance of medical tourism.

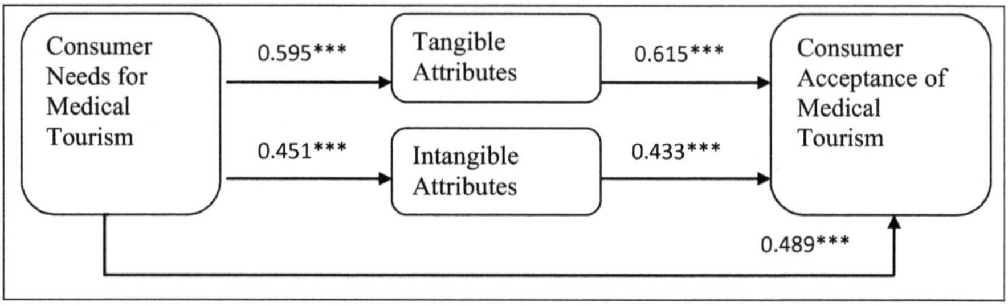

Fig. 11.2 Consumer decision-making process on acceptance of medical tourism
Numeric values over lines are standardized coefficient values
Coefficients are significant at the 0.05 level (*** $p < 0.001$)

11.5 Discussion and Implications for Managerial Action

This study explains the process of consumer decision-making on acceptance of medical tourism. This study finds that (1) consumer needs for medical tourism are the most important factor in the process of accepting medical tourism, (2) consumer needs for medical tourism also serve as an important antecedent in the process of accepting medical tourism, and (3) both tangible and intangible attributes of medical tourism serve as important intermediaries/translators in the process of consumer decision-making on acceptance of medical tourism.

For marketers and practitioners, it is the most important thing to do at first to identify consumer needs for medical tourism, before making any offering for medical tourism destinations and medical tour packages. When they understood consumer needs for medical tourism, then they should promote their offerings and reach out consumers through the two prisms of tangible and intangible attributes of medical tourism. The tangible attributes of medical tourism can be translated through the medical facility and infrastructure of the destination such as medical facilities, hospital equipment, recognition of medical professionals, support services, accreditation of medical service providers, medical tour agencies, medical tour operators, medical tour facilitators, medical tour information centers, medical dispute and arbitration agency, tourist attractions, tour festivals and events, transportation, accommodations, food, leisure and tour facilities, tourist information centers, tourist complaint and arbitration agency, etc. The intangible attributes of medical tourism can be translated and shared through positive feedback, word of mouth, referrals, consultations and information about service quality standards, hospital accreditation, service quality recognition and awards, tour packages, tour programs, travel expenses, visa issues and language services, etc. In this regard, information-rich websites and social networking sites must be created to serve this market as a tool to guide and attract medical tourism consumers. In addition, this study implies that the success of medical tourism facilitators largely depends on how to identify consumer needs for medical tourism, how to translate the tangible and intangible attributes of medical tourism, and how to promote medical tourism destinations and how to reach out to medical tourism consumers.

For researchers it is important to understand what critical factors influence the process of consumer decision-making on acceptance of medical tourism. The acceptance of medical tourism is serious for the consumers who want to meet their specific needs for medical tourism, which are largely different from those of leisure tourists and business travelers and the like. Of course, there are a lot of complexities surrounding this issue, including legal, ethical, and professional issues for medical treatment overseas. This study attempted to understand consumer behavior in a simplified manner in the specific niche market of medical tourism, apart from the existing work in tourism research and the general multigrid framework in textbooks of consumer behavior.

11.6 Conclusion and Future Research

This study proposed a simplified framework to understand consumer behavior in the process of consumer decision-making on acceptance of medical tourism. This study found that (1) consumer needs for medical tourism are the most important factor in the process of accepting medical tourism, (2) consumer needs for medical tourism also serve as an important antecedent in the process of accepting medical tourism, and (3) both tangible and intangible attributes of medical tourism serve as important intermediaries/translators in the process of consumer decision-making on acceptance of medical tourism.

Future research can investigate the further details of consumer needs for medical tourism and the further details of tangible and intangible attributes of medical tourism. In addition, the importance of each attribute in the process of consumer decision-making on acceptance of medical tourism can be weighted. In doing so future research may find some differences in consumer needs for medical tourism and in the translation of tangible and intangible attributes between consumer profiles and cultural profiles.

References

Adams, S. A. (2011). Sourcing the crowd for health services improvement: The reflexive patient and "share-your-experience" websites. *Social Science & Medicine, 72*(7), 1069–1076.

Bartholomew, D. J. (1996). *The statistical approach to environmental measurement*. San Diego: Academic.

Basilevsky, A. (1994). *Statistical factor analysis and related methods*. New York: Wiley.

Chang, G., & Caneday, L. (2011). Web-based GIS in tourism information search: Perceptions, tasks, and trip attributes. *Tourism Management, 32*, 1435–1437.

Chomvilailuk, R., & Srisomyong, N. (2014). Three dimensional perceptions of medical/health travelers and destination brand choices: Cases of Thailand. *Procedia-Social and Behavioral Sciences, 175*, 376–383.

Connell, J. (2006). Medical tourism: Sea, sun, sand and surgery. *Tourism Management, 27*(6), 1093–1100.

Connell, J. (2013). Contemporary medical tourism: Conceptualization, culture and commodification. *Tourism Management, 34*, 1–13.

Cormany, D., & Baloglu, S. (2011). Medical travel facilitator websites: An exploratory study of web page contents and services offered to the prospective medical tourist. *Tourism Management, 32*, 709–716.

Crooks, V. A., Turner, L., Snyder, J., Johnston, R., & Kingsbury, P. (2011). Promoting medical tourism to India: Messages, images, and the marketing of international patient travel. *Social Science & Medicine, 72*(5), 726–732.

Dudensing, R., Hughes, D. W., & Shields, M. (2011). Perceptions of tourism promotion and business challenges: A survey-based comparison of tourism businesses and promotion organizations. *Tourism Management, 32*(6), 1453–1462.

Dunn, G. (1989). *Design and analysis of reliability studies: The statistical evaluation of measurement errors*. London: Edward Arnold Publishers.

Edmonds, A. (2011). "Almost invisible scars": Medical tourism to Brazil. *Signs: Journal of Women in Culture and Society, 36*(2), 297–302.

El-Gohary, H. (2012). Factors affecting e-marketing adoption and implementation in tourism firms: An empirical investigation of Egyptian small tourism organisations. *Tourism Management, 33*(5), 1256–1269.

Enderwick, P., & Nagar, S. (2011). The competitive challenge of emerging markets: The case of medical tourism. *International Journal of Emerging Markets, 6*(4), 329–350.

García-Altés, A. (2005). The development of health tourism services. *Annals of Tourism Research, 32*(1), 262–266.

Heung, V. C. S., Kucukusta, D., & Song, H. (2011). Medical tourism development in Hong Kong: An assessment of the barriers. *Tourism Management, 32*(5), 995–1005.

Horng, J. S., & Tsai, C. T. (2010). Government websites for promoting East Asian culinary tourism: A cross-national analysis. *Tourism Management, 31*(1), 74–85.

Hung, K., & Law, R. (2011). An overview of Internet-based surveys in hospitality and tourism journals. *Tourism Management, 32*(4), 717–724.

Johnson, T. J., & Garman, A. N. (2010). Impact of medical travel on imports and exports of medical services. *Health Policy, 98*(2–3), 171–177.

Kline, R. B. (2005). *Principles and practices of structural equation modeling* (2nd ed.). New York: Guilford Press.

Labonté, R., Runnels, V., Packer, C., & Deonandan, R. (Eds.). (2013). *Travelling well: Essays in medical tourism.* (Transdisciplinary Studies in Population Health Series, 4(1)). Ottawa: Institute of Population Health, University of Ottawa.

Lee, C. G. (2010). Health care and tourism: Evidence from Singapore. *Tourism Management, 31*(4), 486–488.

Lee, S., Jeon, S., & Kim, D. (2011). The impact of tour quality and tourist satisfaction on tourist loyalty: The case of Chinese tourists in Korea. *Tourism Management, 32*(5), 1115–1124.

Lertwannawit, A., & Gulid, N. (2011). International tourists' service quality perception and behavioral loyalty toward medical tourism in Bangkok metropolitan area. *The Journal of Applied Business Research, 27*(6), 1–11.

Li, X., Lai, C., Harrill, R., Kline, S. F., & Wang, L. (2011). When east meets west: An exploratory study on Chinese outbound tourists' travel expectations. *Tourism Management, 32*(4), 741–749.

Lunt, N., & Carrera, P. (2010). Medical tourism: Assessing the evidence on treatment abroad. *Maturitas, 66*(1), 27–32.

Lunt, N., Hardey, M., & Mannion, R. (2010). Nip, tuck and click: Medical tourism and the emergence of web-based health information. *The Open Medical Informatics Journal, 4*, 1–11.

Luo, Q., & Zhong, D. (2015). Using social network analysis to explain communication characteristics of travel-related electronic word-of-mouth on social networking sites. *Tourism Management, 46*, 274–282.

Malhotra, N. K. (2010). *Marketing research: An applied orientation* (6th ed.). Upper Saddle River: Pearson.

Moghimehfar, F., & Nasr-Esfahani, M. H. (2011). Decisive factors in medical tourism destination choice: A case study of Isfahan, Iran and fertility treatments. *Tourism Management, 32*(6), 1431–1434.

Munar, A. M., & Jacobsen, J. K. S. (2014). Motivations for sharing tourism experiences through social media. *Tourism Management, 43*, 46–54.

Ormond, M. E., & Sothern, M. (2012). You, too, can be an international medical traveler: Reading medical travel guidebooks. *Health & Place, 18*(5), 935–941.

Penney, K., Snyder, J., Crooks, V. A., & Johnston, R. (2011). Risk communication and informed consent in the medical tourism industry: A thematic content analysis of Canadian broker websites. *BMC Medical Ethics, 12*(1), 17–26.

Runnels, V., & Carrera, P. M. (2012). Why do patients engage in medical tourism? *Maturitas, 73*(4), 300–304.

Sobo, E. J., Herlihy, E., & Bicker, M. (2011). Selling medical travel to US patient-consumers: The cultural appeal of website marketing messages. *Anthropology & Medicine, 18*(1), 119–136.

Suziedelyte, A. (2012). How does searching for health information on the Internet affect individuals' demand for health care services? *Social Science & Medicine, 75*(10), 1828–1835.

Turner, L. (2011). Canadian medical tourism companies that have exited the marketplace: Content analysis of websites used to market transnational medical travel. *Globalization and Health, 7*, 40–56.

Viladrich, A., & Baron-Faust, R. (2014). Medical tourism in tango paradise: The internet branding of cosmetic surgery in Argentina. *Annals of Tourism Research, 45*(March), 116–131.

Visit Medical Korea. (2016). *Visit medical Korea information*. Available from: http://english.visitmedicalkorea.com. Accessed 18 Jan 2016.

Wang, K. C., Chou, S. H., Su, C. J., & Tsai, H. Y. (2007). More information, stronger effectiveness? Different group package tour advertising components on web page. *Journal of Business Research, 60*(4), 382–387.

Wongkit, M., & McKercher, B. (2013). Toward a typology of medical tourists: A case study of Thailand. *Tourism Management, 38*, 4–12.

Ye, B. H., Qiu, H. Z., & Yuen, P. P. (2011). Motivations and experiences of Mainland Chinese medical tourists in Hong Kong. *Tourism Management, 32*(5), 1125–1127.

Yeoh, E., Othman, K., & Ahmad, H. (2013). Understanding medical tourists: Word-of-mouth and viral marketing as potent marketing tools. *Tourism Management, 34*, 196–201.

Yu, J. Y., & Ko, T. G. (2012). A cross-cultural study of perceptions of medical tourism among Chinese, Japanese and Korean tourists in Korea. *Tourism Management, 33*(1), 80–88.

Zhang, J., Seo, S., & Lee, H. (2013). The impact of psychological distance on Chinese customers when selecting an international healthcare service country. *Tourism Management, 35*, 32–40.

Part III
Future Directions

Chapter 12
Positive and Negative Perceptions of Residents Toward Tourism Development: Formative or Reflective

S. Mostafa Rasoolimanesh, Shuhaida Md. Noor, and Mastura Jaafar

Abstract This paper investigates whether residents' positive and negative perceptions of tourism development, in the Asian context, are reflective or formative in nature. This assessment assumes positive perceptions (PP) and negative perceptions (NP) as both unidimensional and multidimensional constructs inclusive of economic, social, cultural, and environmental components. Data for this study was collected from residents of the Lenggong World Heritage Site in Malaysia and analyzed using partial least squares structural equation modeling (PLS-SEM), confirmatory tetrad analysis (CTA), and other recently developed fit models, such as geodesic discrepancy (d_G), unweighted least squares discrepancy (d_ULS), and standardized root mean square residual (SRMR). The results confirm the formative nature of PP and NP. In addition, the findings indicate that the PP and NP measurement model is best conceptualized as a reflective-formative second-order construct. This study makes a significant theoretical and methodological contribution to the resident perception literature, and the quantitative research in tourism. Future quantitative researches in resident perception field should take into consideration the formative nature NP and PP, when the research framework are conceptualized.

Keywords Residents' perception · Negative perceptions · Positive perceptions · Formative · Reflective · Confirmatory tetrad analysis (CTA) · Model fits for PLS

S. M. Rasoolimanesh (✉)
School of Hospitality, Tourism and Events, Taylor's University, Subang Jaya, Malaysia
e-mail: mostafa.rasoolimanesh@taylors.edu.my

S. Md. Noor
School of Communication, Universiti Sains Malaysia, Penang, Malaysia
e-mail: shuhaida@usm.my

M. Jaafar
School of Housing, Building, and Planning, Universiti Sains Malaysia, Penang, Malaysia
e-mail: masturaj@usm.my

S. Rezaei (ed.), *Quantitative Tourism Research in Asia*, Perspectives on Asian Tourism, https://doi.org/10.1007/978-981-13-2463-5_12

12.1 Introduction

The past two decades have witnessed the publication of a multitude of international studies focused on the perceptions of residents toward tourism development (Andereck et al. 2005; Ap 1992; Gu and Wong 2006; Hall and Page 2014; Huang and Hsu 2005; Jaafar et al. 2015, 2017; Kayat 2002; Kim et al. 2013; Rasoolimanesh et al. 2015, 2017a, b; Zamani-Farahani and Musa, 2008, 2012). Most of these studies, published in refereed journals, employed quantitative methods with multi-item research instruments used for the measurement of positive and negative perceptions (Jaafar et al. 2015, 2017; Kim et al. 2013; Rasoolimanesh et al. 2015, 2017a; Zamani-Farahani and Musa 2008, 2012). In some studies, researchers use items associated with tourism development to make direct measurements of residents' perceptions, while in other studies, researchers might use multidimensional constructs (inclusive of economic, social, and environmental dimensions) to measure these perceptions (Jaafar et al. 2017; Jurowski et al. 1997). Notwithstanding, each of these approaches involves reflective measures of residents' positive and negative perceptions toward the impacts of tourism development. A reflective construct assumes that the measurement items are similar and interchangeable, sharing a common theme and meaning (Hair et al. 2017). Conversely, a formative construct involves numerous indicators that in combination define the construct (Diamantopoulos et al. 2008). Therefore, in a formative construct, the exclusion or inclusion of one or more indicators affects the meaning and content validity of the construct (Chang et al. 2016; Coltman et al. 2008; Diamantopoulos et al. 2008). Items used to measure formative constructs are not interchangeable with one another as each indicator represents a distinct aspect of the construct (Coltman et al. 2008). After having carefully reviewed numerous uni- and multidimensional measurement item constructs used in previous studies to measure positive and negative resident perceptions, we found that studies often tended to neglect the possibility of formative constructs. Residents' perceptions have economic, social, cultural, and environmental dimensions thus allowing them to be understood from different perspectives. These perspectives facilitate a more comprehensive understanding of residents' positive and negative perceptions toward tourism development. This expanded understanding can similarly be observed when the constructs used to measure residents' perceptions are defined using associative items that are subject to direct observation. Items associated with positive and negative perceptions represent different aspects of the perceived impacts of tourism from the perspective of residents. This analysis of previous studies highlights the value of formative constructs of perception and suggests that the constructs used in some studies to measure residents' positive and negative perceptions of the impacts of tourism development may have been poorly conceptualized. Therefore, this study undertakes to reconsider these positive and negative resident perception constructs from this viewpoint.

This study critically reviews the literature pertaining to reflective and formative concepts, while simultaneously exploring the vast body of literature concerning residents' perceptions toward tourism development, focusing particularly on studies conducted in Asian countries. This paper subsequently undertakes to discuss the reflective and formative nature of residents' positive and negative perceptions toward tourism development based on the outcomes of the aforementioned review. To empirically test the reflective-formative nature of the positive and negative constructs of residents' perceptions, we rely on data collected from the Lenggong World Heritage Site (WHS) in Malaysia. We analyze this data by way of confirmatory tetrad analysis in partial least squares (CTA-PLS) path modeling and several newly developed fit indices for PLS. Having conceptualized and tested these constructs empirical, we propose a number of recommendations and new directions for future studies to further expand this field of research. The remainder of this chapter is organized as follows. In the next section, we review the literature on residents' perceptions toward tourism development, measuring residents' perceptions, and understanding reflective and formative concepts. Following the literature review, we outline the research methods and describe the results. Then, we discuss the results and findings of this study. We conclude this chapter by identifying some of the contributions and implications of this study and outline the limitations and a number of suggestions for future research.

12.2 Residents' Perceptions Toward Tourism Development

Several previous studies have explored the perceptions of residents toward the impacts of tourism development on local communities (Andereck et al. 2005; Gu and Wong 2006; Kim et al. 2013; Rasoolimanesh et al. 2015, 2017b; Sharpley 1994, 2014; Vareiro et al. 2013). Several of these studies were conducted in Asian countries (Gu and Wong 2006; Huang and Hsu 2005; Jaafar et al. 2015, 2017; Kayat 2002; Rasoolimanesh et al. 2015, 2017a; Zamani-Farahani and Musa 2008, 2012). Communities are directly affected by the development of local tourism industries and infrastructure and by subsequent interactions with tourists (Rasoolimanesh et al. 2015). These impacts influence community values, behavior patterns, lifestyles, and the quality of life of local community members (Huang and Hsu 2005; Jaafar et al. 2015, 2017; Kayat 2002). The development of tourism can exert a number of economic, sociocultural, and environmental effects on host communities, with these effects being both positive and negative (Rasoolimanesh et al. 2015, 2017a). Positive economic effects include increasing family incomes, improved standards of living, and the creation of new jobs and additional employment opportunities (Choi and Sirakaya 2006; Ko and Stewart 2002; Rasoolimanesh et al. 2017a). The negative economic effects of tourism, as identified in the literature, include a rise in the cost of living, as well as spiraling prices for goods, property, and services, relative poverty through low-paying jobs in the tourism industry, and increasing property taxes (Andereck et al. 2005; Látková and Vogt 2012; Rasoolimanesh et al. 2015).

Several previous studies have investigated the positive and negative impacts of tourism development on local communities from the social and cultural perspective (Deery et al. 2012; Jaafar et al. 2017). The development of tourism can result in recreational and entertainment facilities being improved (Tovar and Lockwood

2008); facilitates an understanding of cultural identity and the preservation and revival of traditional arts, culture, and crafts (Jaafar et al. 2017; Kim 2002; Rasoolimanesh et al. 2017b); and encourages the local community to take pride in their culture (McGehee et al. 2002; Rasoolimanesh et al. 2017b). However, the development of tourism also introduces a number of undesirable social and cultural influences that can adversely affect the value systems of families and family relationships (Jaafar et al. 2017), contribute toward the overcrowding of facilities and traffic congestion on the roads (Ko and Stewart 2002; Látková and Vogt 2012; Rasoolimanesh et al. 2015); result in the increased prevalence of crime, including prostitution (Akama and Kieti 2007; Matarrita-Cascante 2010; Park and Stokowski 2009) and recreational drug use (Deery et al. 2012; Jaafar et al. 2017; Ko and Stewart 2002); and increase both visible litter and public alcoholism (Jaafar et al. 2017; Ko and Stewart 2002; Látková and Vogt 2012). Moreover, from an environmental perspective, the negative impacts of tourism development include damage to the natural environment and ecosystems as well as increased air, water, and other forms of environmental pollution (Ko and Stewart 2002). Therefore, in weighing these negative impacts against the benefits of tourism development, members of affected communities face a critical dilemma (Sharpley and Telfer 2008). While the perceived positive impacts of tourism development might encourage residents to support tourism development, the perceived negative impacts might similarly result in the loss of their support for tourism development (Rasoolimanesh et al. 2015; Sharpley 2014).

12.3 Measuring Residents' Perceptions

Most studies of residents' perceptions, including those conducted in Asian countries, have employed quantitative methods to measure residents' positive and negative perceptions of tourism development using multi-item instruments (Andereck et al. 2005; Gursoy et al. 2002; Jaafar et al. 2015, 2017; Kim et al. 2013; Látková and Vogt 2012; Nunkoo and Ramkissoon 2012; Rasoolimanesh 2017a, 2017b; Vareiro et al. 2013). For this purpose, different types of scales have been used. In some studies, researchers have employed multi-item constructs to measure the perceptions of residents (Ko and Stewart 2002; Látková and Vogt 2012; Rasoolimanesh et al. 2015), whereas in other studies, researchers applied various multi-item constructs to represent the economic, social, cultural, and environmental dimensions of residents' perceptions separately (Jaafar et al. 2017; Jurowski et al. 1997). To the best of our knowledge, all previous quantitative studies have measured residents' positive and negative perceptions as a reflective construct. Moreover, we have not found any studies that have explored the various aspects of residents' perceptions of the impact of tourism development (e.g., economic, social, cultural, and environmental) simultaneously while articulating these positive and negative perceptions as a multidimensional construct. Previous studies have measured residents' positive and negative perceptions either as a reflective multi-item construct or have explored the different aspects of residents' perceptions separately.

12.4 Understanding Reflective and Formative Concepts

There is a long tradition in the social sciences of assessing latent variables (LVs) (Diamantopoulos et al. 2008), which are a type of variable that cannot be measured directly and that are thus measured based some other observable variable (Coltman et al. 2008). That said, the nature of the observable variable can give rise to a number of different approaches to the measurement of the LV. Therefore, there are different types of measurement specifications (Diamantopoulos et al. 2008). According to classical test theory, the observed items are reflections of the LVs, meaning that the LVs actually cause the observed items and that the observed items share a common variance with the LVs (Bollen and Bauldry 2011). Consequently, type of measurement model is called a *reflective measurement model* (Hair et al. 2017). In a reflective measurement model, each item can be described by linear function of its associated LV plus error (Diamantopoulos et al. 2008). A reflective construct assumes that the measurement items are similar and interchangeable, meaning that the items share a common theme and meaning, and that the direction of causality is from construct to items (Coltman et al. 2008).

However, not all constructs used in the social sciences are necessarily amenable to the use of reflective measurement models (Bollen and Bauldry 2011; Diamantopoulos et al. 2008). In some cases, instead of causing the observed items, LVs may actually be a product of the observed items, either alone or in combination (Bollen and Bauldry 2011; Coltman et al. 2008). In this case the direction of the relationship is from items to LV, because the items cause the LV. This type of LV is referred to as a *formative measurement model* (Diamantopoulos et al. 2008). In a formative construct, the indicators must be combined to define the construct; therefore, the exclusion or inclusion of one or more indicators will impact the meaning and content validity of the construct (Coltman et al. 2008). Consequently, the measurement items in a formative construct are not interchangeable, and each indicator represents a different feature of the construct (Fassott and Henseler 2015). Each LV, therefore, can be described as a linear regression of its associated items plus error (Coltman et al. 2008). Items in a formative construct can be either positively or negatively intercorrelated, or they may not be correlated at all (Coltman et al. 2008; Diamantopoulos et al. 2008). In contrast, the items in a reflective construct should be highly correlated (Coltman et al. 2008).

12.5 Research Methods

This study aims to assess and compare the reflective and formative nature of residents' positive and negative perceptions. The residents who provided the data informing this study were from the Lenggong Valley World Heritage Site in Malaysia, which is a rural tourism destination. Data was collected using a self-administered questionnaire. The questionnaire was adapted from similar instruments used in previous resident perception studies, with items describing the positive and negative perceptions of residents toward the economic, social, cultural, and environmental impacts of tourism development (Andereck et al. 2005; Jaafar et al. 2017; Nicholas et al. 2009; Rasoolimanesh et al. 2015, 2017a, b). The final instrument contained 33 items, 14 of which were to measure residents' positive perceptions of various impacts (i.e., 3 economic items, 4 social items, 4 cultural items, and 3 environmental items) and 15 items to measure residents' negative perceptions (i.e., 3 economic items, 4 social items, 4 cultural items, and 4 environmental items). A further four items were used to measure residents' support for tourism development. Question items were answered on a 5-point Likert scale, with 1 indicating *strongly disagree* and 5 indicating *strongly agree*. Data collection was performed over a 1-week period during May of 2017. Respondents were systematically selected from among the residents of several villages within Lenggong Valley. The population of these villages totaled 3,826, spread across 750 houses, 150 of which were selected from data collection purposes.

To identify the reflective or formative nature of residents' perception constructs, the current study has employed partial least squares structural equation modeling (PLS-SEM), with data analysis performed using SmartPLS 3.2.6 (Ringle et al. 2015). The analysis of data using this approach is a two-stage process. In first stage, we analyzed the model assuming residents' positive and negative perceptions were multi-item reflective constructs. We then performed a confirmatory tetrad analysis in PLS (CTA-PLS) to assess the formative nature of the positive and negative perception constructs. CTA-PLS is a method of statistical analysis that facilitates the assessment the formative or reflective nature of a construct (Gudergan et al. 2008; Hair et al. 2018; Rigdon 2015). We then repeated the analysis on the assumption that the resident perception constructs might be formative rather than reflective and assessed the quality of the model based on model fit indices. In the second stage, we assumed that residents' positive and negative perceptions constituted a multidimensional construct inclusive of economic, social, cultural, and environmental dimensions. Therefore, in this stage, we began by establishing residents' positive and negative perceptions as a second-order reflective-reflective construct. Then we performed the CTA-PLS for the model, including second-order constructs, to identify the reflective or formative nature of the constructs. We then changed the reflective second-order constructs to formative constructs and assessed the quality of the model using model fit indices. Therefore, the formative nature of the positive and negative perception constructs was assessed as both multi-item single dimension and multi-item multidimensional constructs, with the results having been compared based on model fit indices.

Using CTA-PLS, we can confirm the inappropriateness of the reflective measurement model and support the use of formative measurements (Gudergan et al. 2008; Hair et al. 2018). To perform CTA-PLS, the pairs of covariances between items should be compared (Hair et al. 2018), which requires the creation of tetrads. A tetrad is the difference between one pair of covariances and another pair. Ideally, all tetrads should be zero to confirm the reflective nature of a construct (Gudergan et al. 2008; Hair et al. 2018). If at least one tetrad in the construct is significantly different from zero, we must reject the assumption of a reflective construct and confirm the formative nature of the measurement model (Hair et al. 2018).

12.6 Results

As mentioned earlier, we conducted the analysis in two stages. Table 12.1 shows the results of the assessment of the measurement model during the first stage. In this stage, we assumed that positive and negative perceptions were multi-item constructs. Table 12.1 shows the outer loadings, composite reliability (CR), and average variance extracted (AVE) for the three constructs involved in the model, namely, positive perceptions (PP), negative perceptions (NP), and support for tourism development (SUP). To establish the reliability and convergent validity of a reflective construct, the outer loadings should be higher than 0.7, CR higher than 0.7, and AVE higher than 0.5 (Hair et al. 2017). However, a loading between 0.4 and 0.7 is acceptable if the CR and AVE meet the threshold (Hair et al. 2017). Table 12.1 shows that some NP items had an outer loading below 0.4, while most PP and NP items loaded between 0.4 and 0.7. Moreover, the AVE of PP and NP was below 0.5. Therefore, the results of the measurement model assessment do not support an acceptable reliability or convergent validity for PP or NP, thus necessitating the removal of a number of items from these constructs to establish reasonable reliability and convergent validity.

The aim of the current study is to compare the various alternatives using reflective and formative PP and NP constructs, as well as first-order and second-order PP and NP. Therefore, in addition to CTA-PLS, we used various model fit indices to augment this comparison. To estimate these model fit indices, the model-implied correlation matrix should be compared with the empirical correlation matrix (Henseler et al. 2016; van Riel et al. 2017). Therefore, in using these model fit indices to compare the models, the model that best fits the empirical data is deemed to be the most representative model (van Riel et al. 2017). Should none of the models conform with the fit indices to an acceptable degree, then we must conclude that the data is perhaps too rich or contains more information than what any of the compared models can represent (Henseler et al. 2016). The Bollen-Stine bootstrapping procedure is the suitable method of assessing the model fit indices in PLS (Dijkstra and Henseler 2015). This approach is used to estimate the significance of the differences between the empirical and model-implied correlation matrix (Dijkstra and Henseler 2015). The two primary approaches to estimating these differences include the geodesic discrepancy (d_G) and unweighted least squares discrepancy (d_ULS) (Henseler et al. 2016). In addition to these model fit indices, an approximate model fit criteria for PLS—standardized root mean square residual (SRMR)—can be used to assess the differences between the model-implied and the empirical correlation matrix (Henseler et al. 2016). To arrive at an acceptable PLS model fit using bootstrapping, the values for SRMR, d_G, and d_ULS should be lower than the upper value of the 95% confidence interval ($CI_{0.95}$) (Henseler et al. 2016).

Table 12.2 shows the $CI_{0.95}$ for SRMR, d_G, and d_ULS following the inclusion of PP and NP as multi-item first-order constructs in the model, according to which the values were higher than their associated $CI_{0.95}$. Therefore, none of model fit indices can support the model.

Table 12.1 Assessment of the reflective measurement model

Construct	Items	Loading	CR	AVE
Positive perceptions			0.884	0.355
	1. Tourism development would create more jobs for my community	0.578		
	2. Tourism development would attract more investment to my community	0.561		
	3. Our standard of living would increase considerably because of tourism	0.530		
	4. Tourism development provides more infrastructures and public facilities like, roads, shopping malls, etc.	0.645		
	5. Tourism development enhances the image of local culture, and residents take pride in their culture	0.645		
	6. Tourism development preserves cultural identity of host residents	0.699		
	7. Tourism development promotes cultural exchange	0.659		
	8. Tourism development facilitates meeting visitors and educational experiences	0.684		
	9. Tourism development increases the recreation facilities and opportunities	0.637		
	10. Tourism development improves quality of fire protection	0.454		
	11. Tourism development improves quality of police protection	0.479		
	12. Tourism development helps to preserve the natural environment	0.534		
	13. Tourism development helps to preserve the historical buildings	0.629		
	14. Tourism development improves the area's appearance	0.543		
Negative perceptions			0.904	0.407
	1. Tourism development would increase the price of goods and services	0.230		
	2. Tourism development would increase the price of land and housing	0.366		
	3. Tourism development would increase the cost of living/property taxes	0.329		
	4. Construction of hotels and other tourist facilities would destroy the environment	0.423		
	5. Tourism development would result in traffic congestion	0.610		
	6. Tourism development would cause overcrowding	0.570		
	7. Tourism development would increase noise pollution	0.642		
	8. Tourism development would increase the rate of crime	0.621		

(continued)

Table 12.1 (continued)

Construct	Items	Loading	CR	AVE
	9. Tourism development increases prostitution	0.757		
	10. Tourism development would increase alcoholism	0.709		
	11. Tourism development increases drug abuse	0.708		
	12. Tourism development would change the lifestyle of local community	0.805		
	13. Tourism development would change the traditional culture, and the Malay culture may disappear slowly	0.834		
	14. Tourism development disturbs the local community routine	0.798		
	15. Tourism development would create phony folk culture	0.757		
Support for tourism development			0.954	0.839
	1. The residents should participate in tourism planning and development	0.872		
	2. I believe that tourism should be actively encouraged in my community	0.922		
	3. I support tourism and would like to see it becomes an important part of my community	0.945		
	4. The local authorities and state government should support the promotion of tourism	0.918		

Table 12.2 Model fit indices for reflective measurement model

	SRMR	5.0%	95.0%	d_ULS	5.0%	95.0%	d_G	5.0%	95.0%
Saturated model	0.164	0.061	0.084	15.156	2.063	3.954	5.176	1.128	2.147
Estimated model	0.164	0.061	0.085	15.156	2.069	4.008	5.176	1.125	2.181

In order to perform CTA, all tetrads should be calculated for the items used to articulate each construct, after which the redundant tetrads are eliminated. In the next step, significance testing is performed using bootstrapping in order to calculate the p-value and the bias-corrected confidence interval. This allows us to assess whether the nonredundant tetrads are significantly different from zero (Hair et al. 2017). If the p-value is less than 0.05 and zero is not within the bias-corrected $CI_{0.05}$ and $CI_{0.95}$, the tetrad has not vanished (i.e., is significantly different from zero) (Gudergan et al. 2008; Hair et al. 2018). Should even one tetrad has not vanished, we can conclude that the construct is not reflective, but formative (Gudergan et al. 2008; Hair et al. 2018). Appendix A.1 and A.2 show the results of CTA-PLS for the PP and NP constructs. The results reveal that several tetrads failed to vanish for both PP and NP. Therefore, we can conclude that neither PP nor NP first-order constructs can be reflective and that PP and NP must be formative in nature.

As both CTA-PLS and the model fit indices indicated that the PP and NP constructs were in fact formative, we reassessed the measurement model, including the two formative constructs (PP and NP) and the reflective construct (SUP). Table 12.3 shows the results of the assessment of the measure model. To assess a formative measurement model, it is necessary to assess the collinearity of the associated items using variance inflation factor (VIF), in which the VIF should be lower than 5 (Hair et al. 2017). Moreover, the outer weight of each item should be significant (Hair et al. 2017). Table 12.3 shows that several items have VIFs greater than 5 and that several PP and NP

Table 12.3 Assessment of formative measurement model

Construct	Items	Weight/loading	*P*-value	VIF
Positive perceptions				
	PP1	0.147	0.041	1.81
	PP2	0.161	0.014	1.92
	PP3	0.031	0.696	2.54
	PP4	0.062	0.377	2.67
	PP5	0.105	0.083	2.13
	PP6	0.093	0.130	2.92
	PP7	0.137	0.017	2.26
	PP8	0.160	0.011	2.15
	PP9	0.209	0.010	1.64
	PP10	0.009	0.906	2.06
	PP11	0.152	0.031	2.07
	PP12	0.099	0.097	2.35
	PP13	0.096	0.149	2.87
	PP14	0.209	0.008	1.61
Negative perceptions				
	NP1	–0.025	0.692	2.46
	NP2	0.091	0.175	6.60
	NP3	0.058	0.345	4.66
	NP4	–0.038	0.632	2.92
	NP5	0.119	0.044	4.51
	NP6	0.044	0.480	4.14
	NP7	0.061	0.306	6.09
	NP8	0.019	0.781	8.14
	NP9	0.087	0.069	7.91
	NP10	0.022	0.733	17.48
	NP11	0.027	0.673	14.41
	NP12	0.301	0.007	2.15
	NP13	0.172	0.002	5.82
	NP14	0.181	0.003	5.42
	NP15	0.251	0.003	4.24
Support for tourism development				
	SUP1	0.858	NA	
	SUP2	0.927	NA	
	SUP3	0.950	NA	
	SUP4	0.923	NA	

Table 12.4 Model fit indices for formative measurement model

	SRMR	5.0%	95.0%	d_ULS	5.0%	95.0%	d_G	5.0%	95.0%
Saturated model	0.087	0.047	0.072	4.207	1.223	2.917	2.055	1.244	2.472
Estimated model	0.087	0.046	0.072	4.207	1.210	2.933	2.055	1.240	2.468

Table 12.5 Assessment of measurement model of first-order constructs (Before establishing second-order PP and NP)

	Cronbach's alpha	Composite reliability	Average variance extracted (AVE)
NP_Cultural	0.890	0.922	0.747
NP_Economic	0.889	0.883	0.724
NP_Environmental	0.895	0.890	0.675
NP_Social	0.964	0.967	0.881
PP_Cultural	0.825	0.882	0.652
PP_Economic	0.740	0.832	0.630
PP_Environmental	0.799	0.869	0.689
PP_Social	0.611	0.746	0.511
SUP	0.935	0.954	0.837

items were weighted nonsignificantly. Moreover, the results of model fit assessment using SRMR, d_ULS, and d_G in Table 12.4 show nonsignificant model fit indices for the model including first-order multi-item formative PP and NP constructs.

To establish PP and NP as second-order constructs, this study used a two-stage approach recommended by several previous studies (Becker et al. 2012; Md Noor et al. 2015; Rasoolimanesh et al. 2016). In first stage, we established economic, social, cultural, and environmental dimensions of positive and negative perceptions of residents toward tourism development (PP_Economic, PP_Social, PP_Cultural, PP_Environmental, NP_Economic, NP_Social, NP_Cultural, and NP_Environmental) as first-order reflective constructs and then assessed the measurement model using these eight reflective constructs as well as SUP. Therefore, the model used in the first stage of assessment included nine reflective first-order constructs. It is necessary to established the reliability, convergent validity, and discriminant validity of these reflective constructs before proceed to the second stage (van Riel et al. 2017). Table 12.5 shows that the CR for all nine reflective constructs exceeded 0.7 and that they had an AVE greater than 0.5. Moreover, the results reveal that the outer loading of associated items of all constructs in this stage were higher than 0.4. Therefore, the results support the acceptable reliability and convergent validity of the nine reflective constructs involved in this stage.

Two criteria were used to establish the discriminant validity of the reflective constructs: the Fornell-Larcker criterion and the heterotrait-monotrait (HTMT) ratio (Henseler et al. 2015; Voorhees et al. 2016). To satisfy the Fornell-Larcker criterion, the square root of the AVE of each LV should be higher than its correlation with other LVs (Hair et al. 2017). Table 12.6 demonstrates acceptable discriminant validity based on the Fornell-Larcker criterion. Moreover, the HTMT ratio should be lower than 0.85 or 0.9 to establish discriminant validity (Henseler et al. 2015).

Table 12.6 Discriminant validity using Fornell-Larcker criterion for first-order constructs (before establishing second-order PP and NP)

	NP_Cultural	NP_Economic	NP_Environmental	NP_Social	PP_Cultural	PP_Economic	PP_Environmental	PP_Social	SUP
NP_Cultural	0.865								
NP_Economic	0.074	0.851							
NP_Environmental	0.399	0.611	0.821						
NP_Social	0.525	0.506	0.789	0.939					
PP_Cultural	0.120	–0.027	0.071	0.086	0.807				
PP_Economic	0.076	0.142	0.227	0.107	0.505	0.794			
PP_Environmental	–0.091	0.161	–0.046	–0.105	0.319	0.176	0.830		
PP_Social	0.114	–0.024	0.057	0.075	0.429	0.222	0.436	0.715	
SUP	0.420	0.147	0.182	0.094	0.217	0.230	0.253	0.200	0.915

Table 12.7 shows that all the LVs had an $HTMT_{.85}$ below 0.85. Therefore, using both the Fornell-Larcker criterion and $HTMT_{0.85}$, the results support the acceptable discriminant validity for the nine reflective measurement models assessed in the first stage. After establishing the reliability, convergent validity, and discriminant validity of the reflective constructs, the PP_Economic, PP_Social, PP_Cultural, PP_Environmental, NP_Economic, NP_Social, NP_Cultural, and NP_Environmental first-order construct scores were obtained to use as second-order PP and NP indicators in the second stage. Therefore, the model in the second stage includes two second-order constructs (i.e., PP and NP) and SUP as a reflective first-order construct.

PP and NP were established as second-order reflective constructs in the first step of second stage. Table 12.8 shows the results of the assessment of the measurement model for the reflective second-order PP and NP constructs. Table 12.8 demonstrates acceptable reliability and convergent validity for these constructs. The CR of PP and NP was higher than 0.7, the AVE was higher than 0.5, and the outer loadings were acceptable. The next step in this process, therefore, is the assessment of model fit indices and CTA-PLS for the second-order PP and NP constructs, this allowing us to confirm whether PP and NP are reflective or formative measurement models.

Table 12.9 shows the model fit indices for the model, including second-order PP and NP constructs. As mentioned earlier, to arrive at an acceptable fit for PLS modeling using bootstrapping, the values for SRMR, d_G, and d_ULS should be lower than the upper value of 95% confidence interval ($CI_{0.95}$) (Henseler et al. 2016). Table 12.9 shows that SRMR, d_G, and d_ULS had higher values than their associated $CI_{0.95}$. Therefore, none of model fit indices can support this model. As such, it is necessary to perform CTA-PLS on the alternative model to check whether the PP and NP constructs were reflective or formative.

To perform the CTA for the second-order PP and NP constructs, it was first necessary to establish each of the tetrads and then to eliminate those that were redundant. Then, using bootstrapping, the *p*-value and the bias-corrected confidence interval were calculated to assess whether the nonredundant tetrads were significantly different from zero (Hair et al. 2017). Tables 12.10 and 12.11 show the results of CTA-PLS for the second-order PP and NP constructs. The results reveal that none of the nonredundant tetrads for both PP and NP had vanished. Therefore, the results of CTA-PLS strongly support the formative nature of the second-order PP and NP constructs.

According to the results of CTA-PLS for these second-order constructs and for the model fit indices, the second-order PP and NP constructs switched from reflective to formative, thus necessitating the reassessment of the measurement model, including the two second-order formative constructs (i.e., PP and NP) and the one reflective construct (SUP). Table 12.12 shows the results of the assessment of the measurement model. The results show that the VIF of all items associated with PP and NP was lower than 5 and acceptable. Moreover, the outer weights of PP_Economic, PP_Social, PP_Cultural, PP_Environmental, NP_Cultural, and NP_Environmental were significant. The results of the assessment of the measurement model identified only nonsignificant outer weights for NP_Economic and

Table 12.7 Discriminant validity using $HTMT_{0.85}$ for first-order constructs (before establishing second-order PP and NP)

	NP_Cultural	NP_Economic	NP_Environmental	NP_Social	PP_Cultural	PP_Economic	PP_Environmental	PP_Social	SUP
NP_Cultural									
NP_Economic	0.138								
NP_Environmental	0.374	0.790							
NP_Social	0.520	0.534	0.898						
PP_Cultural	0.138	0.106	0.110	0.102					
PP_Economic	0.135	0.216	0.260	0.122	0.710				
PP_Environmental	0.122	0.230	0.110	0.179	0.430	0.248			
PP_Social	0.130	0.159	0.101	0.108	0.738	0.637	0.657		
SUP	0.427	0.116	0.125	0.071	0.231	0.240	0.253	0.192	

Table 12.8 Assessment of measurement model of reflective second order

Construct	Items	Loading	CR	AVE
Positive perceptions			0.807	0.568
	PP_ Economic	0.668		
	PP_ Social	0.709		
	PP_ Cultural	0.779		
	PP_ Environmental	0.700		
Negative perceptions			0.836	0.512
	NP_ Economic	0.515		
	NP_ Social	0.816		
	NP_ Cultural	0.849		
	NP_ Environmental	0.787		
Support for Tourism Development			0.953	0.837
	SUP1	0.870		
	SUP2	0.922		
	SUP3	0.946		
	SUP4	0.919		

Table 12.9 Model fit indices for second-order reflective model

	SRMR	5.0%	95.0%	d_ULS	5.0%	95.0%	d_G	5.0%	95.0%
Saturated model	0.114	0.056	0.082	1.011	0.244	0.523	0.510	0.165	0.329
Estimated model	0.114	0.056	0.081	1.011	0.246	0.515	0.510	0.166	0.331

Table 12.10 Results of CTA-PLS for second-order PP

Positive perceptions	Tetrad value	*P*-values	Bias	CI low	CI up	Supported (reflective)
1: PP_Cultural,PP_ Economic,PP_ Environmental,PP_Social	0.149	0.039	–0.004	0.026	0.263	NO
2: PP_Cultural,PP_ Economic,PP_Social,PP_ Environmental	0.145	0.05	–0.005	0.014	0.266	NO

Table 12.11 Results of CTA-PLS for second-order NP

Negative perceptions	Tetrad value	*P*-values	Bias	CI low	CI up	Supported (reflective)
1: NP_Cultural,NP_ Economic,NP_ Environmental,NP_Social	–0.144	0.012	0.004	–0.233	–0.046	NO
2: NP_Cultural,NP_ Economic,NP_Social,NP_ Environmental	–0.262	0.000	0.006	–0.369	–0.143	NO

Table 12.12 Assessment of measurement model of formative second order

Construct	Items	Weights/loadings	*P*-value	VIF
Positive perceptions				
	PP_ Economic	0.358	0.003	1.343
	PP_ Social	0.312	0.006	1.396
	PP_ Cultural	0.339	0.000	1.593
	PP_ Environmental	0.358	0.000	1.268
Negative perceptions				
	NP_ Economic	0.224	0.127	1.728
	NP_ Social	0.144	0.307	3.154
	NP_ Cultural	0.646	0.000	1.492
	NP_ Environmental	0.279	0.017	3.182
Support for tourism development				
	SUP1	0.865		NA
	SUP2	0.925		NA
	SUP3	0.948		NA
	SUP4	0.920		NA

Table 12.13 Model fit indices for second-order formative measurement model

	SRMR	5.0%	95.0%	d_ULS	5.0%	95.0%	d_G	5.0%	95.0%
Saturated model	0.060	0.036	0.074	0.282	0.103	0.423	0.149	0.112	0.158
Estimated model	0.060	0.037	0.074	0.282	0.105	0.423	0.149	0.113	0.158

NP_Social. As such, we must determine the significance of their outer loadings to decide whether they should be removed (Hair et al. 2017). The results indicate significant outer loadings for these two second-order NP items, and for all the formative PP and NP constructs, thus meaning that they can be retained. The results of the model fit assessment using SRMR, d_ULS, and d_G for this alternative model, including formative second-order PP and NP constructs, are presented in Table 12.13. Table 12.13 shows significant model fit indices for the model, including second-order formative PP and NP constructs. The values of SRMR, d_ULS, and d_G for the model, including second-order PP and NP constructs, were lower than the upper value of 95% confidence interval ($CI_{0.95}$) (Henseler et al. 2016), thus meaning that they were acceptable. Therefore, the results allude not only to the formative nature of PP and NP, but also to the necessity of establishing these constructs as second-order reflective-formative constructs.

12.7 Discussion

The current study aimed to determine whether the positive and negative perceptions of residents living in an Asian context toward tourism development were reflective or formative in nature. Earlier studies published in the tourism literature, including studies of Asian countries, articulate residents' positive and negative perceptions as multi-item reflective measurement models (Jaafar et al. 2015, 2017; Kim et al. 2013; Rasoolimanesh et al. 2015, 2017a; Zamani-Farahani and Musa 2008, 2012). Few studies, however, employed measures of residents' perceptions that were inclusive of economic, social, cultural, and environmental dimensions (Jaafar et al. 2017; Jurowski et al. 1997). In this study, we used PLS and CTA to elucidate the formative and/or reflective nature of these constructs. Moreover, this assessment explored two alternatives: (a) that positive and negative perceptions constitute multi-item first-order constructs and (b) that positive and negative perceptions are multidimensional second-order constructs inclusive of economic, social, cultural, and environmental dimensions. In order to compare these alternatives, we used quality criteria to assess the measurement model and model fit indices. For the first alternative, the results of CTA-PLS revealed the formative nature of residents' positive and negative perceptions toward tourism development. Using CTA-PLS, the results indicated that several tetrads were significantly different from zero for the items in the PP and NP constructs, meaning that these results could not support the reflective nature of these constructs. Had at least one tetrad been significantly different from zero after having performed CTA, it would have indicated a low correlation between some items within the construct, indicating that the construct cannot be reflective (i.e., it must be formative) (Gudergan et al. 2008; Hair et al. 2018). In addition, the results of the assessment of the measurement model that assumed a reflective PP and NP showed low loadings for several PP- and NP-related items, as well as low AVE, which might indicate that neither PP nor NP is reflective. Moreover, the values of d_G, d_ULS, and SRMR model fit indices, which were used to assess the differences between the model-implied and the empirical correlation matrix, showed that the data included more information that what the model could represent (Henseler et al. 2016). Then, the direction of arrows were changed from the items to construct in PP and NP, and so PP and NP changed to formative constructs. However, the results of the assessment of the measurement model assuming a formative PP and NP showed high collinearity and VIF for some PP- and NP-related items. One possible reason for this high collinearity among some of the PP and NP items may be their similarity with items used to measure the dimensions of PP and NP. We used these items to measure the economic, social, cultural, and environmental aspects of residents' positive and negative perceptions toward tourism development. Moreover, the results of CTA-PLS showed that only some tetrads differed significantly from zero, which would suggest that not all the PP and NP items were necessarily formative. This finding is consistent with the existing resident perception literature, which suggests that residents might differ, both qualitatively and quantitatively, in their perceptions of the impacts of tourism. For instance, the perception of economic impacts of tourism in a community might be high, while few people might perceive the social or environmental impacts. This would justify going forward with the second alternative and establishing PP and NP as second-order constructs inclusive of perceptions of economic, cultural, social, and environmental tourism impacts.

This study performed CTA-PLS for the second alternative model, assuming PP and NP as multidimensional second-order constructs. The results of CTA-PLS showed that all tetrads were significantly different from zero, thus confirming the formative nature of the second-order PP and NP constructs. Moreover, the results for the fit indices for the reflective second-order PP and NP constructs were not acceptable; but when articulated as formative second-order constructs, the results for PP and NP showed SRMR, d_G, and d_ULS values that were lower than the upper value of $CI_{0.95}$ and acceptable (Henseler et al. 2016). Therefore, according to the results of CTA-PLS analysis, the measurement model criteria, and the model fit indices, the best alternative to measure PP and NP in studies of residents' perceptions is to assume a reflective-formative second-order construct inclusive of perceived economic, social, cultural, and environmental impacts.

12.8 Conclusion

To the best of our knowledge, all previous studies in Asian countries of residents' perceptions toward tourism development have assumed that resident's negative and positive perceptions constitute a reflective construct. However, the results of the current study show that the PP and NP are in fact formative nature. In addition, this study identified that the best alternative with which to conceptualize PP and NP is as a reflective-formative multidimensional second-order construct that is inclusive of perceived economic, social, cultural, and environmental impacts, and not as a formative unidimensional construct. These findings amount to a significant theoretical and methodological contribution to the resident perception literature. However, it should be noted that this study was focused entirely on studies of Asian countries and that the data we analyzed was collected from a single case in Malaysia. As such, this study is not without its limitations, and caution should be exercised before generalizing these findings to other geographic locations and communities. More studies are required, in both the Asian and international contexts, to confirm these findings.

The misconceptualization of a measurement model in a study will almost invariably lead to biased and invalid results for the structural model. Therefore, the findings of this study raise a number of concerns about the validity of the results of previous studies of residents' perceptions. Local authorities and correspondence organizations may wish to reconsider the findings of previous studies by correcting the measurement models to arrive at more acceptable and valid structural-level findings. Further research should be undertaken in the future to understand residents' perceptions, with this research assuming formative instead of reflective measures.

Acknowledgment The authors would like to extend their appreciation to the Malaysian Ministry of Higher Education and Universiti Sains Malaysia (USM) for the grant entitled Small Heritage Tourism Destination: An Explanatory Model of Small Heritage Tourism Experience [Grant No. 203.PCOMM.6711579) that makes this paper possible.

Appendices

Appendix A-1: Results of CTA-PLS for first-order PP

PP	Tetrad value	*P*-values	Bias	CI low	CI up	Supported (reflective)
1: PP1,PP10,PP11,PP12	0.010	0.227	0.000	–0.004	0.024	YES
2: PP1,PP10,PP12,PP11	0.004	0.775	0.000	–0.018	0.025	YES
4: PP1,PP10,PP11,PP13	0.010	0.217	0.000	–0.004	0.023	YES
6: PP1,PP11,PP13,PP10	–0.023	0.042	0.001	–0.040	–0.004	YES
7: PP1,PP10,PP11,PP14	0.006	0.341	0.000	–0.005	0.016	YES
10: PP1,PP10,PP11,PP2	–0.001	0.771	0.000	–0.009	0.006	YES
13: PP1,PP10,PP11,PP3	0.005	0.330	0.000	–0.004	0.014	YES
17: PP1,PP10,PP4,PP11	–0.028	0.154	0.001	–0.060	0.005	YES
20: PP1,PP10,PP5,PP11	–0.043	0.040	0.001	–0.077	–0.007	NO
24: PP1,PP11,PP6,PP10	–0.045	0.048	0.001	–0.082	–0.006	NO
27: PP1,PP11,PP7,PP10	–0.024	0.114	0.001	–0.049	0.002	YES
31: PP1,PP10,PP11,PP9	0.001	0.866	0.000	–0.013	0.016	YES
42: PP1,PP12,PP2,PP10	–0.051	0.040	0.001	–0.091	–0.009	NO
52: PP1,PP10,PP12,PP6	0.003	0.769	0.000	–0.016	0.022	YES
56: PP1,PP10,PP7,PP12	–0.010	0.472	0.000	–0.031	0.012	YES
58: PP1,PP10,PP12,PP8	–0.002	0.808	0.000	–0.012	0.009	YES
59: PP1,PP10,PP8,PP12	–0.035	0.086	0.001	–0.068	–0.001	YES
62: PP1,PP10,PP9,PP12	–0.024	0.169	0.001	–0.053	0.006	YES
66: PP1,PP13,PP14,PP10	–0.002	0.787	0.000	–0.014	0.010	YES
78: PP1,PP13,PP5,PP10	–0.012	0.405	0.000	–0.037	0.013	YES
92: PP1,PP10,PP2,PP14	–0.015	0.575	0.000	–0.058	0.029	YES
98: PP1,PP10,PP4,PP14	0.001	0.967	0.000	–0.026	0.027	YES
100: PP1,PP10,PP14,PP5	0.001	0.898	0.000	–0.013	0.015	YES
104: PP1,PP10,PP6,PP14	–0.004	0.817	0.000	–0.032	0.024	YES
114: PP1,PP14,PP9,PP10	0.010	0.515	0.000	–0.016	0.036	YES
121: PP1,PP10,PP2,PP5	–0.013	0.528	0.000	–0.048	0.021	YES
134: PP1,PP10,PP9,PP2	–0.010	0.468	0.000	–0.033	0.013	YES
144: PP1,PP3,PP6,PP10	0.016	0.417	–0.001	–0.017	0.047	YES
146: PP1,PP10,PP7,PP3	–0.002	0.874	0.000	–0.021	0.017	YES
161: PP1,PP10,PP7,PP4	0.004	0.784	0.000	–0.020	0.026	YES
163: PP1,PP10,PP4,PP8	–0.007	0.524	0.000	–0.024	0.011	YES
166: PP1,PP10,PP4,PP9	–0.030	0.110	0.001	–0.061	0.002	YES
176: PP1,PP10,PP8,PP5	–0.001	0.939	0.000	–0.031	0.027	YES
179: PP1,PP10,PP9,PP5	–0.005	0.699	0.000	–0.024	0.015	YES
181: PP1,PP10,PP6,PP7	–0.001	0.950	0.000	–0.032	0.030	YES
194: PP1,PP10,PP9,PP7	–0.013	0.346	0.000	–0.036	0.010	YES
245: PP1,PP11,PP6,PP13	–0.038	0.062	0.001	–0.071	–0.003	YES

(continued)

PP	Tetrad value	*P*-values	Bias	CI low	CI up	Supported (reflective)
249: PP1,PP13,PP7,PP11	–0.009	0.572	0.000	–0.034	0.017	YES
278: PP1,PP11,PP9,PP14	–0.020	0.195	0.001	–0.045	0.006	YES
284: PP1,PP11,PP4,PP2	0.004	0.726	0.000	–0.015	0.022	YES
291: PP1,PP2,PP6,PP11	0.019	0.282	–0.001	–0.011	0.048	YES
301: PP1,PP11,PP3,PP4	–0.026	0.193	0.001	–0.058	0.007	YES
307: PP1,PP11,PP3,PP6	–0.010	0.450	0.000	–0.032	0.012	YES
330: PP1,PP4,PP8,PP11	–0.019	0.286	0.000	–0.048	0.011	YES
337: PP1,PP11,PP5,PP7	–0.023	0.158	0.000	–0.049	0.004	YES
365: PP1,PP12,PP14,PP13	–0.018	0.351	0.000	–0.050	0.014	YES
391: PP1,PP12,PP14,PP2	–0.002	0.769	0.000	–0.012	0.009	YES
424: PP1,PP12,PP2,PP6	–0.058	0.067	0.001	–0.109	–0.004	YES
452: PP1,PP12,PP9,PP3	–0.014	0.365	0.000	–0.040	0.012	YES
457: PP1,PP12,PP4,PP6	–0.018	0.391	0.000	–0.053	0.017	YES
467: PP1,PP12,PP9,PP4	–0.028	0.202	0.001	–0.062	0.009	YES
537: PP1,PP2,PP7,PP13	0.044	0.065	–0.001	0.004	0.083	YES
544: PP1,PP13,PP3,PP4	–0.002	0.952	0.000	–0.052	0.049	YES
547: PP1,PP13,PP3,PP5	–0.007	0.752	0.000	–0.043	0.029	YES
551: PP1,PP13,PP6,PP3	0.004	0.848	0.000	–0.027	0.034	YES
553: PP1,PP13,PP3,PP7	–0.012	0.577	0.000	–0.049	0.025	YES
557: PP1,PP13,PP8,PP3	–0.010	0.601	0.000	–0.039	0.021	YES
600: PP1,PP7,PP8,PP13	–0.028	0.076	0.001	–0.054	–0.001	YES
644: PP1,PP14,PP9,PP3	0.009	0.484	0.000	–0.012	0.029	YES
678: PP1,PP6,PP8,PP14	–0.011	0.428	0.000	–0.034	0.012	YES
715: PP1,PP2,PP4,PP7	0.066	0.035	–0.001	0.013	0.117	NO
755: PP1,PP3,PP5,PP4	–0.013	0.603	0.000	–0.055	0.029	YES
766: PP1,PP3,PP4,PP9	0.019	0.176	0.000	–0.005	0.043	YES
800: PP1,PP4,PP6,PP5	–0.029	0.113	0.001	–0.058	0.002	YES
848: PP1,PP6,PP8,PP7	–0.030	0.287	0.000	–0.076	0.016	YES
1021: PP10,PP11,PP8,PP9	0.051	0.036	–0.001	0.010	0.090	NO
1361: PP10,PP2,PP7,PP3	–0.018	0.410	0.000	–0.053	0.018	YES
1377: PP10,PP4,PP7,PP2	–0.016	0.170	0.000	–0.036	0.004	YES
1378: PP10,PP2,PP4,PP8	–0.004	0.823	0.000	–0.036	0.027	YES
1464: PP10,PP5,PP7,PP4	–0.024	0.123	0.001	–0.048	0.002	YES
1478: PP10,PP4,PP9,PP6	–0.053	0.072	0.001	–0.099	–0.003	YES
1656: PP11,PP14,PP2,PP13	0.005	0.744	0.000	–0.022	0.031	YES
1819: PP11,PP14,PP5,PP7	0.018	0.201	0.000	–0.006	0.042	YES
1969: PP11,PP4,PP6,PP8	0.036	0.021	–0.001	0.009	0.061	NO
2077: PP12,PP13,PP4,PP5	0.120	0.003	–0.003	0.051	0.184	NO
2153: PP12,PP14,PP7,PP3	0.059	0.010	–0.001	0.021	0.096	NO
2353: PP12,PP5,PP7,PP8	–0.008	0.637	0.000	–0.037	0.021	YES
2562: PP13,PP7,PP9,PP3	–0.018	0.387	0.000	–0.052	0.016	YES

Appendix A-2: Results of CTA-PLS for first-order NP

NP	Tetrad value	*P*-values	Bias	CI low	CI up	Supported (reflective)
1: NP1,NP10,NP11,NP12	0.032	0.418	–0.001	–0.034	0.094	YES
2: NP1,NP10,NP12,NP11	0.122	0.361	–0.005	–0.103	0.338	YES
4: NP1,NP10,NP11,NP13	0.061	0.075	–0.002	0.002	0.115	YES
6: NP1,NP11,NP13,NP10	0.238	0.036	–0.006	0.046	0.419	NO
10: NP1,NP10,NP11,NP15	0.020	0.522	–0.001	–0.032	0.069	YES
13: NP1,NP10,NP11,NP2	–0.007	0.763	0.000	–0.044	0.031	YES
17: NP1,NP10,NP3,NP11	–0.932	0.000	0.023	–1.206	–0.611	NO
20: NP1,NP10,NP4,NP11	–0.548	0.000	0.016	–0.765	–0.299	NO
24: NP1,NP11,NP5,NP10	–0.688	0.000	0.018	–0.935	–0.406	NO
25: NP1,NP10,NP11,NP6	0.017	0.649	–0.001	–0.045	0.077	YES
27: NP1,NP11,NP6,NP10	–0.523	0.000	0.014	–0.756	–0.262	NO
31: NP1,NP10,NP11,NP8	0.035	0.479	–0.001	–0.048	0.114	YES
36: NP1,NP11,NP9,NP10	–0.225	0.049	0.007	–0.407	–0.029	NO
38: NP1,NP10,NP13,NP12	0.347	0.001	–0.010	0.172	0.501	NO
41: NP1,NP10,NP14,NP12	0.258	0.023	–0.008	0.063	0.437	NO
43: NP1,NP10,NP12,NP15	0.234	0.011	–0.008	0.075	0.378	NO
47: NP1,NP10,NP2,NP12	–0.580	0.000	0.017	–0.799	–0.325	NO
50: NP1,NP10,NP3,NP12	–0.523	0.000	0.015	–0.724	–0.293	NO
68: NP1,NP10,NP9,NP12	–0.092	0.069	0.003	–0.172	–0.006	YES
73: NP1,NP10,NP13,NP15	0.473	0.000	–0.013	0.263	0.658	NO
76: NP1,NP10,NP13,NP2	0.014	0.817	0.000	–0.084	0.112	YES
80: NP1,NP10,NP3,NP13	–0.524	0.000	0.013	–0.718	–0.304	NO
88: NP1,NP10,NP13,NP6	0.179	0.045	–0.004	0.028	0.322	NO
93: NP1,NP13,NP7,NP10	–0.301	0.022	0.007	–0.510	–0.077	NO
98: NP1,NP10,NP9,NP13	–0.040	0.159	0.001	–0.085	0.008	YES
110: NP1,NP10,NP4,NP14	–0.259	0.004	0.008	–0.397	–0.105	NO
112: NP1,NP10,NP14,NP5	0.133	0.135	–0.004	–0.017	0.276	YES
115: NP1,NP10,NP14,NP6	0.123	0.253	–0.003	–0.057	0.297	YES
123: NP1,NP14,NP8,NP10	–0.154	0.294	0.004	–0.392	0.091	YES
124: NP1,NP10,NP14,NP9	0.204	0.113	–0.006	–0.013	0.410	YES
130: NP1,NP10,NP15,NP3	0.027	0.754	0.000	–0.113	0.166	YES
134: NP1,NP10,NP4,NP15	–0.297	0.002	0.009	–0.449	–0.127	NO
140: NP1,NP10,NP6,NP15	–0.192	0.036	0.004	–0.339	–0.038	NO
152: NP1,NP10,NP3,NP2	0.002	0.984	–0.001	–0.168	0.171	YES
161: NP1,NP10,NP6,NP2	–0.097	0.108	0.003	–0.194	0.005	YES
175: NP1,NP10,NP3,NP5	–0.386	0.001	0.009	–0.565	–0.187	NO
185: NP1,NP10,NP8,NP3	0.046	0.312	–0.001	–0.030	0.120	YES
186: NP1,NP3,NP8,NP10	0.843	0.000	–0.021	0.535	1.109	NO
192: NP1,NP4,NP5,NP10	–0.060	0.671	0.000	–0.294	0.173	YES

(continued)

NP	Tetrad value	P-values	Bias	CI low	CI up	Supported (reflective)
200: NP1,NP10,NP8,NP4	–0.009	0.876	0.000	–0.099	0.081	YES
202: NP1,NP10,NP4,NP9	–0.477	0.000	0.013	–0.667	–0.261	NO
213: NP1,NP5,NP8,NP10	0.568	0.001	–0.015	0.285	0.822	NO
250: NP1,NP11,NP12,NP4	0.025	0.744	–0.001	–0.102	0.152	YES
270: NP1,NP13,NP14,NP11	–0.032	0.483	0.001	–0.105	0.044	YES
326: NP1,NP11,NP2,NP15	–0.539	0.001	0.014	–0.784	–0.266	NO
333: NP1,NP15,NP4,NP11	–0.485	0.001	0.011	–0.710	–0.238	NO
361: NP1,NP11,NP2,NP7	–0.963	0.000	0.025	–1.263	–0.613	NO
365: NP1,NP11,NP8,NP2	0.027	0.528	0.000	–0.044	0.097	YES
428: NP1,NP11,NP9,NP7	–0.221	0.070	0.007	–0.414	–0.013	YES
433: NP1,NP12,NP13,NP14	0.163	0.082	–0.004	0.005	0.314	YES
470: NP1,NP12,NP3,NP14	–0.589	0.000	0.016	–0.807	–0.339	NO
507: NP1,NP15,NP7,NP12	–0.492	0.000	0.013	–0.709	–0.249	NO
522: NP1,NP2,NP5,NP12	0.179	0.124	–0.006	–0.019	0.364	YES
549: NP1,NP3,NP8,NP12	0.409	0.000	–0.011	0.212	0.584	NO
589: NP1,NP12,NP7,NP8	–0.111	0.444	0.003	–0.348	0.131	YES
636: NP1,NP15,NP5,NP13	–0.909	0.000	0.022	–1.151	–0.624	NO
643: NP1,NP13,NP15,NP8	0.086	0.078	–0.001	0.004	0.165	YES
684: NP1,NP3,NP8,NP13	0.404	0.000	–0.010	0.210	0.579	NO
696: NP1,NP4,NP7,NP13	0.218	0.019	–0.007	0.057	0.364	NO
751: NP1,NP14,NP15,NP8	0.084	0.084	–0.002	0.002	0.161	YES
777: NP1,NP2,NP9,NP14	0.603	0.000	–0.016	0.371	0.803	NO
799: NP1,NP14,NP4,NP6	–0.214	0.070	0.006	–0.403	–0.014	YES
859: NP1,NP15,NP2,NP9	–0.759	0.000	0.018	–0.986	–0.495	NO
861: NP1,NP2,NP9,NP15	0.698	0.000	–0.018	0.447	0.913	NO
867: NP1,NP3,NP5,NP15	0.472	0.000	–0.012	0.266	0.655	NO
921: NP1,NP7,NP9,NP15	0.110	0.173	–0.003	–0.025	0.240	YES
932: NP1,NP2,NP6,NP3	–0.006	0.969	0.000	–0.244	0.233	YES
951: NP1,NP4,NP7,NP2	–0.052	0.455	0.001	–0.166	0.064	YES
960: NP1,NP5,NP6,NP2	–0.090	0.075	0.003	–0.170	–0.004	YES
971: NP1,NP2,NP7,NP6	0.715	0.000	–0.018	0.421	0.973	NO
991: NP1,NP3,NP4,NP6	0.286	0.038	–0.005	0.055	0.507	NO
996: NP1,NP4,NP7,NP3	0.043	0.696	–0.002	–0.142	0.224	YES
1001: NP1,NP3,NP9,NP4	0.257	0.018	–0.005	0.074	0.430	NO
1040: NP1,NP4,NP8,NP5	0.191	0.063	–0.007	0.015	0.354	YES
1055: NP1,NP4,NP8,NP7	0.497	0.001	–0.015	0.232	0.732	NO
1080: NP1,NP8,NP9,NP5	–0.034	0.607	0.001	–0.141	0.076	YES
1086: NP1,NP7,NP9,NP6	–0.092	0.424	0.004	–0.278	0.101	YES
1090: NP1,NP7,NP8,NP9	0.072	0.269	–0.001	–0.037	0.179	YES
1156: NP10,NP11,NP14,NP15	1.349	0.000	–0.033	0.912	1.718	NO
1192: NP10,NP11,NP15,NP5	0.011	0.951	–0.003	–0.289	0.304	YES
1290: NP10,NP8,NP9,NP11	–0.016	0.447	0.001	–0.051	0.020	YES
1576: NP10,NP13,NP6,NP8	0.172	0.014	–0.004	0.053	0.282	NO

(continued)

NP	Tetrad value	*P*-values	Bias	CI low	CI up	Supported (reflective)
1838: NP10,NP2,NP8,NP7	–0.022	0.800	–0.001	–0.168	0.121	YES
1870: NP10,NP3,NP5,NP9	0.050	0.318	–0.001	–0.034	0.131	YES
1929: NP10,NP6,NP9,NP5	–0.521	0.001	0.010	–0.765	–0.257	NO
2506: NP11,NP3,NP4,NP5	–0.254	0.062	0.006	–0.471	–0.024	YES
2665: NP12,NP13,NP2,NP4	0.736	0.000	–0.022	0.420	1.008	NO
2827: NP12,NP14,NP5,NP7	0.430	0.000	–0.012	0.223	0.614	NO
3005: NP12,NP3,NP6,NP4	–0.127	0.344	0.004	–0.345	0.097	YES
3544: NP14,NP15,NP7,NP9	1.019	0.000	–0.024	0.684	1.306	NO

References

Akama, J. S., & Kieti, D. (2007). Tourism and socio-economic development in developing countries: A case study of Mombasa Resort in Kenya. *Journal of Sustainable Tourism, 15*(6), 735–748. https://doi.org/10.2167/jost543.0.

Andereck, K. L., Valentine, K. M., Knopf, R. C., & Vogt, C. A. (2005). Residents' perceptions of community tourism impacts. *Annals of Tourism Research, 32*(4), 1056–1076. https://doi.org/10.1016/j.annals.2005.03.001.

Ap, J. (1992). Residents' perceptions on tourism impacts. *Annals of Tourism Research, 19*(4), 665–690. https://doi.org/10.1016/0160-7383(92)90060-3.

Becker, J. M., Klein, K., & Wetzels, M. (2012). Hierarchical latent variable models in PLS-SEM: Guidelines for using reflective-formative type models. *Long Range Planning, 45*(5), 359–394. https://doi.org/10.1016/j.lrp.2012.10.001.

Bollen, K. A., & Bauldry, S. (2011). Three Cs in measurement models: Causal indicators, composite indicators, and covariates. *Psychological Methods, 16*(3), 265–284. https://doi.org/10.1037/a0024448.

Chang, W., Franke, G. R., & Lee, N. (2016). Comparing reflective and formative measures: New insights from relevant simulations. *Journal of Business Research, 69*(8), 3177–3185. https://doi.org/10.1016/j.jbusres.2015.12.006.

Choi, H. C., & Sirakaya, E. (2006). Sustainability indicators for managing community tourism. *Tourism Management, 27*(6), 1274–1289. https://doi.org/10.1016/j.tourman.2005.05.018.

Coltman, T., Devinney, T. M., Midgley, D. F., & Venaik, S. (2008). Formative versus reflective measurement models: Two applications of formative measurement. *Journal of Business Research, 61*(12), 1250–1262. https://doi.org/10.1016/j.jbusres.2008.01.013.

Deery, M., Jago, L., & Fredline, L. (2012). Rethinking social impacts of tourism research: A new research agenda. *Tourism Management, 33*(1), 64–73. https://doi.org/10.1016/j.tourman.2011.01.026.

Diamantopoulos, A., Riefler, P., & Roth, K. P. (2008). Advancing formative measurement models. *Journal of Business Research, 61*(12), 1203–1218. https://doi.org/10.1016/j.jbusres.2008.01.009.

Dijkstra, T. K., & Henseler, J. (2015). Consistent partial least squares path modeling. *MIS Quarterly, 39*(2), 297–316.

Fassott, G., & Henseler, J. (2015). Formative (Measurement). In C. L. Cooper (Ed.), *Wiley Encyclopedia of Management* (Vol. 9, 3rd ed., pp. 1–4). Chichester: Wiley. https://doi.org/10.1002/9781118785317.weom090113.

Gu, M., & Wong, P. (2006). Residents' perception of tourism impacts: A case study of homestay operators in Dachangshan Dao, north-east China. *Tourism Geographies, 8*(3), 253–273. https://doi.org/10.1080/14616680600765222.

Gudergan, S. P., Ringle, C. M., Wende, S., & Will, A. (2008). Confirmatory tetrad analysis in PLS path modeling. *Journal of Business Research, 61*(12), 1238–1249. https://doi.org/10.1016/j.jbusres.2008.01.012.

Gursoy, D., Jurowski, C., & Uysal, M. (2002). Resident attitudes. *Annals of Tourism Research, 29*(1), 79–105. https://doi.org/10.1016/S0160-7383(01)00028-7.

Hair, J. F., Hult, G. T. M., Ringle, C., & Sarstedt, M. (2017). *A primer on partial least squares structural equation modeling (PLS-SEM)* (2nd ed.). Los Angeles: Sage.

Hair, J. F., Sarstedt, M., Ringle, C. M., & Gudergan, S. P. (2018). *Advanced issues in partial least squares structural equation modeling*. Los Angeles: SAGE.

Hall, C. M., & Page, S. (2014). *The geography of tourism and recreation: Environment, place and space* (4th ed.). New York: Routledge.

Henseler, J., Ringle, C. M., & Sarstedt, M. (2015). A new criterion for assessing discriminant validity in variance-based structural equation modeling. *Journal of the Academy of Marketing Science, 43*(1), 115–135. https://doi.org/10.1007/s11747-014-0403-8.

Henseler, J., Hubona, G., & Ray, P. A. (2016). Using PLS path modeling in new technology research: Updated guidelines. *Industrial Management & Data Systems, 116*(1), 2–20. https://doi.org/10.1108/IMDS-09-2015-0382.

Huang, S., & Hsu, C. H. C. (2005). Mainland Chinese residents' perceptions and motivations of visiting Hong Kong: Evidence from focus group interviews. *Asia Pacific Journal of Tourism Research, 10*(2), 191–205. https://doi.org/10.1080/10941660500135977.

Jaafar, M., Noor, S. M., & Rasoolimanesh, S. M. (2015). Perception of young local residents toward sustainable conservation programmes: A case study of the Lenggong World Cultural Heritage Site. *Tourism Management, 48*, 154–163. https://doi.org/10.1016/j.tourman.2014.10.018.

Jaafar, M., Rasoolimanesh, S. M., & Ismail, S. (2017). Perceived sociocultural impacts of tourism and community participation: A case study of Langkawi Island. *Tourism and Hospitality Research, 17*(2), 123–134. https://doi.org/10.1177/1467358415610373.

Jurowski, C., Uysal, M., & Williams, D. R. (1997). A theoretical analysis of host community resident reactions to tourism. *Journal of Travel Research, 36*(2), 3–11. https://doi.org/10.1177/004728759703600202.

Kayat, K. (2002). Power, social exchanges and tourism in Langkawi: Rethinking resident perceptions. *International Journal of Tourism Research, 4*(3), 171–191 https://doi.org/10.1002/jtr.375.

Kim, K. (2002). *The effects of tourism impacts upon quality of life of residents in the community* (Unpublished doctoral dissertation). Virginia Polytechnic Institute and State University, Blacksburg, VA. Retrieved from http://theses.lib.vt.edu/theses/available/etd-12062002-123337/unrestricted/Title_and_Text.pdf

Kim, K., Uysal, M., & Sirgy, M. J. (2013). How does tourism in a community impact the quality of life of community residents? *Tourism Management, 36*, 527–540. https://doi.org/10.1016/j.tourman.2012.09.005.

Ko, D.-W., & Stewart, W. P. (2002). A structural equation model of residents' attitudes for tourism development. *Tourism Management, 23*(5), 521–530. https://doi.org/10.1016/S0261-5177(02)00006-7.

Látková, P., & Vogt, C. A. (2012). Residents' attitudes toward existing and future tourism development in rural communities. *Journal of Travel Research, 51*(1), 50–67. https://doi.org/10.1177/0047287510394193.

Matarrita-Cascante, D. (2010). Beyond growth. *Annals of Tourism Research, 37*(4), 1141–1163. https://doi.org/10.1016/j.annals.2010.05.004.

McGehee, N. G., Andereck, K. L., & Vogt, C. A. (2002). *An examination of factors influencing resident attitudes toward tourism in twelve Arizona communities*. Presented at the 33rd Annual Travel and Tourism Research Association Conference, Arlington, VA.

Md Noor, S., Rasoolimanesh, S. M., Ganesan, V., & Jaafar, M. (2015). Effective interpretation using various media toward mindfulness: A case study of Malacca. *Journal of Heritage Tourism, 10*(3), 263–279. https://doi.org/10.1080/1743873X.2015.1005625.
Nicholas, L. N., Thapa, B., & Ko, Y. J. (2009). Residents' perspectives of a world heritage site. *Annals of Tourism Research, 36*(3), 390–412. https://doi.org/10.1016/j.annals.2009.03.005.
Nunkoo, R., & Ramkissoon, H. (2012). Power, trust, social exchange and community support. *Annals of Tourism Research, 39*(2), 997–1023. https://doi.org/10.1016/j.annals.2011.11.017.
Park, M., & Stokowski, P. A. (2009). Social disruption theory and crime in rural communities: Comparisons across three levels of tourism growth. *Tourism Management, 30*(6), 905–915. https://doi.org/10.1016/j.tourman.2008.11.015.
Rasoolimanesh, S. M., Jaafar, M., Kock, N., & Ramayah, T. (2015). A revised framework of social exchange theory to investigate the factors influencing residents' perceptions. *Tourism Management Perspectives, 16*, 335–345. https://doi.org/10.1016/j.tmp.2015.10.001.
Rasoolimanesh, S. M., Dahalan, N., & Jaafar, M. (2016). Tourists' perceived value and satisfaction in a community-based homestay in the Lenggong Valley World Heritage Site. *Journal of Hospitality and Tourism Management, 26*, 72–81. https://doi.org/10.1016/j.jhtm.2016.01.005.
Rasoolimanesh, S. M., Jaafar, M., & Barghi, R. (2017a). Effects of motivation, knowledge and perceived power on residents' perceptions: Application of Weber's theory in world heritage site destinations. *International Journal of Tourism Research, 19*(1), 68–79. https://doi.org/10.1002/jtr.2085.
Rasoolimanesh, S. M., Jaafar, M., Kock, N., & Ahmad, A. G. (2017b). The effects of community factors on residents' perceptions toward World Heritage Site inscription and sustainable tourism development. *Journal of Sustainable Tourism, 25*(2), 198–216. https://doi.org/10.1080/09669582.2016.1195836.
Rigdon, E. E. (2015). Structural equation modeling: Nontraditional alternatives. In N. Balakrishnan, T. Colton, B. Everitt, W. Piegorsch, F. Ruggeri, & J. L. Teugels (Eds.), *Wiley StatsRef: Statistics reference online* (pp. 1–12). Chichester: Wiley. https://doi.org/10.1002/9781118445112.stat06528.pub2.
Ringle, C., Wende, S., & Becker, J. (2015). *SmartPLS 3 (Version 3.2.6)*. Boenningstedt: SmartPLS GmbH.
Sharpley, R. (1994). *Tourism, tourists and society* (3rd ed.). Huntingdon: Elm Publications.
Sharpley, R. (2014). Host perceptions of tourism: A review of the research. *Tourism Management, 42*, 37–49. https://doi.org/10.1016/j.tourman.2013.10.007.
Sharpley, R., & Telfer, D. J. (2008). *Tourism and development in the developing world*. Abingdon: Routledge.
Tovar, C., & Lockwood, M. (2008). Social impacts of tourism: An Australian regional case study. *International Journal of Tourism Research, 10*(4), 365–378 https://doi.org/10.1002/jtr.667.
van Riel, A. C. R., Henseler, J., Keméņy, I., & Sasovova, Z. (2017). Estimating hierarchical constructs using consistent partial least squares: The case of second-order composites of common factors. *Industrial Management & Data Systems, 117*(3), 459–477. https://doi.org/10.1108/IMDS-07-2016-0286.
Vareiro, L. M. d. C., Remoaldo, P. C., & Ribeiro, J. A. C. (2013). Residents' perceptions of tourism impacts in Guimarães (Portugal): A cluster analysis. *Current Issues in Tourism, 16*(6), 535–551. https://doi.org/10.1080/13683500.2012.707175.
Voorhees, C. M., Brady, M. K., Calantone, R., & Ramirez, E. (2016). Discriminant validity testing in marketing: An analysis, causes for concern, and proposed remedies. *Journal of the Academy of Marketing Science, 44*(1), 119–134. https://doi.org/10.1007/s11747-015-0455-4.
Zamani-Farahani, H., & Musa, G. (2008). Residents' attitudes and perception towards tourism development: A case study of Masooleh, Iran. *Tourism Management, 29*(6), 1233–1236. https://doi.org/10.1016/j.tourman.2008.02.008.
Zamani-Farahani, H., & Musa, G. (2012). The relationship between Islamic religiosity and residents' perceptions of socio-cultural impacts of tourism in Iran: Case studies of Sare'in and Masooleh. *Tourism Management, 33*(4), 802–814. https://doi.org/10.1016/j.tourman.2011.09.003.

Chapter 13
Vector Autoregressive Models with Multivariate Time Series: Spillovers of International Tourist Arrivals on a Local Economy

Jung Wan Lee and Michael Kwag

Abstract The chapter demonstrates an application of vector autoregressive models with time series in tourism research. The chapter describes how to apply vector autoregression with time series in a multivariate setting to estimate the short-run and long-run effects of international tourist arrivals on a local economy and associated economic spillovers in the local economy in Asia. The study employs unit root tests (i.e., the augmented Dickey and Fuller test, the Phillips and Perron test, the Ng and Perron test, and the Kwiatkowski et al. test), cointegration test (i.e., the Johansen cointegration test), and vector autoregressive models with multivariate time series (i.e., vector autoregression analysis, model diagnostic tests, residual diagnostic tests, and impulse response tests) using quarterly time series data, which are available from the first quarter of 1970 to the third quarter of 2012 (172 observations) retrieved from the Bank of Korea.

Keywords Vector autoregressive model · Time series · Unit root test · Cointegration test · Impulse response function · International tourist arrivals · Tourism · Economic spillover · Tourism research · Korea

J. W. Lee (✉)
Administrative Sciences Department, Metropolitan College, Boston University, Boston, MA, USA
e-mail: jwlee119@bu.edu

M. Kwag
School of Hospitality Administration, Boston University, Boston, MA, USA
e-mail: mkwag@bu.edu

S. Rezaei (ed.), *Quantitative Tourism Research in Asia*, Perspectives on Asian Tourism, https://doi.org/10.1007/978-981-13-2463-5_13

13.1 Introduction of Methodology

Chapter Objectives

- To provide researchers an application of vector autoregressive models with time series in tourism research
- To introduce tools for testing the stationarity of time series and the number of unit roots
- To introduce tools for testing the presence of cointegrating relationships among several non-stationary variables
- To introduce vector autoregression and vector error correction models as a powerful technique for analyzing the association between a time series dependent variable and one or more independent time series
- To explain the nature and methods of vector autoregression and vector error correction models and describe the general model, estimation of parameters, standardized regression coefficient, significance testing, residual analysis, and the meaning of regression coefficients, including short-run and long-run dynamics and elasticities

13.1.1 Time Series Modeling

The structural approach to time series modeling uses economic theory to model the relationship among the variables of interest. Unfortunately, economic theory is often not rich enough to provide a dynamic specification that identifies all of these relationships. Furthermore, estimation and inference are complicated by the fact that endogenous variables may appear on both the left and right sides of equations. These problems lead to alternative, nonstructural approaches to modeling the relationship among several variables. This chapter describes the estimation and analysis of vector autoregression (VAR) and the vector error correction (VEC) models. This chapter also describes tools for testing the stationarity of time series, the number of unit roots, and the presence of cointegrating relationships among several non-stationary variables.

13.1.2 Unit Root Test

It is well known in the literature that the data generating process for many economic variables are characterized by stochastic trends that might result in spurious inference if the time series properties are not carefully investigated. The theory behind autoregressive moving average (ARMA) estimation is based on stationary time series. A series is said to be covariance stationary if the mean and autocovariances of the series do not depend on time. Any series that is not stationary is said to be non-stationary. A common example of a non-stationary series is the random walk, which is a difference stationary series since the first difference of the time series is stationary. The series has a constant forecast value, conditional on, and the variance is increasing over time. A difference stationary series is said to be integrated and is denoted as I(r) where r is the order of integration. The order of integration is the number of unit roots contained in the series or the number of differencing operations it takes to make the series stationary. Standard inference procedures do not apply to regressions, which contain an integrated dependent variable or integrated regressors. Therefore, it is important to check whether a series is stationary or not before using it in a time series regression.

The formal method to test the stationarity of a series is the unit root test. There are several well-known tests for this purpose based on individual time series. They are the augmented Dickey and Fuller (1979, 1981), the Phillips and Perron (1988), the GLS-detrended Dickey and Fuller (Elliott et al. 1996), the Kwiatkowski et al. (1992), the Elliott, Rothenberg, and Stock (1996) point optimal, and the Ng and Perron (2001) unit root tests.

Dickey and Fuller (1979) show that under the null hypothesis of a unit root, this statistic does not follow the conventional Student's t-distribution, and they derive asymptotic results and simulate critical values for various test and sample sizes. The simple Dickey-Fuller unit root test is valid only if the series is an AR(1) process. If the series is correlated at higher-order lags, the assumption of white noise disturbances is violated. The augmented Dickey-Fuller (ADF) test constructs a parametric correction for higher-order correlation by assuming that the series follows an AR(p) process and adding lagged difference terms of the dependent variable to the right-hand side of the test regression. Said and Dickey (1984) demonstrate that the ADF test is asymptotically valid in the presence of a moving average (MA) component, provided that sufficient lagged difference terms are included in the test regression.

Elliott et al. (1996) propose a simple modification of the ADF tests in which the data are detrended so that explanatory variables are taken out of the data prior to running the test regression. While the Dickey-Fuller test with GLS detrending (DFGLS) t-ratio follows a Dickey-Fuller distribution in the constant only case, the asymptotic distribution differs when it includes both a constant and trend.

Phillips and Perron (1988) propose an alternative (nonparametric) method of controlling for serial correlation when testing for a unit root. The Phillips and Perron (PP) method estimates the non-augmented DF test and modifies the t-ratio of the coefficient so that serial correlation does not affect the asymptotic distribution of the test statistic. The asymptotic distribution of the PP modified t-ratio is the same as that of the ADF statistic. As Enders (1995) indicates, the ADF test assumes the errors to be independent and that they have constant variance, while the PP test allows for fairly mild assumptions about the distribution of errors.

Elliott et al. (1996) propose the Elliott, Rothenberg and Stock (ERS) point optimal test, which is based on the quasi-differencing regression. The ERS point optimal test statistic considers the null against the alternative, where is the predicted value from the OLS estimation of the null model. The two statistics of the ERS point optimal test and the DFGLS test are provided by the ERS test, which provide a class of efficient unit root tests, in the sense that they reduce the size distortion and improve the power compared with traditional unit root tests such as augmented

Dickey-Fuller and Phillips-Perron tests. Four different test statistics, including the two in the ERS test and two other tests (see Bhargava, 1986), the modified PP test and the modified point optimal test, are reported by Ng and Perron (2001) test.

Many of the unit root tests described above require a consistent estimate of the residual spectrum at frequency zero. The kernel-based estimator of the frequency zero spectrum is based on a weighted sum of the autocovariances, with the weights which are defined by a kernel function. The entire unit root tests described above test the following null hypothesis: a series has a unit root (non-stationary), while the alternative hypothesis assumes that the series has no unit root (stationary).

Kwiatkowski et al. (1992) propose a different approach from the other unit root tests described above in which the series is assumed to be stationary under the null. The KPSS statistic is based on the residuals from the OLS regression of on the exogenous variables. The estimator of used in this calculation differs from the estimator for used by GLS detrending since it is based on a regression involving the original data and not on the quasi-differenced data. The KPSS statistic tests the following null hypothesis: a series has no unit root (stationary), while the alternative hypothesis assumes that the series has a unit root (non-stationary).

13.1.3 Cointegration Test

The finding that many macro time series may contain a unit root has spurred the development of the theory of non-stationary time series analysis. Engle and Granger (1987) pointed out that a linear combination of two or more non-stationary series may be stationary. If such a stationary linear combination exists, the non-stationary time series are said to be cointegrated. The stationary linear combination is called the cointegrating equation and may be interpreted as a long-run equilibrium relationship among the variables. In other words, if the variables are cointegrated, then there is a long-run relationship, and a force exists, which converges into a long-run equilibrium. There are several tools for testing for the presence of cointegrating relationships among non-stationary variables in a multivariate setting. They are the Johansen (1991, 1995) cointegration test and the Engle and Granger (1987) residual-based cointegration test.

The Engle and Granger (1987) single-equation residual-based cointegration test is a simple unit root test applied to the residuals obtained from a static OLS cointegrating regression. Under the assumption that the series are not cointegrated, the residuals are unit root non-stationary. Therefore, a test of the null hypothesis of no cointegration against the alternative of cointegration may be constructed by computing a unit root test of the null of residual non-stationarity against the alternative of residual stationarity. The regression equations are tested by the method of least squares. The optimal lags are automatically selected for the ADF test based on the Schwarz info criterion. The Engle-Granger test obtains only one single cointegration relationship based on OLS, whereas it is possible to obtain more than one cointegration relationship with the Johansen test, which is a maximum likelihood-based test that requires a large sample. Cheung and Lai (1993) reported that the Johansen approach is more efficient than the Engle-Granger method, because the maximum likelihood procedure has significantly large and finite sample properties.

Johansen (1988, 1991, 1995) considers a simple case where Y_t is integrated of order one *I(1)*, such that the first difference of Y_t is stationary. Suppose the process Y_t is defined by an unrestricted VAR system of order ($n \times 1$). The Johansen approach derives maximum likelihood estimators of the cointegrating vectors for an autoregressive process with independent errors. The Johansen method maximizes the likelihood function for Y_t conditional on any given α, using standard least squares formulae for the regression of ΔY_t on the lagged differences ΔY_{t-1}, ΔY_{t-2}, ... ΔY_{t-j}. The Johansen cointegration test models each variable, which are assumed to be jointly endogenous, as a function of all the lagged endogenous variables in the system. To illustrate the unrestricted VAR cointegration test of Johansen, consider a VAR model written in the error correction form with Gaussian errors as shown in Eq. 13.1.

$$\Delta Y_t = \alpha + \sum_{j=1}^{n-1} \beta_j \Delta Y_{t-j} + \gamma Y_{t-n} + \theta \mathrm{ECT}_t + \varepsilon_t \tag{13.1}$$

where Δ is the difference operator; α is the deterministic component; β, γ, and θ are the parameters to be estimated; ε_t is assumed to be stationary random errors with mean zero, that is, white noise; j is the lag length; t represents 1, 2, 3, ... and n observation; Y_t is the k vector of non-stationary variable and is considered fixed, and the likelihood function is calculated for a given value of this; and ECT_t is the error correction term obtained from the cointegrating vectors.

The Johansen procedure uses two ratio tests: (a) a trace test and (b) a maximum eigenvalue test, to test for a number of cointegration relationships. Both can be used to determine the number of cointegrating vectors present, although they do not always indicate the same number of cointegrating vectors. The trace test is a joint test, the null hypothesis is that the number of cointegrating vectors is less than or equal to r, against a general alternative hypothesis that there are more than r. The maximum eigenvalue test conducts separate tests on each eigenvalue. The null hypothesis is that there are r cointegrating vectors present against the alternative that there are (r + 1) present. The order of r is determined by using the two likelihood ratio test statistics: the trace statistic, λ_{trace}, and the maximum eigenvalue statistic, $\lambda_{\max}$, as shown in Eqs. 13.2 and 13.3:

$$\lambda_{\text{trace}} = -T \sum_{j=r+1}^{n} \mathrm{In}\left(1-\hat{\lambda}_j\right) \tag{13.2}$$

$$\lambda_{\max} = -T \mathrm{In}\left(1-\hat{\lambda}_{r+1}\right) \tag{13.3}$$

where r is the hypothesized number of cointegrating equation, $\hat{\lambda}_j$ is the jth largest estimated eigenvalue of the coefficient matrix, and T is the sample size, the number of observation used for estimation. The result can be sensitive to the number of lags included in the test and the presence of autocorrelation.

13.1.4 Vector Autoregression (VAR) and Vector Error Correction Models (VECM)

The conventional approaches to modeling the relationship among several variables employ the estimation and analysis of vector autoregression (VAR) and the vector error correction models (VECM). The vector autoregression (VAR) is commonly used for forecasting systems of interrelated time series and for analyzing the dynamic impact of random disturbances on the system of variables. The VAR approach sidesteps the need for structural modeling by treating every endogenous variable in the system as a function of the lagged values of all of the endogenous variables in the system.

The mathematical representation of a VAR is:

$$Y_t = \beta_1 Y_{t-1} + \ldots + \beta_p Y_{t-p} + \gamma X_t + \varepsilon_t \tag{13.4}$$

where Y_t is a k vector of endogenous variables, X_t is a d vector of exogenous variables, $\beta_1 \ldots \beta_p$ and γ are matrices of coefficients to be estimated, and ε_t is a vector of innovations that may be contemporaneously correlated but are uncorrelated with their own lagged values and uncorrelated with all of the right-hand side variables. Since only lagged values of the endogenous variables appear on the right-hand side of the equations, simultaneity is not an issue, and OLS yields consistent estimates. Moreover, even though the innovations may be contemporaneously correlated, OLS is efficient and equivalent to GLS since all equations have identical regressors.

Engle and Granger (1987) and Granger (1988) report that if two or more variables are cointegrated, there always exists a corresponding error correction representation in which the short-run dynamics of the variables in the system are influenced by the deviation from equilibrium. The VECM is a technique that facilitates to capture both the dynamic and interdependent relationships of regressors and is a special type of restricted VAR to correct a disequilibrium that may shock the whole system. The VECM implies that changes in one variable are a function of the level of disequilibrium in the cointegrating relationship, as well as changes in the other explanatory variable.

The mathematical representation of a VECM is:

$$Y_t = \beta_1 Y_{t-1} + \ldots + \beta_p Y_{t-p} + \gamma X_t + \theta_1 \mathrm{ECT}_{t-1} + \theta_2 \mathrm{ECT}_{t-2} + \varepsilon_t \tag{13.5}$$

where Y_t is a k vector of endogenous variables, X_t is a d vector of exogenous variables, and $\beta_1 \ldots \beta_p$, γ and θ are matrices of coefficients to be estimated, ECT_t is the error correction term obtained from the cointegrating vectors, and ε_t is assumed to be stationary random errors with mean zero, that is, white noise.

The VECM has cointegration relations built into the specification so that it restricts the long-run behavior of the endogenous variables to converge to their cointegrating relationships while allowing for short-run adjustment dynamics. The cointegration term is known as the error correction term since the deviation from long-run equilibrium is corrected gradually through a series of partial short-run adjustments. The VECM can distinguish between the short-run and long-run dynamics because it can capture both the short-run dynamics between the time series and their long-run equilibrium relationship. The long-run dynamics is implied through the significance of the t-statistics of the lagged error correction terms (i.e., by testing H0: $\theta_1 = 0$). In this case, it estimates the asymptotic variance of the estimator, and then the t-statistics will have asymptotically the standard normal distribution. Therefore, asymptotic t-statistics in this test can be interpreted in the same way as t-statistics, which are used to interpret the statistical significance of coefficients of the lagged error correction terms, which contain the long-run information because it is derived from the long-run cointegrating relationship.

13.2 Literature Review and Hypotheses

13.2.1 Overview of International Tourist Arrivals to South Korea

For the past four decades, South Korea has built up a remarkable record of economic growth and integration in the information and knowledge economy: a competitive education system, a highly skilled and dedicated workforce, and advances in information technology are widely acclaimed as key factors driving the dynamic economy. In recent years, however, rapidly aging population, international pressures regarding environmental issues, and widening gaps between and within industries are increasing economic hardship apparent. Finding the best solutions to these problems is one of the greatest challenges faced by policymakers of South Korea today.

Although Korea's rapid economic growth in the past four decades has been acclaimed as a result of Korea's export-driven economy, the tourism industry could be considered as another contributing sector toward Korea's recent healthy economy. According to the tourism research data of the World Travel and Tourism Council (2011), the tourism industry in Korea contributed about US$63.1 billion on the GDP, which was 7.6% of the total economy in 2010. The tourism industry in Korea also supports about 8% of the total employment. Visitor exports, including expenditures by international visitors on goods and services within the economy, reached US$13.8 billion, which was 23.4% of the total exports in 2010. International visitors' spending includes travel costs and travel services covering tour, business, education, and diplomatic arrivals, among others.

Given the aforementioned statistics, Korea has been eager to promote her tourism sector. The Korea Culture and Tourism Institute and the Korea Tourism Organization publish statistics of short-term international tourist arrivals on a monthly basis. According to the statistics of the Korea Culture and Tourism Institute (2014) and the Korea Tourism Organization (2014), the total number of international arrivals grew by 58% from December 2005 to December 2013. International tourists primarily came from neighboring countries in Asia. Roughly 75% of the total number of international tourists derives from Japan, China, Hong Kong, and Taiwan. In addition, the recent popularity of Korean culture (so-called *HanRyu*) in these countries is one of the factors that increased the number of tourist arrivals. Table 13.1 reports descriptive statistics of international tourist arrivals to South Korea in 2013.

International tourists incur costs while visiting their tourism destinations, which include accommodations, food, tours, and shopping. When choosing a destination, tourists decide whether to visit the destination or not based on comparison of the relative cost of living between the chosen destination and their home country. Economic growth is the increasing ability of a nation to produce more goods and services. Growth can occur in many different ways such as the increased use of land, labor, capital and business resources, and increased productivity of existing

Table 13.1 Numbers of visitor arrivals by purpose and nationality in 2013 (unit: person)

Region	Tour	Business	Official	Others	Total	% of total
Asia	7,779,999	234,375	12,517	1,951,696	9,978,587	82
Americas	655,387	13,085	15,627	231,523	915,622	7.5
Europe	493,105	35,596	1,293	238,191	768,185	6.3
The Pacific	126,366	5,442	527	27,712	160,047	1.3
Africa	20,699	8,530	858	13,327	43,414	0.4
Other areas	132	41	1	26	200	0
Overseas Korean	0	0	0	309,495	309,495	2.5
Total	9,075,688	297,069	30,823	2,771,970	12,175,550	100
Nationality	Tour	Business	Official	Others	Total	% of Total
Japan	2,633,959	27,292	1,753	84,746	2,747,750	22.6
Twain	523,806	5,108	61	15,687	544,662	4.5
Hong Kong	388,483	919	5	11,028	400,435	3.3
Thailand	295,556	552	1216	75,554	372,878	3.1
Malaysia	182,611	1522	66	23,528	207,727	1.7
Philippines	106,512	2889	839	290,446	400,686	3.3
Indonesia	96,631	4794	348	87,416	189,189	1.6
China	3,139,867	126,011	3162	1,057,829	4,326,869	35.5
Maximum	3,139,867	126,011	3162	1,057,829	4,326,869	35.5
Minimum	96,631	552	5	11,028	189,189	1.7
Mean	920,928	21,136	931	205,779	1,148,775	9

resources. Applying information technology to the tourism industry provides a framework for the efficient distribution of goods and services and improves communication between firms and customers.

This study aims to examine direct and indirect economic spillover effects from international tourist arrivals to local economic growth of Korea. Specifically, this study aims to answer the following two questions: First, how does the indirect spillover from international tourist arrivals to other industries affect economic growth in the short term? Second, is there a long-run equilibrium relationship between international tourist arrivals, tourism industry, information communications technology, retail industry, and overall economic growth? If we explicitly disentangle direct and indirect economic spillover effects from international tourist arrivals on local economic growth, policy decisions will be more informative and effective.

13.2.2 Direct Effects of Tourism on the Local Economy

Because the economic effects of tourism have been to create income, taxes, hard currency, and jobs, tourism has made a significant contribution to the local economy of many countries (Choi and Sirakaya 2006; Dwyer and Forsyth 2008). Several studies show that the expansion of tourism activity positively influences economic growth in developed countries (e.g., Cárdenas-García et al. 2015) and especially in developing economies (e.g., Sahli and Nowak 2007). Castro-Nuño et al. (2013) report the effect of tourism on economic growth shows a positive elasticity between GDP and tourism. Massidda and Mattana (2013) report long-run and short-run relationships per capita across international tourism arrivals. Real GDP and total international commercial transactions have been beneficial for the Italian economy. Much literature reports that tourism positively affects economic growth in African countries (Fayissa et al. 2008), the same in Mexico (Sánchez-Carrera et al. 2008), the same in Singapore (Lee 2008), the same in Taiwan (Chen and Chiou-Wei 2009), and the same in Spain (Balaguer and Cantavella-Jordá 2002). Many researchers (e.g., Dritsakis 2004; Gunduz and Hatemi-J 2005) have also proposed that tourism growth is a major factor of overall economic growth in the long run. These empirical studies present strong evidence of a positive relationship between tourism growth and economic growth in several countries.

In this regard, Sahli and Nowak (2007) report that many governments have engaged in tourism growth for the purpose of local economic growth. A study reports that tourism development positively influences economic growth and poverty reduction in the Nicaraguan economy (Croes and Vanegas 2008). Mihalic (2002) reports several advantages of tourism growth, such as foreign exchange earnings and generating export revenues, as a growth strategy compared to the exports of goods and services. Felsenstein and Fleischer (2003) report that the contribution of local festivals shows modest but positive local economic growth, suggesting some justification for public policy measures for local festivals as a tourism strategy. It is said that tourism not only stimulates the growth of the industry but also triggers local economic growth of the countries (Lee and Chang 2008). In summary, as evidenced by a wide range of aforementioned studies, tourism plays an important role in local economic growth. Hence, the use of tourism in many countries continues to attract a large interest from their governments' viewpoint of its potential spillover effects on local economic growth.

13.2.3 Indirect Effects of Tourism on the Local Economy

According to indirect effects of tourism on the local economy, it can be found from several types of analyses, such as input output analysis, linkage analysis, and computable general equilibrium analysis. According to Pratt (2015), tourism plays a significant role of driving the economy of small island developing states. These small islands have little other economic activities due to its small size, remoteness, environmental vulnerability, and socioeconomic factors. The tourism impacts the economies of these small islands in several ways, such as infrastructure improvements (roads, airport, and other utilities), foreign investments, and employment opportunity. Schubert et al. (2011) support aforementioned analysis that tourism affects the improvement of infrastructure and generate employment. Furthermore, its spillover stimulates other industries. For example, considering countries that have small economies and they are dependent on tourism, tourism allows these countries to expand their scale of productions, which benefit their economies of scale. In addition, Snieska et al. (2014) claim that indirect effects of tourism on the local economy in rural area are also significant. Beside of foreign investments, its spillover can be seen through decreasing unemployment rate, job creations, and impacts on the green economy. This also helps improve the quality of life for both rural residents and visitors as well as preserve and develop local cultures and landscape.

Considering the relationship between tourism and urban economic growth, a study measures the spillover effects of tourism in China by using spatial economic analysis and finds that the economic growth in one city will affect in other nearby cities as well (Ma et al. 2015). The spillover from one city to another city can generate more job opportunities and decrease both information communication costs and transportation costs. In addition, the spillover effects also affect the economic gap between cities, which results in economic growth between groups. For example, after facing global financial crisis in 2008 in Taiwan, which previously relied on promoting exports, the Taiwanese government reconsidered the national economic development policy by redirecting her tourism policies (Chien et al. 2014). The government allows Chinese tourists to visit Taiwan. This benefits Taiwan economy in terms of job creation, gross added value, and employment income. Besides, the indirect effects of tourism can stimulate the national economy in terms of foreign investments as a source of income and foreign exchange (Chindris-Vasioiu and Tocan 2015). Furthermore, a study observes that the indirect effects of tourism on the local economy can be seen in terms of poverty reduction in Central America

(Vanegas 2014). Focusing on a tourism-led economy allows several countries in Central America to fight against poverty, which eventually affects local economic growth among the regions. A study explains the relationship between tourism and local labor markets by providing some evidence from Canada: every new casino opens in an area, and there are job creations in the gambling industry, which lasts for around 5 years (Humphreys and Marchand 2013).

13.2.4 Tourism and Its Spillover on Other Business Sectors

It is said that investments in tourism and of itself may appear to be insufficient for local economic growth (Du et al. 2016). Instead, tourism's contribution to the long-term growth of an economy comes through its role as an integral part of a broader development strategy. In this regard, we find many valuable spillovers from the expansion of tourism. Travel agencies, tour operators, hotels, and airlines use Internet-enabled business models and information management systems. In addition, they increasingly use social networking sites and Internet websites such as global distribution systems (i.e., Amadeus, Sabre, Worldspan). According to a survey from AirPlus Community (2010), corporate travel professions are significantly more active on social networking sites, which indicate increased digital opportunities for tourism-related recommendations and services. The results of a survey indicate that online travel reservation (66%) is one of the main activities over the Internet, which is becoming increasingly common across the global tourism industry (Zickuhr 2010). A study reports that travelers are beginning to adopt alternative channels in search of more authentic experiences through the Internet and tourism-related websites (Xiang et al. 2015).

Shopping tourism has been acknowledged as a primary travel motive. Wong and Wan (2013) identify a four-dimensional construct that reflects tourists' satisfaction of a service, product, environment, merchandise value, and service differentiation during their shopping excursion (see also Dolnicar et al. 2015; Yuan and Jang 2008). A study reports that physical infrastructure and tourist attractions are significant determinants of inbound tourism flows (Yang and Wong 2012). Lim and McAleer (2005) report specific trends exist in terms of international tourist arrivals, for example, from Japan to Australia from 1976 to 2000. However, they generally overlook the direct and indirect spillover effects of international tourist arrivals on other industries of the economy. To bridge the gap, this study aims to examine direct and indirect spillover effects of international tourist arrivals on other industries and on the overall health of the Korean economy. In an empirical model, additional independent variables, such as information communications technology and the wholesale and retail industry, are included to enhance predictability of the model and to address omitted variable bias. To the best of the authors' knowledge, these endogenous variables have not been included in the existing empirical models. Based on the discussion raised above, the following hypotheses are constructed:

Hypothesis 1: A long-run equilibrium relationship is likely to exist among international tourist arrivals, economic growth, tourism, and other industries.

Hypothesis 2: The direct spillover of international tourist arrivals is likely to be related to local economic growth in the short term, ceteris paribus.

Hypothesis 3: The indirect spillover of international tourist arrivals is likely to be related to other industries of the economy in the short term.

13.3 Data and Research Methods

13.3.1 Data

This section describes data and outlines the methodology used in the selection of indicators and the normalization of data. The normalization of the data is necessary to transform the values to the similar unit of measurement. Therefore, all series were transformed into a natural log that mitigates possible distortions of dynamic properties of the series. Log transformation is a preferred method since each resulting coefficient in the regression equation represents the ratio of the incremental change in the logarithm function with respect to the incremental change in the logarithm argument. Table 13.2 reports the measurement items, descriptions of data, and sources.

Table 13.2 Measurement items and data sources

Measurement items	Descriptions of data and sources
Economic growth (GDP)	GDP is used as a proxy for local economic growth, which is measured by percentage change in gross domestic product (GDP) in market prices. GDP is the sum of gross value added by all resident producers in the economy, plus any product taxes and minus any subsidies not included in the value of the products. It is calculated without making deductions for depreciation of fabricated assets or for depletion and degradation of natural resources
International tourist arrivals (VISIT)	VISIT is used as a proxy for international tourist arrivals. International tourist arrivals represent the tourism receipts of international tourist arrivals and their expenditures in Korea. These statistics are collected and published monthly by the Bank of Korea
Tourism (TOUR)	TOUR is used as a proxy for the economic output of hospitality and tourism industry. It is the combined output of the hospitality and tourism industry, including that of restaurants and hotels, and culture and entertainment services, as well as leisure industries dealing directly with tourists
Information communications technology (ICT)	ICT represents the economic output of information and communications technology industry at market prices, including communication and publishing, broadcasting, film, and information services
Wholesale and retail (RETAIL)	RETAIL represents the economic output of the wholesale and retail industry at market prices, including the wholesale and retail trade, transport, and storage

Table 13.3 Results of correlation analysis

	GDP	VISIT	TOUR	ICT
VISIT	0.987*** [78.188]			
TOUR	0.986*** [75.799]	0.994*** [123.768]		
ICT	0.993*** [111.834]	0.987*** [78.253]	0.987*** [81.031]	
RETAIL	0.993*** [107.431]	0.978*** [59.955]	0.978*** [59.608]	0.988*** [81.364]

The numeric values in square brackets [] are t-statistics. Probabilities for rejection of the null hypothesis are employed at the 0.05 significance level (***, $p < 0.01$)

All of these economic time series data are collected and retrieved from the Bank of Korea Economic Statistics System database. The sample is restricted to those periods for which quarterly data are available from the first quarter of 1970 to the third quarter of 2012 (172 observations). Table 13.3 reports the results of correlation analysis.

13.3.2 Empirical Model

Engle and Granger (1987) specify that if two or more variables are cointegrated, a corresponding error correction term exists in which the short-run dynamics of the variables in the system are influenced by the deviation from equilibrium. In this case, a vector error correction model (VECM) is formulated to reintroduce the information lost in the differencing process, thereby capturing long-run as well as short-run dynamics.

The following equation illustrates a multivariate vector autoregressive model with the error correction term. This equation will be used to test long-run and short-run spillover effects from tourism to other industries and to overall local economic growth in Korea.

$$
\begin{aligned}
\Delta \text{lnGDP}_t = {} & \alpha_1 + \sum_{j=1}^{n-1} \beta_{1j} \Delta ln\text{GDP}_{t-j} + \sum_{j=1}^{n} \beta_{2j} \Delta ln\text{VISIT}_{t-j} + \sum_{j=1}^{n} \beta_{3j} \Delta ln\text{TOUR}_{t-j} \\
& + \sum_{j=1}^{n} \beta_{4j} \Delta ln\text{ICT}_{t-j} + \sum_{j=1}^{n} \beta_{5j} \Delta ln\text{RETAIL}_{t-j} + \sum_{j=1}^{n} \beta_{6j} \left\{ \Delta ln\text{VISIT}_{t-j} * \Delta ln\text{TOUR}_{t-j} \right\} \\
& + \sum_{j=1}^{n} \beta_{7j} \left\{ \Delta ln\text{VISIT}_{t-j} * \Delta ln\text{ICT}_{t-j} \right\} + \sum_{j=1}^{n} \beta_{8j} \left\{ \Delta ln\text{VISIT}_{t-j} * \Delta ln\text{RETAIL}_{t-j} \right\} \\
& + \zeta_1 \text{ECT}_{t-1} + \varepsilon_{1t}
\end{aligned}
$$

where:

Δ indicates the difference operator.
α is the deterministic component (constant).
β and ς denote the parameters to be estimated.

ECT_{t-1} is the error correction term obtained from the cointegrating vectors deriving from the long-run cointegrating relationship.
$\mathcal{E}_t$ follows stationary random errors with mean zero, that is, white noise.
t represents 1, 2, 3… and *n* observations, and *j* is the lag length.
GDP refers to economic growth.
VISIT represents the tourism receipt of international tourist arrivals.
TOUR refers to the economic output of the tourism and hospitality industry.
ICT represents the economic output of the information communications technology industry.
RETAIL represents the economic output of the wholesale and retail industry.

The VECM is useful for capturing both the long-run and the short-run dynamics when the variables are cointegrated. The long-run dynamic is indicated by the significance of t-statistics of the lagged error correction term, i.e., by testing null hypothesis: $\varsigma = 0$. The short-run dynamics, i.e., Granger causality in the VEC model, can be tested by the Wald test. The block exogeneity Wald test in the VEC model provides Chi-squared statistics of coefficients on the lagged endogenous variables, which are used to interpret the statistical significant of coefficients of the regressors.

13.3.3 Unit Root Test

Table 13.4 reports the result of the unit root tests. All test equations were tested by the least squares method. The optimal lag in the tests was automatically selected based on the Schwarz information criterion. The bandwidth for the tests was selected

Table 13.4 Results of unit root test

	GDP	VISIT	TOUR	ICT	RETAIL
	Level	Level	Level	Level	Level
Methods	1st difference	1st difference	1st difference	1st difference	1st difference
ADF test	−2.206	−2.323	−1.040	−2.457	−1.574
	−3.576***	−20.228***	−5.450***	−3.931***	−4.872***
PP test	−2.379	−2.501	−1.185	−2.453	−2.508
	−48.222***	−30.669***	−26.850***	−16.560***	−23.498***
NP test	3.764	−0.924	4.268	6.543	3.243
	−2.632**	−7.334***	−3.158***	−4.896***	−2.280**
KPSS test	1.571***	1.514***	1.579***	1.576***	1.576***
	0.331	0.518	0.150	0.336	0.347

ADF test refers to the augmented Dickey and Fuller unit root test, *PP* test refers to the Phillips and Perron unit root test, *NP* test refers to the Ng and Perron unit root test, *KPSS* test refers to the Kwiatkowski et al. unit root test
Probability values for rejection of the null hypothesis are based on the 0.05 level (**, $p < 0.05$; ***, $p < 0.01$)

based on the Newey-West estimator using the Bartlett kernel function. The null hypothesis of a unit root cannot be rejected in the level of the series, but all null hypotheses of a unit root are rejected in the first difference of the series. The results in Table 13.4 confirm that all series are integrated in the order of one.

13.4 Empirical Results

13.4.1 Results of Cointegration Test

The Johansen cointegration test represents each variable as a function of all lagged endogenous variables in the system. It uses two ratio tests: a tract test and a maximum eigenvalue test, to examine the number of cointegration relationships. Both tests can be used to determine the number of cointegrating vectors present although they do not always indicate the same number of cointegrating vectors. If trace statistics and maximum eigenvalue statistics yield different results, the result of the maximum eigenvalue test is preferred because of the benefit of carrying out separate tests on each eigenvalue.

Table 13.5 reports the results of the Johansen cointegration test. For the Johansen cointegration test, the assumptions of cointegration tests allow for individual effects but no individual linear trends in a vector autoregressive model. The results in Table 13.5 seem to conflict with each other statistic. The trace test results indicate at least three cointegrating vectors at the 0.05 level, while the maximum eigenvalue test results indicate no cointegrating vector at the 0.05 level. The conflicting results of the Johansen cointegration test indicate that hypothesis 1 cannot be rejected, which meant that there is no long-run equilibrium relationship between economic growth, international tourist arrivals, tourism, information communications technology, and retail industry in Korea at the 0.05 significance level (H1: trace statistics <0.05, but maximum eigenvalue statistics >0.05).

Table 13.5 Results of Johansen cointegration test

Regression model	GDP = *f*(VISIT, TOUR, ICT, RETAIL)	
Number of cointegration (r)	Trace statistic	Maximum eigenvalue statistic
$r = 0$	79.389***	29.420
$r \leq 1$	49.968**	19.841
$r \leq 2$	30.126**	14.876
$r \leq 3$	14.179	10.718
$r \leq 4$	1.356	1.356

Probability values for rejection of the null hypothesis are based on the 0.05 level (**, $p < 0.05$ and ***, $p < 0.01$)

13.4.2 Results of Vector Autoregressive Models with Time Series

Statistical inference is sensitive to parameter instability, serial correlation in residuals, and residual skewness. Skewness of the series is not significantly different from a normal distribution. Histogram normality Jarque-Bera test (i.e., null hypothesis: residuals are multivariate normal) is not rejected. Breusch-Godfrey serial correlation Lagrange multiplier or LM test (i.e., null hypothesis: no serial correlation at lag order 2) is not rejected. Heteroskedasticity test (i.e., null hypothesis: no autoregressive conditional heteroskedasticity or ARCH effect at lag order 1) is not rejected. Thus, this model yields acceptable results to employ VAR models. There are considerably fewer outliers, and the fluctuation bands are smaller (see Fig. 13.1).

As such Table 13.6 reports the results of vector autoregressive (VAR) estimates, model diagnostic tests, and residual diagnostic tests. Table 13.6 indicates the results of VAR regression analysis with the coefficients of regressors representing elasticities in the short run.

In testing hypothesis 2 that the direct spillover of international tourist arrivals is likely to be related to local economic growth in the short run, Table 13.6 shows that international tourist arrivals and tourism have a positive impact on local economic growth and are statistically significant at the 0.05 level (H2: $p < 0.01$). The results indicate that international tourist arrivals and tourism have positive and direct effects on the Korean economy.

In testing hypothesis 3 that the indirect spillover of international tourist arrivals is likely to be related to other industries of the economy, Table 13.6 indicates that the indirect spillover of international tourist arrivals is positively associated with the wholesale and retail industry of the Korean economy in the short run and is statistically significant at the 0.05 level (H3: $p < 0.01$). The results indicate that interna-

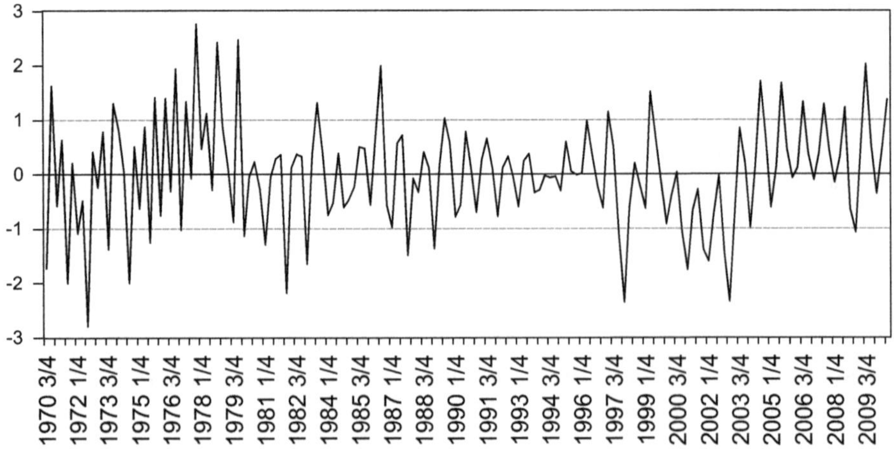

Fig. 13.1 Graph of standardized residuals

Table 13.6 Results of vector autoregressive models with time series

Endogenous variables	Model 1: coefficient	Model 2: coefficient
VISIT	0.950 [9.590]***	0.933 [9.736]***
TOUR	1.784 [9.480]***	1.562 [8.213]***
ICT	0.163 [6.634]***	0.189 [3.732]***
RETAIL	0.743 [18.260]***	0.581 [11.412]***
VISIT*TOUR		0.005 [2.787]**
VISIT*ICT		0.004 [1.501]
VISIT*RETAIL		0.021 [4.746]***
Constant	4.211 [15.600]	5.184 [14.879]
R-squared	0.995	0.996
Adjusted R-squared	0.995	0.996
F-statistic	8666.781***	5709.156***
Durbin-Watson statistic	2.308 (obs. = 162)	1.991(obs. = 161)
Normality test[a]	14.412***	1.112
Serial correlation LM test[a]	7.996*** F(2, 155)	0.056 F(2, 151)
Heteroskedasticity test[a]	0.136 F(1, 159)	1.441 F(1, 158)

VAR model 1: GDP = *f*(VISIT, TOUR, ICT, RETAIL)
VAR model 2: GDP = *f*(VISIT, TOUR, ICT, RETAIL, VISIT*TOUR, VISIT*ICT, VISIT*RETAIL)
[a]Residual diagnostics: (1) Histogram normality test: the numeric values are Jarque-Bera statistics. (2) Breusch-Godfrey serial correlation LM test: the numeric values are F- statistics. (3) Heteroskedasticity test (ARCH): the numeric values are F- statistics
The numeric values in square brackets [] are t-statistics. Probabilities for rejection of the null hypothesis are employed at the 0.05 significance level (**, $p < 0.05$ and ***, $p < 0.01$)

tional tourist arrivals have contributed to the output of the tourism and hospitality industry, including restaurants and hotels, and culture and entertainment services, as well as leisure industries dealing with tourists. In addition, international tourist arrivals have contributed to the economic output of the Korean wholesale and retail industry, including the wholesale and retail trade, transport, and storage services in Korea.

In addition to the results in Table 13.6, it may be worthy of note that the impulse response function, which is a graphical representation of the responding shock of the dependent variable to shocks of the error term, traces out the effect of the dynamics of the dependent variable on all the endogenous variables in the VAR model. The impulse response function captures the effect of a one-time shock to one of the innovations in future values of the endogenous variables. To interpret the impulse responses, the Cholesky transformation method is used. The results in Fig. 13.2 are computed for a 36-period horizon in the VAR model. The confidence bands are computed using the 95% confidence intervals.

Figure 13.2 represents the results of the impulse responses of economic growth to a Cholesky shock at one standard deviation for each endogenous variable. The graphs show that the impulse responses of economic growth to international tourist arrivals, tourism, and retail industry shocks start to have a positive impact to some extent in the short term and the impact declines drastically after 10 months, before

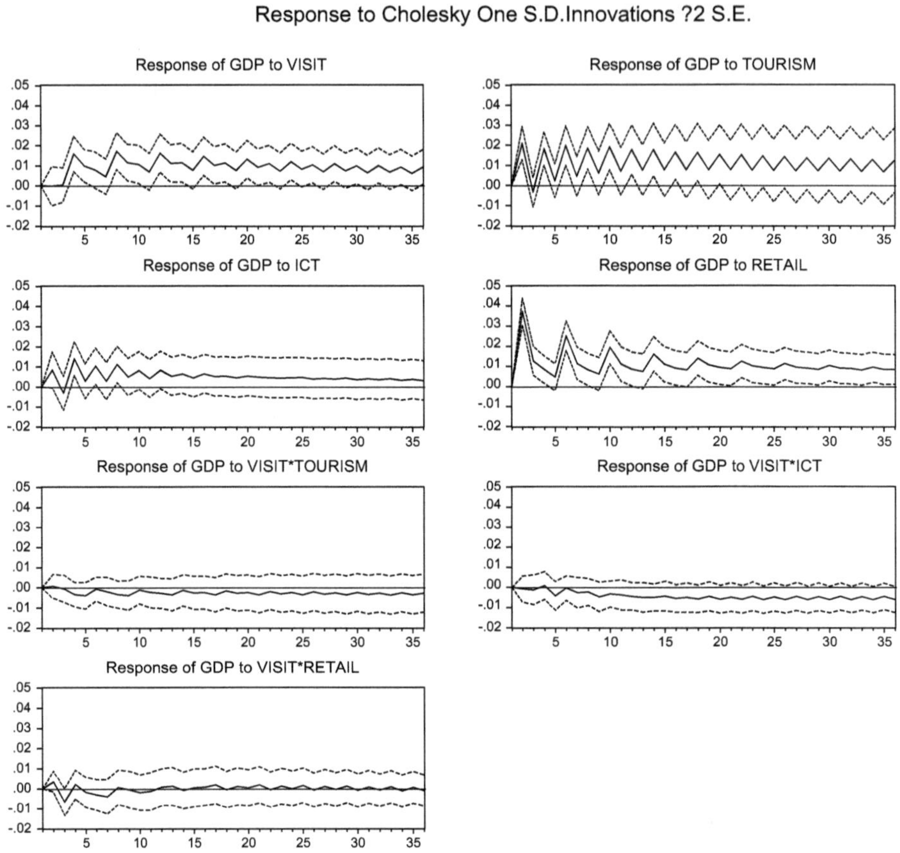

Fig. 13.2 Impulse responses of GDP to innovations of each endogenous variable
Note: The dotted lines in the graphs show the 95% confidence bands that are computed using the 95% confidence intervals

finally diminishing after 15 months. After 15 months, the impulse responses of economic growth to the shocks of all the endogenous variables in the VAR model are not significant, and the effect of each shock on economic growth will be asymptotic to some constant or reduced almost to zero.

13.5 Discussion and Policy Implications

This section provides policy implications with inferences based on the results from Table 13.5 and Fig. 13.3. Growth policies may emphasize the following for implementation and alteration with respect to magnitude of elasticity shown in the two bullet points discussed below.

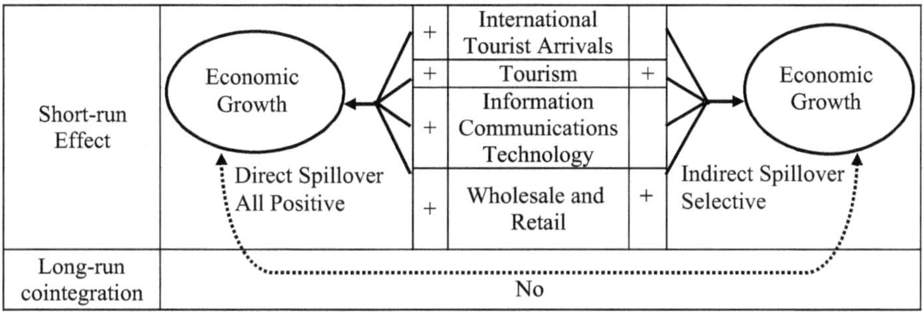

Fig. 13.3 Summary of the findings

- ***A 1% growth in international tourist arrivals increases local economic growth directly by 0.933% in the short run.***

According to the finding on Table 13.5, the economic growth would seem not to be affected by international tourist arrivals in the long run. However, the economic growth has been directly affected by international tourist arrivals in the short run. The findings support the strong positive direct effects of international tourist arrivals on local economic growth, which have played a critical role in boosting local economic growth in Korea. Considering the direct effect of international tourist arrivals, the findings support that the national economic growth has been enhanced by the income received from international tourist arrivals. Their associated spillovers may include government policy adjustments, business activities, cultural creations, and other tourism policies. The best practices may include international marketing promotion for Korean culture, e.g., *Hanryu* and *K-Pop*, open-door electronic visa policy (e-visa), promotion for medical tourism, and promotion for foreign investments, especially in hotel and resort developments.

- ***A 1% growth in the indirect spillover from international tourist arrivals to the wholesale and retail industry increases overall economic growth by 0.021%.***

A positive impact of the indirect spillover from international tourist arrivals to the wholesale and retail industry of the Korean economy indicates that the two industries are at the whim of the sustainable economic growth: that is, mechanisms to promote the spillover from international tourist arrivals to the wholesale and retail industry could be successful to the overall economy. By promoting the spillover from international tourist arrivals to the retail industry, the overall economy of the nation could be benefited from the two industries, which can strengthen the pull factors of the economy.

Besides indirect effects from international tourist arrivals to the retail industry, the spillover effect has affected other industries in the supply chains of tourism. For example, international tourist arrivals affect developments of information communications technology, transformations of production, distribution and consumption of goods, and other business services. In addition, international tourist arrivals also create job opportunities for local labor. Therefore, the government should focus on promoting tourism toward goodwill service connections between international tourist arrivals and other industries in the supply chains, because the spillover to improve the productivity in each industry is still substantial. For instance, the nation has benefited greatly from strengthening the relationship between international tourist arrivals and the retail industry, which would become a solid foundation for future sustainable economic growth in Korea.

Considering both direct and indirect effects, the spillover from international tourist arrivals has played an important role in the economic growth of Korea. Thus, policymakers should assess and understand the spillover of tourism toward other industries and overall economy. The government should acknowledge changes in tourism trends and, accordingly, adjust tourism policies, which should allow the flow of supply chains between industries. For example, policymakers should focus on those countries that have a large number of tourists visiting Korea such as Japan, China, and some Southeast Asian countries. In other words, policymakers should encourage international tourist arrivals from aforementioned countries to visit Korea by increasing cultural and business activities across the country, as well as reduce the remained obstacles to visit Korea. For example, the government should provide reasonable and sufficient air links, favorable visa requirements, and convenient language services as well as by promoting the awareness of national brand image, so-called, *Hanryu* and *K-pop*.

In addition, to strengthen the image of the nation as one of the best destinations of tourism, policymakers should concern about the uneven economies in each country, which might affect the tourists' decision either to visit Korea or an alternative. Implementation of such policies must be efficient when providing economic incentives to tourism sectors. The implementation will also give positive spillover effects in transportation, accommodation, and tourism operations, which will improve tourism firms' economic returns to investments. This improves productivity of its own industry and spreads spillover effects while reducing unnecessary expenses, thereby, giving the tourism business the edge they need to stay ahead of the competition. As such the spillover effects will enhance the national economy as a whole.

13.6 Conclusion

The significant direct and indirect spillovers of international tourist arrivals to economic growth, tourism, and the wholesale and retail industry are empirically documented in this study. The results of this study also indicate that the economic spillover from international tourist arrivals to the retail industry can serve as a leading driver for sustainable economic growth of the nation. Economic opportunities, entrepreneurship, and innovation in conjunction with the sustainable economic growth can arise in all sectors: therefore, simple assessments based on high-growth industries may mislead the spillover effect of certain industries.

References

AirPlus Community. (2010). *Travel management pros see real value in social media tools*. Available from: http://www.airpluscommunity.com/2010/06/28/. Accessed 18 Jan 2015.

Balaguer, J., & Cantavella-Jordá, M. (2002). Tourism as a long-run economic growth factor: The Spanish case. *Applied Economics, 34*(7), 877–884.

Bhargava, A. (1986). On the theory of testing for unit roots in observed time series. *Review of Economic Studies, 53*, 369–384.

Cárdenas-García, P. J., Sánchez-Rivero, M., & Pulido-Fernández, J. I. (2015). Does tourism growth influence economic development? *Journal of Travel Research, 54*(2), 206–221.

Castro-Nuño, M., Molina-Toucedo, J. A., & Pablo-Romero, M. P. (2013). Tourism and GDP: A meta-analysis of panel data studies. *Journal of Travel Research, 52*(6), 745–758.

Chen, C. F., & Chiou-Wei, S. Z. (2009). Tourism expansion, tourism uncertainty and economic growth: New evidence from Taiwan and Korea. *Tourism Management, 30*(6), 812–818.

Cheung, Y. W., & Lai, K. S. (1993). Finite-sample sizes of Johansen's likelihood ratio tests for co-integration. *Oxford Bulletin of Economics and Statistics, 55*, 313–328.

Chien, C., Hong, C., & Li, J. (2014). Spillover effects of tourism policy changes on hospitality industries. *Global Journal of Business Research, 8*(2), 111–124.

Chindris-Vasioiu, O., & Tocan, M. (2015). Sustainable development in tourism – Factor of economic growth. *Knowledge Horizons-Economics, 7*(2), 160–164.

Choi, H. C., & Sirakaya, E. (2006). Sustainability indicators for managing community tourism. *Tourism Management, 27*(6), 1274–1289.

Croes, R., & Vanegas, M., Sr. (2008). Cointegration and causality between tourism and poverty reduction. *Journal of Travel Research, 47*(1), 94–103.

Dickey, D. A., & Fuller, W. A. (1979). Distribution of the estimators for autoregressive time series with a unit root. *Journal of the American Statistical Association, 74*(366), 427–431.

Dickey, D. A., & Fuller, W. A. (1981). Likelihood ratio statistics for autoregressive time series with a unit root. *Econometrica, 49*(4), 1057–1071.

Dolnicar, S., Coltman, T., & Sharma, R. (2015). Do satisfied tourists really intend to come back? Three concerns with empirical studies of the link between satisfaction and behavioral intention. *Journal of Travel Research, 54*(2), 152–178.

Dritsakis, N. (2004). Tourism as a long-run economic growth factor: An empirical investigation for Greece using causality analysis. *Tourism Economics, 10*(3), 305–316.

Du, D., Lew, A. A., & Ng, P. T. (2016). Tourism and economic growth. *Journal of Travel Research, 55*(4), 454–464.

Dwyer, L., & Forsyth, P. (2008). Economic measures of tourism yield: What markets to target? *International Journal of Tourism Research, 10*(2), 155–168.

Elliott, G., Rothenberg, T. J., & Stock, J. H. (1996). Efficient tests for an autoregressive unit root. *Econometrica, 64*, 813–836.

Enders, W. (1995). *Applied econometric time series*. New York: Wiley.

Engle, R. F., & Granger, C. W. J. (1987). Co-integration and error correction: Representation, estimation, and testing. *Econometrica, 55*(2), 251–276.

Fayissa, B., Nsiah, C., & Tadasse, B. (2008). Impact of tourism on economic growth and development in Africa. *Tourism Economics, 14*(4), 807–818.

Felsenstein, D., & Fleischer, A. (2003). Local festivals and tourism promotion: The role of public assistance and visitor expenditure. *Journal of Travel Research, 41*(4), 385–392.

Granger, C. W. J. (1988). Some recent developments in a concept of causality. *Journal of Econometrics, 39*, 199–211.

Gunduz, L., & Hatemi-J, A. (2005). Is the tourism-led growth hypothesis valid for Turkey? *Applied Economics Letters, 12*(8), 499–504.

Humphreys, B. R., & Marchand, J. (2013). New casinos and local labor markets: Evidence from Canada. *Labour Economics, 24*, 151–160.

Johansen, S. (1988). Statistical analysis of cointegration vectors. *Journal of Economic Dynamics and Control, 12*(2–3), 231–254.
Johansen, S. (1991). Estimation and hypothesis testing of co-integration vectors in Gaussian vector autoregressive models. *Econometrica, 59*, 1551–1580.
Johansen, S. (1995). *Likelihood-based inference in cointegrated vector autoregressive models*. Oxford: Oxford University Press.
Korea Culture and Tourism Institute. (2014). *Tourism statistics*. Available from: http://stat.tour.go.kr/. Accessed 18 Jan 2015.
Korea Tourism Organization. (2014). *Tourism statistics*. Available from: http://kto.visitkorea.or.kr/eng/tourismStatics/keyFacts/KoreaMonthlyStatistics.ktov. Accessed 18 Jan 2015.
Kwiatkowski, D., Phillips, P., Schmidt, P., & Shin, Y. (1992). Testing the null hypothesis of stationary against the alternative of a unit root: How sure are we that economic time series have a unit root? *Journal of Econometrics, 54*(1–3), 159–178.
Lee, C. G. (2008). Tourism and economic growth: The case of Singapore. *Regional and Sectoral Economic Studies. Euro-American Association of Economic Development, 8*(1), 89–98.
Lee, C. C., & Chang, C. P. (2008). Tourism development and economic growth: A closer look at panels. *Tourism Management, 29*(1), 180–192.
Lim, C., & McAleer, M. (2005). Analyzing the behavioral trends in tourist arrivals from Japan to Australia. *Journal of Travel Research, 43*(4), 414–421.
Ma, T., Hong, T., & Zhang, H. (2015). Tourism spatial spillover effects and urban economic growth. *Journal of Business Research, 68*(1), 74–80.
Massidda, C., & Mattana, P. (2013). A SVECM analysis of the relationship between international tourism arrivals, GDP and trade in Italy. *Journal of Travel Research, 52*(1), 93–105.
Mihalic, T. (2002). Tourism and economic development issues. In *Tourism and development: Concepts and issues* (pp. 81–111). Clevedon: Channel View Publications.
Ng, S., & Perron, P. (2001). Lag length selection and the construction of unit root tests with good size and power. *Econometrica, 69*(6), 1519–1554.
Phillips, P. C. B., & Perron, P. (1988). Testing for a unit root in time series regression. *Biometrica, 75*(2), 335–346.
Pratt, S. (2015). The economic impact of tourism in SIDS. *Annals of Tourism Research, 52*, 148–160.
Sahli, M., & Nowak, J.-J. (2007). Does inbound tourism benefit developing countries? A trade theoretic approach. *Journal of Travel Research, 45*(4), 426–434.
Said, S. E., & Dickey, D. A. (1984). Testing for unit roots in autoregressive moving average models of unknown order. *Biometrika, 71*, 599–607.
Sánchez-Carrera, E. J., Brida, J. G., & Risso, W. A. (2008). Tourism impact on long-run Mexican economic growth. *Economic Bulletin, 23*(21), 1–8.
Schubert, S. F., Brida, J. G., & Risso, W. A. (2011). The impacts of international tourism demand on economic growth of small economies dependent on tourism. *Tourism Management, 32*, 377–385.
Snieska, V., Barkauskiene, K., & Barkauskas, V. (2014). The impact of economic factors on the development of rural tourism: Lithuanian case. *Procedia-Social and Behavioral Sciences, 156*, 280–285.
Vanegas, M. (2014). The triangles of poverty, economic growth, and inequality in Central America: Does tourism matter? *Worldwide Hospitality and Tourism Themes, 6*(3), 277–292.
Wong, I. A., & Wan, Y. K. P. (2013). A systematic approach to scale development in tourist shopping satisfaction: Linking destination attributes and shopping experience. *Journal of Travel Research, 52*(1), 29–41.
World Travel and Tourism Council. (2011). *Travel and tourism economic impact 2011*. London: World Travel and Tourism Council.
Xiang, Z., Wang, D., O'Leary, J. T., & Fesenmaier, D. R. (2015). Adapting to the internet: Trends in travelers' use of the web for trip planning. *Journal of Travel Research, 54*(4), 511–527.

Yang, Y., & Wong, K. K. F. (2012). A spatial econometric approach to model spillover effects in tourism flows. *Journal of Travel Research, 51*(6), 768–778.

Yuan, J. J., & Jang, S. S. (2008). The effects of quality and satisfaction on awareness and behavioral intentions: Exploring the role of a wine festival. *Journal of Travel Research, 46*(3), 279–288.

Zickuhr, K. (2010). *Generations 2010*. Available from: http://www.pewinternet.org/Reports/2010/Generations-2010.aspx. Accessed 18 Jan 2015.

Chapter 14
Beyond Structural Equation Modelling in Tourism Research: Fuzzy Set/Qualitative Comparative Analysis (fs/QCA) and Data Envelopment Analysis (DEA)

Naser Valaei, Sajad Rezaei, Ree C. Ho, and Fevzi Okumus

Abstract This chapter discusses the methods of applying both the data envelopment analysis (DEA) and fuzzy set/qualitative comparative analysis (fs/QCA) in tourism research. Unlike conventional quantitative methods in social sciences, research such as system of regression and multivariate procedures are mostly based on frequency and consistency thresholds. The basis of fuzzy set analysis is the fact that there is no "single correct answer". Indeed, fuzzy sets fill the gap between qualitative and quantitative methods of measurement, and QCA is one of the few methods that cover "limited diversity". This method addresses the shortcoming of most traditional methods which presume that causal conditions are "independent" constructs and the impact on the outcome variable are both additive and linear. In addition, DEA is a nonparametric quantitative data analysis method that is used to examine the relationship between inputs to a production process and the outputs of that process. It acts as a mathematical programming technique to develop and provide the best possible solutions. This chapter shows that application of fuzzy set/QCA (fs/QCA) and DEA method in Asian tourism research would yield a fruitful contribution to the literature.

Keywords Quantitative methods · Fuzzy set/qualitative comparative analysis (fs/QCA) · Data envelopment analysis (DEA) · Tourism research

---

N. Valaei
KEDGE Business School, Talence, France

S. Rezaei
Universität Hamburg, Hamburg, Germany

R. C. Ho (✉)
Taylor's University, Subang Jaya, Selangor, Malaysia

F. Okumus
University of Central Florida, Orlando, FL, USA

S. Rezaei (ed.), *Quantitative Tourism Research in Asia*, Perspectives on Asian Tourism, https://doi.org/10.1007/978-981-13-2463-5_14

14.1 Introduction

QCA is rooted in the comparative case-based reasoning which is heavily used by sociologists and was later transformed into logical methodology by Charles C Ragin et al. (2008). One of the core concepts in QCA is configuration which indicates the holistic view towards interconnected elements (Fiss et al. 2013). Configurational theory has been hampered by the inconformity between its method and theory. As configurational theory is intertwined with the concepts of "synergistic effects, non-linearity, and equifinality", empirical/correlational/traditional/conventional research has interlaced with the concepts of "additive effects, linearity, and unifinality" (Fiss et al. 2013). A certain casual inference has to meet four theories of causality such as "manipulation theory mechanisms and capacities, counterfactual theory, neo-humean regularity theory" (Brady 2002). According to Charles C Ragin (2006), one of the questionable practices in current social science is to reach theoretical conclusions based on correlational hypotheses.

Moreover, data envelopment analysis (DEA) is a nonparametric quantitative data analysis method that is used to examine the relationship between inputs to a production process and the outputs of that process. DEA was used by Charnes et al. (1978) as the measurement tool in evaluating the efficiency and productivity of production unit. In contrast to regression models that produce average measurement for all units involved, DEA acts as mathematical programming technique to develop and provide the best possible solutions (Donthu et al. 2005). Furthermore, the best units can be selected and used as benchmark to assess the performance in sales and marketing activities (Donthu et al. 2005).

Although DEA is known as a useful and rigorous quantitative method in many disciplines (Donthu et al. 2005), a few tourism studies have used it as a methodological choice. One of the main reasons for the paucity of the studies using DEA in tourism studies is due to the absence of recognized systematic procedures and evaluation instruments in the field (Donthu et al. 2005). This chapter is the first attempt that intends to provide researchers with the necessary information and guidance to apply DEA in tourism research by using examples from the Asian context. Therefore, this chapter aims at providing a critical review on QCA, fs/QCA, their fundamental terminologies, and the comparison between conventional/traditional methods (e.g. regression analysis, correlational analysis, and SEM) and fs/QCA. This chapter indicates that applying fs/QCA would improve our understanding of tourist destination image compared with hypothetical results and offers methodological bases to researchers in conducting fs/QCA and DEA in a tourism context.

14.1.1 QCA and fs/QCA

To substantiate an argument and stressing a mixture of causal conditions, researchers need to juxtapose cases that are thoroughly related with each other, and the optimal juxtaposition is between combinations of cases that are dissimilar to merely one causal condition (Ragin and Sonnett 2005; Woodside 2014). QCA was introduced by Ragin (1987, 2003) to analyse the arrangements of crisp set memberships, namely, conventional Boolean sets. QCA was initially used to serve as the midpoint in segregating "case-oriented" and "variable-oriented" methods. Its simulated structure combines the principal attributes in the "case-oriented" qualitative method and principal attributes found in the "variable-oriented" quantitative method (Fiss et al. 2013). The configurative nature of QCA structure is suitable in explicating the causes and effects of the interrelationships among the constructs in social science studies (Ragin and Sonnett 2005). The logic of the comparative "case-oriented" methodology provides the accuracy needed in explaining the precision of the intersectional relationship and conditions required. This type of precision is missing in several quantitative methods that similar cases would attenuate degrees of freedom and statistical power (Ragin and Sonnett 2005; Woodside 2013).

Studies on configurational theory have debatably been hampered by a lack of congruence between investigation methods and theoretical backgrounds. Although theoretical examinations of configurational theory as a consequence indicate equifinality, nonlinearity, and simulated impacts, empirical research still applies econometric methods to a great extent that by their essence are inclined to entail unifinality, additive results, and linearity (Fiss et al. 2013). In addition, regularity theory distinguishes a required state to mark out causation. Due to the fact that interconnection is not causation. No justification exists on why entirely logical limitations on lawlike expressions should be enough to describe the distinctive nature of causal relationships (Brady 2002; Valaei et al. 2017). In social science research, several theories are imprecise at which they try to identify how causal conditions merge to provide results. This causes researchers to make a number of possible causal constructs from a list of theoretical prospects related to the doubtful result (Ragin and Sonnett 2005). Having an easy situation to apply set-theoretic logic to circumstances that differ in degree or level is an advantage of fuzzy sets (Ragin 2006).

Fuzzy sets have several latent applications in the social sciences that need to be discovered. These new fuzzy set approaches are established based on arguments shown in recent studies on fuzzy logic method of investigation. They additionally create a rigorous association to the crisp set fundamental features shown in "the comparative method", particularly those in connection with the problem of restricted variety (Ragin 2003; Valaei et al. 2017).

The inference from fuzzy set results is powerful, rigorous, and effective to a great extent because they provide scholars to carefully adjust the fractional in the set members by applying the numerical value of "0" (known as non-membership) or "1" (known as full membership), respectively (Ragin 2007a). The union of these two produces fs/QCA, a group of approaches which provides social scientists with an unconventional method as a substitute to the traditional quantitative methods that are mainly formed on correlational reasoning (Ragin 2007b).

14.1.2 SEM vs. fs/QCA

QCA neither make causal conclusions from sample nor does it derive population attributes from a sample. The purpose of QCA is to support the causal inference in tandem with the case-based knowledge (Ragin 2007a). Unlike correlational method, QCA is designed as a set of **theoretic** approach (Ragin 2007b). Researchers begun suspecting the conventional quantitative regression-based methods, as they work best with large sample size (Ragin and Sonnett 2005). In addition to support or reject a causal relationship, they draw their theoretical conclusions mainly by large sample size (Fearon 1991). Furthermore, the traditional methods can also be argued in terms of testing the significance of a relationship based on cut-off measures of P-value (P-value <0.05, with 95% of probability, and P-value <0.01, with 99% of probability).

Unlike regression-based methods, QCA has a complex viewpoint rather than simple, and it is one of the few methods that cover "limited diversity". Researchers in traditional quantitative approach use models and techniques to base their reasoning in terms of essence of causation. They also claim that the nature of causation, also called "independent variables", has additive and linear effect that works best to explain the phenomenon (outcome condition) (Ragin and Sonnett 2005; Valaei et al. 2017). Table 14.1 shows the terminologies and differences in regression-based methods (symmetric modelling) and fs/QCA (configurational or asymmetric modelling).

An advantage of QCA method is its capability to enable tools in determining the intermediate solutions in addition to its capability to enable tools for inferring the low to high membership continuum (Ragin and Sonnett 2005). Opposed to regression-based methods, another merit of QCA is its ability to work with small number of cases (even between 10 and 20), and providing other approaches in examining causal relationships (Fearon 1991). In addition, opposed to regression-based and multivariate methods, QCA does not draw conclusion based on a "single correct answer" (e.g. a positive/negative relationship between two variables). Rather, it provides different decisions (solutions) based on consistency and frequency thresholds (Ragin 2007a).

Uncertainty exists in most of social science theories, as they fail to determine how causal conditions are incorporated to generate outcomes. In this stream, researchers extract variables from several theories related to the phenomenon they are investigating to develop their theoretical model (Ragin and Sonnett 2005). Another merit of QCA over multivariate method is the configurational structure of QCA that allows "counterfactuals" (positive or negative hypothetical matched cases of different variables) to be found for a proposed relationship (Ragin and Sonnett 2005).

Table 14.1 General terms for calibration and counterfactual analysis in fs/QCA

Symbol	=	Source
Upper-case letter	"The presence of a condition"	Ragin and Sonnett (2005)
Lower-case letter	"The absence of a condition"	Ragin and Sonnett (2005)
Multiplication (·)	"Combined conditions (logical *and*)"	Ragin and Sonnett (2005)
Addition (+)	"Alternate combinations of conditions (logical *or*)"	Ragin and Sonnett (2005)
"→"	"Sufficient for"	Ragin and Sonnett (2005
For four causal conditions	"Number of combinations = 2 k, where k is the number of causal conditions"	Ragin and Sonnett (2005)
Crisp set analysis	"Consistency scores = 1.0 = > perfect consistency"	Ragin (2006)
Fuzzy set analysis	"Scores close to 0.0 indicate that a case is mostly out (membership = 0.0)"	Ragin (2006)
	"Scores close to 0.2 indicate that a case is mostly but not fully out of the target set (membership = 0.2)"	
	"Scores close to 0.0 indicate that a case is more out than in the target set (membership = 0.4)"	
	"Scores greater than 0.5 = a case is more in than out (membership = 0.5)"	
	"Scores close to 0.6 indicate that a case is more in than out of the target set (membership = 0.6)"	
	"Scores close to 0.8 indicate that a case is mostly but not fully in the target set (membership = 0.8)"	
	"Scores close to 1.0 = a case is mostly in (membership = 1.0)"	
	"Coverage >1 or 2 (for small sample size, the coverage should be 1, and for large sample size, the coverage should be 2)"	
	Consistency >0.8 or 0.85 (for small sample size, the threshold for the consistency is 0.8, and for large sample size, the threshold for the consistency is 0.85)"	
	"Full membership = 1.0"	Ragin (2003, 2006)
	"Full non-membership = 0.0"	
	"If membership in X is consistently less than or equal to membership in Y = all the cases will fall on or above the main diagonal of the plot, yielding a consistency score of 1.0 (or 100% consistent)"	
	"If membership in X is consistently larger than 80% or falls above the main diagonal of the plot, then it 'almost always' yields sufficiency in Y"	

14.1.3 Truth Table

QCA follows the idea of complicated combinations and an identical outcome may be resulted from different positive and negative combinations of conditions shown in "truth tables" (Valaei et al. 2017). Truth tables include the reasonably feasible amalgamations of causal recipes/solutions (analogous to net effect in symmetric modelling) and the outcome (analogous to dependent variable in symmetric modelling) related to each amalgamation (Ragin and Sonnett 2005). When the truth table (analogous to correlation matrix in symmetric modelling) is established, "standard analysis" method is constructed for the truth table. Standard analyses spontaneously enable the parsimonious, complex, and intermediate solutions. According to Ragin et al. (2008), standard analysis is the suitable method because it provides an exclusive way to obtain the intermediate solution.

In fs/QCA, the attainment of causal recipes (solutions) is the main purpose and can be achieved by culminating the membership value close to "1". The solution/complex recipe of the truth table gives a line for each path (causal relationship) to the outcomes. As mentioned previously, one of the merits of QCA method is its capability to enable tools in determining the intermediate solutions in addition to its capability to enable tools for inferring the endpoints of the complexity continuum (Charles C Ragin et al. 2008). In choosing the causal recipes, researchers need to consider those solutions that have the highest consistency and coverage.

14.1.4 Crisp Set Analysis and Calibration to Fuzzy Set

Fuzzy set theory was originally proposed by Zadeh (1965). In traditional methods, measures are questions that shape research constructs and are parts of theoretical frameworks that are deduced from theories. According to C. C. Ragin (2007b), quantitative method of measurement is strongly theory-centric. Further, in qualitative method of measurement, the meaning is appended to the particular measurements. However, fuzzy sets fill the gap between qualitative and quantitative methods of measurement. C.C. Ragin (1987) integrated the QCA with the fuzzy set. In order to use the crisp values, they need to be calibrated. Calibration is the transformation of the crisp values (e.g. 7-point Likert scale) to fuzzy forms (Ragin 2007b). Crisp set has binary values, and fuzzy set uses an interval of "0.0 to 1.0" for set membership (Ragin 2006).

Fuzzy sets expand crisp values by allowing membership scores that can vary based on the crisp values. For instance, the use of Likert scale of "5 point" for measurement, i.e. "5" for "strongly agree" (crisp values between 1 and 5), "4" for "agree", "3" for "neutral", "2" for "disagree", and "1" for "strongly disagree". These values are calibrated into fuzzy sets ranging from "strongly agree" (full membership) for value "1", "agree" for value "0.75", "neutral" (crossover point) for value "0.50", "disagree" for value "0.25", and "strongly disagree" (full non-membership) for value "0". The interpretation of the fuzzy sets for this calibration is as follows: "1" indicates that the strongly satisfied respondents are fully member of the satisfied respondents. "0" indicates that the strongly dissatisfied respondents are fully member of dissatisfied respondents or fully non-member of satisfied respondents. "0.75" indicates that respondents are satisfied with 75 percent and dissatisfied with 25%. According to C. Ragin (2007a), values of "0.2" and "0.3" show that objects are weak members of the set. On the other hand, values of "0.8" and "0.9" show that objects are strong members of the set but not full membership. More detailed symbols and definitions on calibration from crisp values to fs/QCA are provided in Table 14.1.

14.1.5 Data Envelopment Analysis (DEA)

DEA was initially used as the assessment instrument to evaluate and compare the efficiency of the decision-making units by Charnes et al. (1978). Donthu et al. (2005) refer to DEA as a rigorous quantitative approach. In addition, it can identify the most effective unit and subsequently employ it as the benchmark to assess the performance in sales and marketing activities (Donthu et al. 2005). The productivity achieved is often the key focus for measuring the marketing operations (Donthu et al. 2005). The primary reasons for the paucity of the studies using DEA in marketing are due to the absence of recognized systematic procedures and evaluation instruments made available for marketers (Donthu et al. 2005).

14.1.6 Relative Efficiency

The required productivity can be obtained by evaluating both the input and output data with DEA method (Donthu et al. 2005). Relative efficiency is another term used to describe the ratio obtained comparing the inputs and outputs involved. A decision-making unit is considered "efficient" when a ratio of 1 is attained. If the ratio attained scored below 1, then it is regarded as "less efficient". The aim is to obtain the highest ratio and subsequently match it with other ratio computed for other efficient units. Hence, the attained productivity is often regarded as relative efficiency (Donthu et al. 2005).

14.1.7 Comparisons Between DEA and Regression Models

Although regression analysis is the more popular and rigorous quantitative method, it is plagued by its inability to deal with a list of input and output due to its reliance on the dependent construct (Donthu et al. 2005). It is only served as estimation

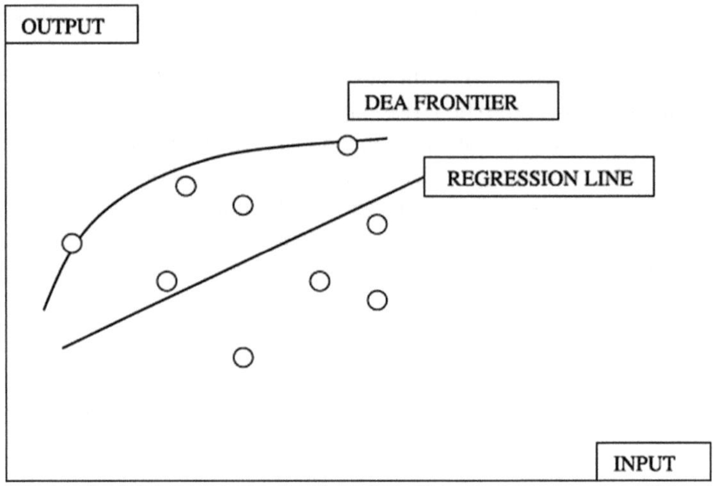

Fig. 14.1 DEA vs. regression (Donthu et al. 2005)

model and provides no avenue for feedback needed for further improvement. Furthermore, it has to depend upon the data restriction that relates directly to the list of hypotheses. In contrast, a list of efficient units can be obtained and compared to the most efficient unit for the future improvement. Hence, regression analysis provides the average estimation for the decision-making unit, but DEA can be better used to evaluate the most efficient units with a list of other possibilities. The most efficient decision-making unit could be obtained on the efficient frontier, and it may not necessarily show graphically above the regression line. In addition, DEA can detect the selected facets of the efficiency frontier as shown in Fig. 14.1.

14.1.8 DEA: Advantages and Disadvantages

Major benefit of DEA lies with its capability to include an array of independent variables as well as dependent variables. In addition, it also can be used to measure the outcomes in accordance with their magnitudes (Berg 2010). There is no requirement for DEA to state the mathematic formula involved in deriving the functions. Hence, this is advantageous for researchers to unravel theoretical relationships which are hard to achieve by regression modelling (Donthu et al. 2005). Delving further, it is also ready to manage several variables in dealing with the outcome measurement. This leads to better analysis in determining the cause of the inefficiency of the intended studies.

One of the major disadvantages of DEA is its model specifications. It has to include or exclude the variable and subsequently leads to the outcomes deviated from the actual results (Berg 2010). This is due to the sensitivity on the decision in choosing the needed variables. Therefore, it is hard to identify the finest specifications. This often resulted in the expansion of endogenous and exogenous variables.

Furthermore, Donthu et al. (2005) listed three main weaknesses of the DEA technique as follow:

1. The impact of omitting important variables
2. The impact of outliers
3. The impact of missing observations

14.1.9 Application in Tourism and Hospitality

With the development of tourism and hospitality industry worldwide (Bihamta et al. 2017; Mohseni et al. 2018; Rezaei et al. 2018), the wide applications of DEA in other industries manifested its practicality and relevancy. Given the fact that efficient frontier was rather popular in other industries, it was only until recently that DEA has been considered by tourism scholars. A number of tourism studies have employed it to measure and evaluate the performance of the industry such as Benito et al. (2014), Köksal and Aksu (2007), and Rotoli et al. (2015). DEA was proven to be useful in managing several variables in achieving efficiency as well as the ability to identify the benchmark needed for comparison.

In general, DEA was used to measure the tourism efficiency. Tourism efficiency analyses have been conducted in subcategories of tourism industry like hotel industry, travel agencies, and restaurants in the extant literature (Brockett and Golany 1996; Dabestani et al. 2016; Ramírez-Hurtado and Contreras 2017). Likewise, economic efficiency assessment also has been considered in tourism industry. DEA evaluates tourism efficiency at the international level utilizing fundamental description of the indicators such as total labour, expenditures, and arrival and departure time.

Hwang and Chang (2003) investigated the variance in efficiency due to the changes of the customer management process in hotel industry with DEA technique. In their study, they managed to evaluate the performance of the hotel managers and the variance in efficiency. Economic efficiency of the hotels in Portugal was investigated with the use of DEA technique. The result of the efficiency was divided into technical efficiency and allocative efficiency based on prices of both the inputs and outputs collected. The findings were useful in providing insights to further improve the hotels in terms of the economic as well as management perspective.

Brown and Ragsdale (2002) focused on what marketability of hotels' brand equity could be developed further for hotel industry. They applied DEA method to examine the competitiveness of market efficiency against the rival companies, based on the customer value and satisfaction level of customer. The findings indicated that brand names were not influential as expected in determining the desired customer satisfaction and customer value. In their study, brand equity was compared to the nearest competitor. Specifically, the highly inefficient brand as determined by the customers contributed to the negative consequences. More complaints, negative feedback, and poor staff evaluation were just some of the indicators in lower rating of the hotels. Hence, DEA method is a valuable tool which can be applied to further improve the market efficiency of the hotel brands.

14.1.10 Application in Travel Agency and Restaurant

Ramírez-Hurtado and Contreras (2017) applied DEA method to evaluate the franchise efficiency under the situation of scarce resource management. The main objective was to manage these resources to avoid unnecessary wastage. Their sample was the franchise travel agencies in Spain. Their study found that the half of the travel agencies were efficient. Hence, it showed that a great number of franchisees were considered as inefficient. Hence, it was concluded that Tobit model can be applied to improve the inefficiency found in the franchise system.

A study on developing the operation efficiency strategy for airports was conducted by Liu (2016). It aimed to measure the total efficiency level as well as the operational level of airport-related services. The study investigated ten selected airport firms in East Asia by using network data envelopment analysis (NDEA) method. Furthermore, it also applied panel data model to find out the major drivers for the efficiency at subprocess levels. There was only one single airport that attained the overall and subprocess at the stage 1 of the study. In general, the other nine airports did not achieve the required efficiencies. Stage 2 of the study was conducted with the use of regression model. The results showed that the efficient aeronautical service was highly depended on the number of airlines served and the number of destinations. It implied that airline service efficiency can be achieved with service quality and unrelated to direct airline revenue. Similarly, Köksal and Aksu (2007) also examined the efficiency of travel agency operations in Turkey with DEA method. The finding found no interrelationship between operating efficiency and groups of travel agencies.

DEA was used by Reynolds and Thompson (2007) to examine the restaurant efficiency level. The study centres on the cause-effect relationship between productivity of the restaurant and a list of uncontrollable variables. A total of 62 restaurants were selected. The uncontrollable variables identified were hourly wages, seat number, and whether they are stand-alone type of restaurant. The study further validated the appropriateness of DEA to measure the efficiency of the restaurants.

Therefore, researchers in tourism studies should move towards more complex methods of analysis to reap more knowledge from data. Even though there is a controversy on the use of qualitative vs. quantitative approaches, QCA method bridges the gap between these schools of thought. However, it is recommended that tourism researchers start to apply QCA, fs/QCA, and necessary condition analysis rather than conventional approaches, i.e. regression analysis, SEM, and correlational analysis. It is necessary to consider contrary conditions in tourism research. For instance, chances are that there is a disparate segment such as luxury tourism market, which is different from the whole population. Drawing conclusions by excluding the viewpoints of this segment will result in wrong decision-making. In addition, considering the multi-cultural aspect of Asian countries, researchers should provide complex casual recipes for their outcome variables. This is due to the high degree of complexity in Asian markets, namely, cultural, technological, economical, and political factors that influence the outcome condition (e.g. tourist loyalty).

14.2 Conclusion

This chapter mainly discusses how we can apply fs/QCA and data envelopment analysis (DEA) methods in tourism research. Tourism researchers have started to apply advanced methods in their studies such as SEM (Rezaei et al. 2016, 2017; Rezaei and Valaei 2017a, b; Rich 2013; Shahijan et al. 2015). It is likely that a hypothesis gets rejected in traditional methods but it is supported or conditionally supported in QCA and fs/QCA methods (Valaei et al. 2017). Conventional quantitative methods in social sciences, research such as system of regression, and multivariate procedure are based on frequency and consistency thresholds. However, the basis of fuzzy set analysis depends on the notion of "no single correct answer" to derive the conclusion from data analysis reasonably, and the disparate results are traced by disparate decisions (Ragin 2007a, b).

Indeed, fuzzy sets fill the gap between qualitative and quantitative methods of measurement (Ragin 2007a, b). QCA is one of the few methods that cover "limited diversity" (Ragin and Sonnett 2005). This method addresses the shortcoming of most traditional methods which presume that causal conditions are "independent" constructs whose impacts on the outcome variable are both additive and linear (Ragin and Sonnett 2005). This chapter shows that application of fuzzy set/QCA (fs/QCA) method in Asian tourism research would yield a fruitful contribution to the literature. This chapter further indicates that applying fs/QCA would improve our understanding of tourist destination image compared with hypothetical results. This chapter offers methodological bases to researchers in conducting fs/QCA and DEA in a tourism context.

It is necessary for researchers to draw accurate conclusions based on data analysis, and mistakenly rejecting a hypothesis will result in wrong decision-making at all levels. Even though there are advancements in conventional methods both in factor-based models (examples for software performing covariance-based SEM are Amos, MPlus, EQS, and Lisrel) and composite-based models (examples for software performing variance-based SEM are SmartPLS, WarpPLS, and PLS-GUI), tourism researchers need to apply complex methods due to the nature of complex latent concepts. This chapter suggested that tourism researchers should consider complex methods (necessary condition analysis, QCA, and fs/QCA) to provide complex causal recipes for managers and decision-makers. One of the implications of QCA method is INUS conditions (Braumoeller 2017). INUS conditions are neither sufficient nor necessary conditions which are distinction from other sufficient and necessary conditions (Horsten and Weber 2014). For instance, tourist satisfaction can be a combination of several INUS conditions (causal recipes with logical "and", logical "or", and logical "not"), which provide researchers with more complex perspective towards the outcome condition (tourist satisfaction).

References

Benito, B., Solana, J., & López, P. (2014). Determinants of Spanish regions' tourism performance: A two-stage, double-bootstrap data envelopment analysis. *Tourism Economics, 20*(5), 987–1012.

Berg, S. (2010). *Water utility benchmarking: Measurement. Methodology and performance incentives*. London: IWA Publishing.

Bihamta, H., Jayashree, S., Rezaei, S., Okumus, F., & Rahimi, R. (2017). Dual pillars of hotel restaurant food quality satisfaction and brand loyalty. *British Food Journal, 119*(12), 2597–2609. https://doi.org/10.1108/BFJ-07-2016-0344.

Brady, H. E. (2002). *Models of causal inference: Going beyond the Neyman-Rubin-Holland theory*. Paper presented at the Annual Meetings of the Political Methodology Group.

Braumoeller, B. F. (2017). Causal complexity and the study of politics. *Political Analysis, 11*(3), 209–233. https://doi.org/10.1093/pan/mpg012.

Brockett, P. L., & Golany, B. (1996). Using rank statistics for determining programmatic efficiency differences in data envelopment analysis. *Management Science, 42*(3), 466–472.

Brown, J. R., & Ragsdale, C. T. (2002). The competitive market efficiency of hotel brands: An application of data envelopment analysis. *Journal of Hospitality & Tourism Research, 26*(4), 332–360.

Charnes, A., Cooper, W. W., & Rhodes, E. (1978). Measuring the efficiency of decision making units. *European Journal of Operational Research, 2*(6), 429–444.

Dabestani, R., Shahin, A., Saljoughian, M., & Shirouyehzad, H. (2016). Importance-performance analysis of service quality dimensions for the customer groups segmented by DEA: The case of four star hotels. *International Journal of Quality & Reliability Management, 33*(2), 160–177.

Donthu, N., Hershberger, E. K., & Osmonbekov, T. (2005). Benchmarking marketing productivity using data envelopment analysis. *Journal of Business Research, 58*(11), 1474–1482.

Fearon, J. D. (1991). Counterfactuals and hypothesis testing in political science. *World Politics, 43*(02), 169–195.

Fiss, P. C., Marx, A., & Cambré, B. (2013). Configurational theory and methods in organizational research: Introduction. *Research in the Sociology of Organizations, 38*, 1–22.

Horsten, L., & Weber, E. (2014). INUS conditions. In *Wiley StatsRef: Statistics reference online*. Hoboken: Wiley.

Hwang, S.-N., & Chang, T.-Y. (2003). Using data envelopment analysis to measure hotel managerial efficiency change in Taiwan. *Tourism Management, 24*(4), 357–369.

Köksal, C. D., & Aksu, A. A. (2007). Efficiency evaluation of A-group travel agencies with data envelopment analysis (DEA): A case study in the Antalya region, Turkey. *Tourism Management, 28*(3), 830–834.

Liu, D. (2016). Measuring aeronautical service efficiency and commercial service efficiency of East Asia airport companies: An application of network data envelopment analysis. *Journal of Air Transport Management, 52*, 11–22.

Mohseni, S., Jayashree, S., Rezaei, S., Kasim, A., & Okumus, F. (2018). Attracting tourists to travel companies' websites: The structural relationship between website brand, personal value, shopping experience, perceived risk and purchase intention. *Current Issues in Tourism, 21*(6), 616–645. https://doi.org/10.1080/13683500.2016.1200539.

Ragin, C. C. (1987). *The comparative method: Moving beyond qualitative and quantitative strategies*. Oakland: University of California Press.

Ragin, C. C. (2003). *Recent advances in fuzzy-set methods and their application to policy questions*. http://www.compasss.org/Ragin2003.pdf

Ragin, C. C. (2006). Set relations in social research: Evaluating their consistency and coverage. *Political Analysis, 14*(3), 291–310.

Ragin, C. (2007a). Chapter 5: Qualitative comparative analysis using fuzzy sets (fsQCA). Forthcoming in N. Rihoux, & C. Ragin (Eds.), *Configurational comparative analysis*, London: Sage Publications.

Ragin, C. C. (2007b). Fuzzy sets: Calibration versus measurement. In D. Collier, H. Brady, & J. M. Box-Steffensmeier (Eds.), *Methodology: Oxford handbooks of political science*. Oxford: Oxford University Press.

Ragin, C. C., & Sonnett, J. (2005). Between complexity and parsimony: Limited diversity, counterfactual cases, and comparative analysis. In *Vergleichen in der Politikwissenschaft* (pp. 180–197). Heidelberg: Springer.

Ragin, C. C., Rubinson, C., Schaefer, D., Anderson, S., Williams, E., & Giesel, H. (2008). *User's guide to fuzzy-set/qualitative comparative analysis*. University of Arizona, p. 87.

Ramírez-Hurtado, J. M., & Contreras, I. (2017). Efficiency of travel agency franchises: A study in Spain. *Service Business, 11*(4), 717–739.

Reynolds, D., & Thompson, G. M. (2007). Multiunit restaurant productivity assessment using three-phase data envelopment analysis. *International Journal of Hospitality Management, 26*(1), 20–32.

Rezaei, S., & Valaei, N. (2017a). Branding in a multichannel retail environment: Online stores vs app stores and the effect of product type. *Information Technology and People, 30*(4), 853–886. https://doi.org/10.1108/ITP-12-2015-0308.

Rezaei, S., & Valaei, N. (2017b). Crafting experiential value via smartphone apps channel. *Marketing Intelligence and Planning, 35*(5), 688–702. https://doi.org/10.1108/MIP-08-2016-0141.

Rezaei, S., Ali, F., Amin, M., & Jayashree, S. (2016). Online impulse buying of tourism products: The role of web site personality, utilitarian and hedonic web browsing. *Journal of Hospitality and Tourism Technology, 7*(1), 60–83. https://doi.org/10.1108/JHTT-03-2015-0018.

Rezaei, S., Mazaheri, E., & Azadavar, R. (2017). Determinants of experienced tourists' satisfaction and actual spending behavior: A PLS path modelling approach. *International Journal of Culture, Tourism, and Hospitality Research, 11*(2), 157–181. https://doi.org/10.1108/IJCTHR-09-2015-0107.

Rezaei, S., Shahijan, M. K., Valaei, N., Rahimi, R., & Ismail, W. K. W. (2018). Experienced international business traveller's behaviour in Iran: A partial least squares path modelling analysis. *Tourism and Hospitality Research, 18*(2), 163–190. https://doi.org/10.1177/1467358416636930.

Rich, S. R. (2013). Handbook of research methods in tourism: Quantitative and qualitative approaches. *Journal of Tourism and Cultural Change, 11*(1–2), 135–136. https://doi.org/10.1080/14766825.2013.764112.

Rotoli, F., Cawood, E. N., & Christidis, P. (2015). A data envelopment analysis approach for accessibility measures: Simulating operational enhancement scenarios for railway across Europe. *European Transport Research Review, 7*(2), 18.

Shahijan, M. K., Rezaei, S., Preece, C. N., & Ismail, W. K. W. (2015). International medical travelers' behavioral intention: An empirical study in Iran. *Journal of Travel and Tourism Marketing, 32*(5), 475–502. https://doi.org/10.1080/10548408.2014.916248.

Valaei, N., Rezaei, S., & Ismail, W. K. W. (2017). Examining learning strategies, creativity, and innovation at SMEs using fuzzy set qualitative comparative analysis and PLS path modeling. *Journal of Business Research, 70*, 224–233.

Woodside, A. G. (2013). Moving beyond multiple regression analysis to algorithms: Calling for adoption of a paradigm shift from symmetric to asymmetric thinking in data analysis and crafting theory. *Journal of Business Research, 66*(4), 463–472.

Woodside, A. G. (2014). Embrace• perform• model: Complexity theory, contrarian case analysis, and multiple realities. *Journal of Business Research, 67*(12), 2495–2503.

Zadeh, L. A. (1965). Fuzzy sets. *Information and Control, 8*(3), 338–353.

Chapter 15
Ethical Considerations in Quantitative Tourism and Hospitality Researches

Maryam Emami, Sajad Rezaei, Behrooz Sangani, and See-Kwong Goh

Abstract Research ethics is an extremely important part of every study. Tourism scholars receive guidance on ethical issues in research from different sources throughout their careers such as experiences from participation in research activities, quantitative data analysis or research methods methodology courses, short trainings, workshops, professional readings, and involving in peer-review processes. However, in a structured quantitative context, such as tourism and hospitality, the acquaintances and relationships between the research philosophy, design, and analysis choices related to research ethics are largely implicit and informal. That is, methodology scholars and research advisors might convey ethical principles and conducts about research design and data analysis, by reporting experientially and by modeling appropriate professional behavior. Thus, this chapter aims to provide a practical review of ethical issues in quantitative methods in tourism and hospitality researches. More specifically, this chapter focuses on several potential issues that might emerge in conducting research methods in tourism context. Data collection method and questionnaire design in this regard are considered as practical examples. Moreover, based on the principles of ethical issues in behavioral research, this chapter proposes and offers substantial guides in anticipating ethics in tourism and hospitality research.

Keywords Quantitative research · Business ethic · Tourism researchers

M. Emami (✉)
Enterprise Change, Infosys Consulting, Houston, TX, USA

S. Rezaei
Universität Hamburg, Hamburg, Germany
e-mail: sajad.rezaei@uni-hamburg.de

B. Sangani
Department of Management, Payame Noor University, Tehran, Iran

S.-K. Goh
Taylor's Business School, Taylor's University, Subang Jaya, Malaysia
e-mail: SeeKwong.Goh@taylors.edu.my

S. Rezaei (ed.), *Quantitative Tourism Research in Asia*, Perspectives on Asian Tourism, https://doi.org/10.1007/978-981-13-2463-5_15

15.1 Introduction

The hospitality and tourism industry is one of the greatest global industries and represents a field of study in which many issues have not been uncovered (Ariffin 2013; Hobson and Teaff 1994). It is one of the world's greatest industries in terms of employment and is considered to be an important contributor to the economic development worldwide (Teng 2011) and a major influence on social, environmental, and financial results (Chung and Parker 2010; Wilkins et al. 2009). In fact, "the hospitality sector that provides lodging, meals and related services for travelers" (Chris Paxson 2009, p. 89) is classified as one of the services (Ariffin and Maghzi 2012; Chung and Parker 2010). Although research in the hospitality context has grown dramatically, there is still a lack of consensus concerning how to perfectly illustrate hospitality domain. This is also prominent in understanding ethical considerations in quantitative research.

Ethics has a prominent role in every aspect of human life, including the business world. Principles for ethical considerations in the business world have been discussed in many researches. Talking about this subject, ethics come into the picture when researchers involve human subjects in their work. Ethics' definition among a society or profession members by Douglas C. Frechtling (2017) includes the principles, rules, and standards, which dictate the conduct that should be right, good, and authentic. Code of ethics is described as a set of principles or standards that are systematized. They believe that an appropriate ethical behavior in a professional setting is characterized by code of ethics. "Professional codes of ethics" as the term conveys, refer to these rules and principles which act as a method of which associations, business firms, nonprofit organizations, and governments prefer. They need to be related to the organizational values and be meaningful to members of the business practices (Douglas C. Frechtling 2017).

Ethics in tourism research have been quite an interesting topic for many researchers. Following the codes of ethics in tourism research is a crucial factor, but as it is observed by Douglas C. Frechtling (2017), many fail to do so. They incorporated probability sample survey method in parallel with ethical reporting rules based on code of ethics, that is obtained from the World Association for Public Opinion Research (WAPOR) in their latest work on tourism research articles. They found out that most articles did not meet the criteria of these rules, while they failed to fulfill the value of transparency in the research, that acts as a foundation for analysis within the realm of social science studies. There are various parts of ethics in tourism studies that need further investigations.

In addition, ethics are required to be taken seriously in the Internet research. There is a need for a process approach when it comes to ethics. Ethical issues should be addressed and be resolved at any given stage of a research or a project. Instead of simply having a set of ethical practices, there should be a spectrum of characteristics within ethical questions. These questions should be deliberately considered by both researchers and observers of ethics to ensure the quality of ethical issues in the realm of research in such ongoing, revolutionized world (Markham and Buchanan 2015). Indeed, ethical principles and materials have been embedded mostly in qualitative topics, and most of the ethical topics in quantitative data analysis on specific areas of concern to tourism researchers (e.g., measurement, assumptions, sampling, design, model specification, reliability and validity assessment, and model fitting evaluation) are vague and disseminated, while they lack an ethical imperative. Consequently, this chapter aims to provide a practical review of ethical issues in quantitative methods in tourism and hospitality researches and focuses on several potential issues that might emerge in conducting quantitative research methods in tourism context. Moreover, based on the principles of ethical issues in business research, this chapter proposes and offers guidelines in anticipating ethical problems once they emerged.

15.2 Business Ethic

Business ethics has gained quite a reputation in the research world for more than two decades now. Looking at the ethics, previous researchers (Behi and Nolan 1995) believe that confidentiality and/or anonymity should be guaranteed in the subjects, while consent is being explicitly explained. Another important factor is to maintain dignity and to provide a balance for both individuals and society to gain benefits from ethics and to be immune to harm. Business ethics have also touched various arenas. Practicing the business ethics is considered a crucial part of conducting a business, and having adequate research and information on it will help the practitioners in this area. However, a review of tourism and hospitality domain shows that studies were mostly conducted using qualitative methods. Since effective codes of ethics are the concern of many researchers (MacCannell 2012), there is a gap in tourism research which needs to be addressed in this issue. It is of high importance to define ethical codes in tourism, that are clear and to evaluate those ethics to see how effective they are. The concern of lack of research in the tourism industry was voiced by researchers like Moscardo (2010), stating that there is a need for critical approaches that are reflective of awareness in this realm, since there are many difficulties and challenging issues in tourism ethical questions. Lack of empirical studies, in the hospitality industry, and observing ethics in it, have been stated by many scholars. Another problem is a very limited source of theoretical background which investigated ethics in tourism and hospitality.

Ethics in business research carry a significant load as well. It is a complicated concept that comes with a lot of challenges. One of the biggest difficulties to conduct a science of human behavior is the matter of ethics. The hardest part is not to select sample size or to make the right measurement that is the perfect fit, but is doing the right thing that is ethical. As a result of being ethical, one can live with the consequences that their actions bring (Bernard 2017, p. 25). As an instance, the

Internet is one of the places which is desperately trying to practice business ethics. Issues like privacy concerns, informed consent of the subjects, protecting confidentiality, and public and private data definition are all within the realm of ethics. The main issue that has been the focus of ethic researchers is that they face hardship in defining the spaces between public and private domains (Eysenbach and Till 2001). The vitality of ethics in the business research is blatant, and it will pave the way towards establishing a more ethical business practices. One of the factors that need to be taken into consideration in research ethics is research methodology. Ryan (2005) states that categorizing a research work as good might refer to the fact that is has a database, which is rich with large samples of thoughtful and articulate respondents. On the other hand, Freese (2007) believes that other researchers are not allowed to replicate the results of a survey, unless the research methodology and its details are already published. He also notes that the techniques of the research can be duplicated for additional population only if the work has been previously published in details. As a result, transparency in research ethics plays an essential role from publication point of view.

15.3 Types of Ethical Consideration

Regarding ethics and its impacts (Panter and Sterba 2011, p. 19), three types are addressed by Frankel (1989) as educational, aspirational, and regulatory. Firstly, tourism researchers should consider educational type. In specific occupations, disciplines, or any other companies and institutions, people face various issues; thus, it is the job of educational ethics to address such issues. Educational issues outline certain different behaviors, which are favored by individuals. If there are any variances from these ethical issues that are significant, the guidelines will be suspected, from ethical perspective. Regarding quantitative professions, ASA's Ethical Guidelines for Statistical Practice (1999) and the Association for Computing Machinery's Code of Ethics and Professional Conduct (1992), are the two instances of documents on ethics. The notion is not a set of comprehensive rules, which provide people with instructions on how to behave or to react to a situation. In fact, this acts as an inspiration for people to be more competent, honest, collegial, respectful, loyal, congenial, and trustworthy. The assumption that embodies this type is the person has already learned how to be ethical. Another type of ethics is regulatory, that should be considered by researchers. They are more on the legal side, and they use verbs like "shall" to convey ethical obligations. Evidence and proof within a structural standard in a specific period of time should be presented; Therefore, in case of any violations, there will be certain punishments. Like the National Society of Professional Engineers' Code of Ethics for Engineers (2007), such ethical codes are exhaustive, meaning that a penalty will not apply to someone unless a violation is proved according to a defined set of rules (Wilkinson 1999).

In any given research domain, ethics should be reflected within its own principles and the subject matter at hand; otherwise, the research will end up being a failure (MacCannell 2012). In the realm of public opinion, this code describes professional practices along with ethics. Furthermore, there is a list of disclosure items that is published by AAPOR (2015), which ensures that publishing qualitative research and papers is transparent when the content analysis is employed. In addition, the World Association for Public Opinion Research Code of Ethics created a guidance based on ethical practice principles that acts as a framework of professional standards for its members. This guideline also ought to be acceptable principles for public as well as users of research (WAPOR 2011). In order to measure how compliant a research is to the ethical standard, code of ethics is being utilized. Within a professional setting or discipline, code of ethics will help to determine whether or not, a work product or even an action is ethical (D. A. Fennell and Malloy 2007).

Another arena that attracts attention from business ethic research is tourism and hospitality studies. Köseoglu et al. (2016) have found out the significance of business ethics articles in their recent bibliometric study. They have identified a huge increase in the number of articles in business ethics, that contain tourism or hospitality management areas. There are abundant numbers of publications (e.g., Duffy and Smith 2004; D. A. Fennell and Malloy 2007; Lovelock and Lovelock 2013; MacCannell 2012; McLean 1993), that looked deeply into ethics and its dimension in tourists, visitors, and tourism industry in general, but looking at tourism and ethics in the research of tourism, it can be observed that there is a blatant lack of studies (Douglas C. Frechtling 2017). In the hospitality industry, understanding ethics is regarded as a challenging work (Lee and Tsang 2013), and there is a desperate need to understand the moral and perception of the ethics of stakeholders in this industry. Interviewing hotel managers in a research by Yeh (2012) has demonstrated that business ethics should receive more emphasis from hospitality properties: proper trainings and education will assist both students and employees to learn how to deal with ethical issues.

According to MacCannell (2012), areas like defining clear ethical codes for tourism research and evaluating how effective those codes of ethics are, need more attention and research. There are some complaints that the existing codes of ethics in tourism have not gone through effective evaluation system and that is why they are not creating the expected behaviors (Lovelock and Lovelock 2013; Malloy and Fennell 1998). One of the observed shortcomings in the tourism research derives from research methodology. Most published work relied on qualitative data collection methodology (Douglas C. Frechtling 2017), and none of these works looked into reporting ethics using a distinct data collection methodology that is quantitative in the realm of tourism journals and their research manuscripts. Looking at tourism journals, Douglas Frechtling (2017) asserts that the articles which follow the probability sample surveys are in line with a recognized international code of ethics. This code of ethics is widely popular, and such articles are in compliance with transparency codes and their value in such publications in the research zone.

15.4 Data Collection Method and Questionnaire Design

In tourism and hospitality research, the primary data are collected to examine the relationship that might exist between the research constructs via online medium. Data collection methods and approaches are critical and important to the efficiency of the population undertaken, in which the communication within society makes the online data collection method to be considered as a natural development of methodology (Turunen et al. 2010). The fast adoption of Internet technology and the expansion of Internet users make the online data collection a robust and powerful method in today's survey research (Albaum et al. 2011; Allen et al. 2006; Davidov and Depner 2011; Malhotra and Krosnick 2007; Payne and Barnfather 2012; Sills and Song 2002; Willis 2012). Furthermore, the Internet makes it possible for researchers to access different cohorts of users for the purpose of gathering information, transaction, and data (Boyer et al. 2002; Eusebio et al. 2004; McMellon and Schiffman 2001; Shiu and Dawson 2004; Wharton et al. 2003). Therefore, the online marketing research and web technology attract marketing managers and researchers alike (James et al. 2009; Sethuraman et al. 2005; Stieger and Reips 2010), to perform marketing research via online instruments.

Literature discussed the several advantages of online data collection over offline data collection methods. The online data collection method is advantageous due to its convenience, low cost, verifiability, return system, easy access, and feedback mechanism over paper-and-pencil data collection method (Kaye and Johnson 1999). Accordingly, the Internet surveys might replace traditional methods and become a conventional method and tool of data collection approach in research (James et al. 2009; Martin and Dirk 2009) as it is an advantageous method of data collection (Heath and Stewart 2003; Stieger and Reips 2010). Researchers (Albaum et al. 2011; Davidov and Depner 2011; Eusebio et al. 2004; Wharton et al. 2003; Yetter and Capaccioli 2010) claim that compared to paper-and-pencil data collection method, online data collection method has several advantages which are as follows:

(i) Lower total costs
(ii) Greater speed in questionnaire distribution and collection
(iii) Greater response rates
(iv) Less response errors
(v) Greater completed total responses
(vi) Access to wide social distance

However, despite the advantages and critics of the Internet as online instrument compared to traditional paper-and-pencil method, researches declare that there is no difference in terms of accuracy of the information gathered through online data collection method (Wharton et al. 2003). Technically, Sethuraman et al. (2005) found that the Internet data collection approach is better in dealing with the "basis of internal consistency and predictive (face) validity." Previously, researchers (McMellon and

Schiffman 2001) criticized online data collection approach over paper-and-pencil method because of certain problems including the following items:

(i) Self-selection bias
(ii) Sampling bias
(iii) Costs of random sampling
(iv) Technical problems such as incompatibility of server formats
(v) Non-transferability of word processor codes
(vi) Length of the survey
(vii) Challenge in dealing with the less computer literate respondent's inability to upload and download questionnaire files

E-mailed and online-based questionnaires are generally two types of data collection approach through online instrument methodology (Wharton et al. 2003). These types of methods also pose their own limitations over each other. An online data collection method provides a simple and powerful mode of dispersion and collection of response that can save time and money (Eusebio et al. 2004; Wharton et al. 2003). Turunen et al. (2010) suggest performing the option for an online response to an online questionnaire along with the traditional paper-and-pencil methods and form. Several associations and firms offer online research as a subject such as the Association of Internet Researchers (AoIR), Interactive Marketing Research Organization (IMRO), and Deutsche Gesellschaft für Online-Forschung (DGOF) (Martin and Dirk 2009). Park et al. (2012) and Thirumalai and Sinha (2011) collected their research data via www.bizrate.com. The data required to examine customer satisfaction with their online purchase patterns was collected from BizRate.com, which is a publicly available data source. BizRate's combined customer ratings of online retailers are considered to be among the credible indicators of online customer satisfaction and loyalty (Thirumalai and Sinha 2011). Finn et al. (2009), for their study, used data that were provided by Web Mystery Shoppers International Inc. (webmysteryshoppers.com), which is a market research company that conducts online mystery shopping studies. Hencehence, tourism and hospitality researchers should report the advantages and disadvantages of their online data collection approach, as an ethical verification.

15.5 Research Publication as an Ethical Concern

Another subject that we need to pay attention to is how much of tourism research has been shared with people in the tourism industry. Unfortunately, owners, tourism managers, and other people involved in this industry have little knowledge being transferred to them from published materials from tourism journals (Douglas C. Frechtling 2004). It will be beneficial for people in the industry to have access to such research knowledge, so they can have a chance to develop awareness about the challenges and find suitable solutions. One step to be taken is from practitioners in the hotel industry to develop ethics in this realm. They accentuate the importance of ethics via creating school programs and practices in the industry to develop and to implement hospitality ethics (Dopson and Nelson 2003; Yeh 2012). That might act as an additional way to actually practice ethics, instead of just talking about it in the hotel and tourism industry. Training and courses in ethics are in high demand since the desperate need for them is felt from the employees. There is a report by 81% of managers, who articulated the fact that there are no courses or workshops on ethical issues offered by their organizations (Knani 2014; Yeh 2012). That might be an effective approach for developing awareness on ethical practices in the hospitality industry.

As far as universities and publications on top tourism journals are concerned, D. Fennell (2013) made a point that ethical dimensions play a prominent role in these researches. Such pressures will bring about venomous effects on the freedom of academia and growing knowledge in such environments. He further explored the hardships in the tourism industry that derive from the audit culture, which is pretty new. The latest audit system has powerful influence on knowledge creation. Research ethics in tourism have proved to have a substantial weight. In addition, setting high research KPIs in some universities has resulted in dishonesty among academics. As argued by Fennell (2013), when academics are coerced into publishing in high-impact and high-quality journals with unreasonable KPIs, scholars are most likely not contributing to the development of the research paper. As a matter of fact, some academics might not have participated in the research and have not even read the article, but just have their names inserted into the manuscript, in order to meet their KPIs and to secure their tenures.

Another ethical area with regards to dishonesty is about manipulation and falsification of data and results. Despite the fact that there are research ethic committees and codes of conduct to govern these areas, intentional manipulation and falsification are hard to audit, and if one wants to avoid this, it requires high self-integrity. For instance, a researcher deliberately removed some data in order to gain a more favorable results or alter the results to achieve the intended outcome. These actions will dilute the value of academic research and further damage the trust between academia and the wider society (Befring 2015; Wasserman 2013). As such, researchers should be transparent and honest in reporting the methods and findings, which permits various stakeholders to challenge, verify, and replicate the study (Douglas C. Frechtling 2017; Wasserman 2013).

15.6 Conclusion

Research ethics is an extremely important part of every study, but tourism scholars receive guidance on ethical issues in research from different sources throughout their careers, such as experiences from participation in research activities, quantitative data analysis or research method courses (Aiken et al. 2008), short trainings, workshops, professional reading, and involving in peer review processes. Research methodology and ethics are intertwined. Quantitative methodology and the way it is reflected in the APA (American Psychological Association, 2002), ethics codes are the topics of many researches. They are also interested in discovering the place that the null hypothesis significance testing has in quantitative research approaches in the recent years. Additionally, dilemmas like extensive data retention and its popularity as a cost and benefit tradeoff, along with practices of archiving, that are needed to create a balance for reviewers or even consultees, need attention.

In tourism research, like any other arenas, ethics has a noteworthy presence. Acting with integrity and honesty is an essential responsibility of researchers in tourism studies, while they are pursuing justice (Ryan 2005), and in most institutions these days, researchers ought to meet the requirements that are articulated by ethic committees. An interesting point that is overlooked by researchers in tourism is the opinion of children in this industry as well. Since it is a difficult task to collect the data from this age range (researchers will need parents or the government's permission for data collection), there is a scarcity of information from children. Quantitative researches can be conducted to understand the travel experiences of kids, which might lead to generating additional theories in the tourism industry (Poria and Timothy 2014).

However, according to what Panter and Sterba (2011) stated in page one of their work, even in structured quantitative contexts, "the connections between design/analysis decisions and research ethics are largely implicit and informal. That is, methodology professors and research advisors convey ethical principles about research design, data analysis, and reporting experientially and by modeling appropriate professional behavior—without necessarily being cognizant that they are doing so and without labeling their behavior as such." There is a pressure to run simple analysis, while in reality, there are available options for more complex data analysis, which are more appropriate in statistical terms (Panter and Sterba 2011). As it can be concluded, ethics and its role in the business studies are priority since they are making sure that business practices and people involve in them have the ample information. It will help them to go through suitable trainings, using that information. Indeed, ethical principles and materials have been embedded mostly in qualitative topics, and most of the ethical topics in quantitative data analysis on specific areas of concern to tourism researchers (e.g., measurement, assumptions, sampling, design, model specification, reliability and validity assessment, and model fitting evaluation) are vague and disseminated, while they lack an ethical imperative (Panter and Sterba 2011). Tourism and hospitality suffered from lack of ethical research and that acts as a hindrance to managers and employees in both arenas. Conducting enough qualitative research, along with quantitative ones, is of top importance. They will provide the opportunity for people in these businesses to be aware of ethical codes to practice in the tourism and hospitality. Thus, this chapter aims to provide a practical review of ethical issues in quantitative methodology in tourism research.

References

AAPOR. (2015, November). American Association for Public Opinion Research code of ethics. Retrieved from https://www.aapor.org/Standards-Ethics/AAPOR-Code-of-Ethics.aspx

Aiken, L. S., West, S. G., & Millsap, R. E. (2008). Doctoral training in statistics, measurement, and methodology in psychology: Replication and extension of Aiken, West, Sechrest, and Reno's (1990) survey of PhD programs in North America. *American Psychologist, 63*(1), 32.

Albaum, G., Wiley, J., Roster, C., & Smith, S. M. (2011). Visiting item non-responses in internet survey data collection. *International Journal of Market Research, 53*(5), 687–703.

Allen, G. N., Burk, D. L., & Davis, G. B. (2006). Academic data collection in electronic environments: Defining acceptable use of internet resources. *MIS Quarterly, 30*(3), 599–610.

Ariffin, A. A. M. (2013). Generic dimensionality of hospitality in the hotel industry: A host–guest relationship perspective. *International Journal of Hospitality Management, 35*(0), 171–179. https://doi.org/10.1016/j.ijhm.2013.06.002.

Ariffin, A. A. M., & Maghzi, A. (2012). A preliminary study on customer expectations of hotel hospitality: Influences of personal and hotel factors. *International Journal of Hospitality Management, 31*(1), 191–198. https://doi.org/10.1016/j.ijhm.2011.04.012.

Befring, E. (2015, November 03). *Quantitative methods*. Retrieved from https://www.etikkom.no/en/library/introduction/methods-and-approaches/quantitative-methods/

Behi, R., & Nolan, M. (1995). Ethical issues in research. *British Journal of Nursing, 4*(12), 712–716.

Bernard, H. R. (2017). *Research methods in anthropology: Qualitative and quantitative approaches*. Lanham: Rowman & Littlefield.

Boyer, K. K., Olson, J. R., Calantone, R. J., & Jackson, E. C. (2002). Print versus electronic surveys: A comparison of two data collection methodologies. *Journal of Operations Management, 20*(4), 357–373. https://doi.org/10.1016/s0272-6963(02)00004-9.

Chris Paxson, M. (2009). Boomer boom for hospitality: Opportunities and challenges. *Journal of Hospitality Marketing & Management, 18*(1), 89–98. https://doi.org/10.1080/19368620801989360.

Chung, L. H., & Parker, L. D. (2010). Managing social and environmental action and accountability in the hospitality industry: A Singapore perspective. *Accounting Forum, 34*(1), 46–53. https://doi.org/10.1016/j.accfor.2009.10.003.

Davidov, E., & Depner, F. (2011). Testing for measurement equivalence of human values across online and paper-and-pencil surveys. *Quality & Quantity, 45*(2), 375–390. https://doi.org/10.1007/s11135-009-9297-9.

Dopson, L. R., & Nelson, A. A. (2003). Future of hotel education: Required program content areas for graduates of US hospitality programs beyond the year 2000–part two. *Journal of Hospitality & Tourism Education, 15*(3), 11–17.

Duffy, R., & Smith, M. (2004). *The ethics of tourism development*. London: Routledge.

Eusebio, S., Jr., Joao Luiz, B., & Stuart, J. B. (2004). Developing automated e-survey and control tools: An application in industrial management. *Industrial Management & Data Systems, 104*(3), 189–200.

Eysenbach, G., & Till, J. E. (2001). Ethical issues in qualitative research on internet communities. *BMJ, 323*(7321), 1103–1105.

Fennell, D. (2013). The ethics of excellence in tourism research. *Journal of Travel Research, 52*(4), 417–425.

Fennell, D. A., & Malloy, D. (2007). *Codes of ethics in tourism: Practice, theory, synthesis* (Vol. 33). Clevedon: Channel View Publications.

Finn, A., Wang, L., & Frank, T. (2009). Attribute perceptions, customer satisfaction and intention to recommend E-services. *Journal of Interactive Marketing, 23*(3), 209–220. https://doi.org/10.1016/j.intmar.2009.04.006.

Frankel, M. S. (1989). Professional codes: Why, how, and with what impact? *Journal of Business Ethics, 8*(2–3), 109–115.

Frechtling, D. C. (2004). Assessment of tourism/hospitality journals' role in knowledge transfer: An exploratory study. *Journal of Travel Research, 43*(2), 100–107.
Frechtling, D. C. (2017). On the ethics of tourism research. *Journal of Travel Research*, 004728751772975. https://doi.org/10.1177/0047287517729756.
Freese, J. (2007). Replication standards for quantitative social science: Why not sociology? *Sociological Methods & Research, 36*(2), 153–172.
Heath, M., & Stewart, A. (2003). A comparison of online and postal data collection methods in marketing research. *Marketing Intelligence & Planning, 21*(2), 85–95.
Hobson, J. S. P., & Teaff, J. D. (1994). Hospitality and leisure/recreation. *Journal of Hospitality & Leisure Marketing, 2*(1), 43–54. https://doi.org/10.1300/J150v02n01_05.
James, B. W., Vallen, H., Gerald, A., & Peter, T. (2009). Selecting techniques for use in an internet survey. *Asia Pacific Journal of Marketing and Logistics, 21*(4), 455–474.
Kaye, B. K., & Johnson, T. J. (1999). Research methodology: Taming the cyber frontier. *Social Science Computer Review, 17*(3), 323–337. https://doi.org/10.1177/089443939901700307.
Knani, M. (2014). Ethics in the hospitality industry: Review and research agenda. *International Journal of Business and Management, 9*(3), 1.
Köseoglu, M. A., Sehitoglu, Y., Ross, G., & Parnell, J. A. (2016). The evolution of business ethics research in the realm of tourism and hospitality: A bibliometric analysis. *International Journal of Contemporary Hospitality Management, 28*(8), 1598–1621.
Lee, L. Y.-S., & Tsang, N. K. (2013). Perceptions of tourism and hotel management students on ethics in the workplace. *Journal of teaching in travel & tourism, 13*(3), 228–250.
Lovelock, B., & Lovelock, K. (2013). *The ethics of tourism: Critical and applied perspectives*. London: Routledge.
MacCannell, D. (2012). On the ethical stake in tourism research. *Tourism Geographies, 14*(1), 183–194.
Malhotra, N., & Krosnick, J. A. (2007). The effect of survey mode and sampling on inferences about political attitudes and behavior: Comparing the 2000 and 2004 ANES to internet surveys with nonprobability samples. *Political Analysis, 15*(3), 286–323. https://doi.org/10.1093/pan/mpm003.
Malloy, D. C., & Fennell, D. A. (1998). Codes of ethics and tourism: An exploratory content analysis. *Tourism Management, 19*(5), 453–461.
Markham, A., & Buchanan, E. (2015). *Ethical decision-making and internet research: Recommendations from the Aoir ethics working committee (version 2.0)*. Association of Internet Researchers, 2012.
Martin, L., & Dirk, J. S. (2009). Internet research differs from research on internet users: Some methodological insights into online travel research. *Tourism Review, 64*(1), 32–47.
McLean, D. (1993). Ethics in hospitality management: Edited by Stephen SJ Hall. Educational Institute of the American Hotel and Motel Association (PO Box 1240, East Lansing MI 48826, USA) ISBN 0-86612-067-x, 1992, viii+ 285 pp.(exhibits, bibliography) $27.96. *Annals of Tourism Research, 20*(2), 384–388.
McMellon, C. A., & Schiffman, L. G. (2001). Cybersenior research: A practical approach to data collection. *Journal of Interactive Marketing, 15*(4), 46–55. https://doi.org/10.1002/dir.1022.
Moscardo, G. (2010). *Tourism research ethics: Current considerations and future options*. Oxford: Goodfellow Publishers.
Panter, A. T., & Sterba, S. K. (2011). *Handbook of ethics in quantitative methodology*. New York: Routledge.
Park, I., Cho, J., & Rao, H. R. (2012). The effect of pre- and post-service performance on consumer evaluation of online retailers. *Decision Support Systems, 52*(2), 415–426. https://doi.org/10.1016/j.dss.2011.10.001.
Payne, J., & Barnfather, N. (2012). Online data collection in developing nations: An investigation into sample Bias in a sample of south African university students. *Social Science Computer Review, 30*(3), 389–397. https://doi.org/10.1177/0894439311407419.

Poria, Y., & Timothy, D. J. (2014). Where are the children in tourism research? *Annals of Tourism Research, 47*, 93–95.
Ryan, C. (2005). Ethics in tourism research: Objectivities and personal perspectives. In *Tourism research methods: Integrating theory with practice* (pp. 9–19). Cambridge, MA: CABI.
Sethuraman, R., Kerin, R. A., & Cron, W. L. (2005). A field study comparing online and offline data collection methods for identifying product attribute preferences using conjoint analysis. *Journal of Business Research, 58*(5), 602–610. https://doi.org/10.1016/j.jbusres.2003.09.009.
Shiu, E. C. C., & Dawson, J. A. (2004). Comparing the impacts of internet technology and national culture on online usage and purchase from a four-country perspective. *Journal of Retailing and Consumer Services, 11*(6), 385–394. https://doi.org/10.1016/j.jretconser.2003.12.004.
Sills, S. J., & Song, C. (2002). Innovations in survey research. *Social Science Computer Review, 20*(1), 22–30. https://doi.org/10.1177/089443930202000103.
Stieger, S., & Reips, U.-D. (2010). What are participants doing while filling in an online questionnaire: A paradata collection tool and an empirical study. *Computers in Human Behavior, 26*(6), 1488–1495. https://doi.org/10.1016/j.chb.2010.05.013.
Teng, C.-C. (2011). Commercial hospitality in restaurants and tourist accommodation: Perspectives from international consumer experience in Scotland. *International Journal of Hospitality Management, 30*(4), 866–874. https://doi.org/10.1016/j.ijhm.2011.01.007.
Thirumalai, S., & Sinha, K. K. (2011). Customization of the online purchase process in electronic retailing and customer satisfaction: An online field study. *Journal of Operations Management, 29*(5), 477–487. https://doi.org/10.1016/j.jom.2010.11.009.
Turunen, M., Paanala, A., Villman, J., Nevalainen, A., & Haverinen-Shaughnessy, U. (2010). Evaluating housing quality, health and safety using an internet-based data collection and response system: A cross-sectional study. *Environmental Health: A Global Access Science Source, 9*, 69–69.
WAPOR. (2011, December 1). *World Association for Public Opinion Research code of ethics*. Retrieved from https://wapor.org/about-wapor/code-of-ethics/
Wasserman, R. (2013). Ethical issues and guidelines for conducting data analysis in psychological research. *Ethics & Behavior, 23*(1), 3–15.
Wharton, C. M., Hampl, J. S., Hall, R., & Winham, D. M. (2003). PCs or paper-and-pencil: Online surveys for data collection. *Journal of the American Dietetic Association, 103*(11), 1458–1459. https://doi.org/10.1016/j.jada.2003.09.004.
Wilkins, H., Merrilees, B., & Herington, C. (2009). The determinants of loyalty in hotels. *Journal of Hospitality Marketing & Management, 19*(1), 1–21. https://doi.org/10.1080/19368620903327626.
Wilkinson, L. (1999). Statistical methods in psychology journals: Guidelines and explanations. *American Psychologist, 54*(8), 594.
Willis, P. (2012). Talking sexuality online – Technical, methodological and ethical considerations of online research with sexual minority youth. *Qualitative Social Work, 11*(2), 141–155. https://doi.org/10.1177/1473325011400488.
Yeh, R. (2012). Hotel general managers' perceptions of business ethics education: Implications for hospitality educators, professionals, and students. *Journal of Human Resources in Hospitality & Tourism, 11*(1), 72–86.
Yetter, G., & Capaccioli, K. (2010). Differences in responses to web and paper surveys among school professionals. *Behavior Research Methods, 42*(1), 266–272.

MIX
Papier aus verantwortungsvollen Quellen
Paper from responsible sources
FSC® C105338

If you have any concerns about our products,
you can contact us on
ProductSafety@springernature.com

In case Publisher is established outside the EU,
the EU authorized representative is:
Springer Nature Customer Service Center GmbH
Europaplatz 3, 69115 Heidelberg, Germany

Printed by Libri Plureos GmbH
in Hamburg, Germany